# Climate Change Management

**Series Editor**

Walter Leal Filho, International Climate Change Information and Research Programme, Hamburg University of Applied Sciences, Hamburg, Germany

The aim of this book series is to provide an authoritative source of information on climate change management, with an emphasis on projects, case studies and practical initiatives – all of which may help to address a problem with a global scope, but the impacts of which are mostly local. As the world actively seeks ways to cope with the effects of climate change and global warming, such as floods, droughts, rising sea levels and landscape changes, there is a vital need for reliable information and data to support the efforts pursued by local governments, NGOs and other organizations to address the problems associated with climate change. This series welcomes monographs and contributed volumes written for an academic and professional audience, as well as peer-reviewed conference proceedings. Relevant topics include but are not limited to water conservation, disaster prevention and management, and agriculture, as well as regional studies and documentation of trends. Thanks to its interdisciplinary focus, the series aims to concretely contribute to a better understanding of the state-of-the-art of climate change adaptation, and of the tools with which it can be implemented on the ground.

**Notes on the quality assurance and peer review of this publication**

Prior to publication, the quality of the works published in this series is double blind reviewed by external referees appointed by the editor. The referees are not aware of the author's name when performing the review; the referees' names are not disclosed.

Walter Leal Filho · Franziska Wolf
Editors

# Climate Change and Health: Perspectives from Developing Countries

Springer

*Editors*
Walter Leal Filho
Hamburg University of Applied Sciences
Hamburg, Germany

Franziska Wolf
Hamburg University of Applied Sciences
Hamburg, Germany

ISSN 1610-2002 ISSN 1610-2010 (electronic)
Climate Change Management
ISBN 978-3-031-93176-5 ISBN 978-3-031-93177-2 (eBook)
https://doi.org/10.1007/978-3-031-93177-2



This Springer imprint is published by the registered company Springer Nature Switzerland AG
The registered company address is: Gewerbestrasse 11, 6330 Cham, Switzerland

# Contents

# Chapter 1
# Effects of Ambient Temperature and Its Effect Modifiers on Malaria in Bamako, Mali

**Adama Boua Doumbia, Konin Pierre-Claver Kakou, Walter Leal Filho, Hamadoun Sangho, and Amoro Coulibaly**

**Abstract** *Background.* The existence of previous studies on the effects of various meteorological factors related to malaria is limited in Mali. This study tries to address this gap, by investigating the effects of ambient temperature and its effect modifiers on malaria in Bamako. Specifically, it sought to analyse temporal trends in weather conditions in connection with malaria cases, to identify patterns, seasonality, and potential long-term trends, and to correlate specific weather events that may increase incidences of malaria. *Methods.* Daily data on malaria cases and meteorological factors from 2012 to 2022 were collected. Then, we quantified relationships between meteorological variables and malaria case outcomes. The association between meteorological factors and malaria was explored using Regression models, the Pearson

A. B. Doumbia (✉)
Doctoral Research Programme on Climate Change and Education, West African Science Service Centre on Climate Change and Adapted Land Use (WASCAL), School of Agriculture and Environmental Sciences, University of the Gambia, P.O. Box 3530, Kanifing, Banjul, The Gambia
e-mail: ada.bouadoumbia@gmail.com; adama.doumbia@haw-hamburg.de

K. P.-C. Kakou
Doctoral Research Programme on Climate Change and Energy, West African Science Service Centre on Climate Change and Adapted Land Use (WASCAL), University of Abdou Mounouni, Niger, The Gambia
e-mail: koninclaver@gmail.com

W. Leal Filho
Research and Transfer Centre "Sustainable Development and Climate Change Management", Hamburg University of Applied Sciences, Ulmenliet 20, 21033 Hamburg, Germany
e-mail: walter.leal2@haw-hamburg.de

H. Sangho
Faculté de Médicine et Odontostomatologie, Université des Sciences des Techniques et des Technologies de Bamako, Bamako, Mali
e-mail: prsangho@gmail.com

A. Coulibaly
Département Science et Technique Agricole, Institut Polytechnique Rural de Formation et de Recherche Appliquée de Katibougou, Katibougou, Mali
e-mail: coulibalya2002@yahoo.fr

 
W. Leal Filho and F. Wolf (eds.), *Climate Change and Health: Perspectives from Developing Countries*, Climate Change Management,
https://doi.org/10.1007/978-3-031-93177-2_1

Chi-Square Test, and the Generalized Linear Model (GLM) with Poisson Distribution. *Results.* The findings suggest that malaria increases with rising levels of rainfall and humidity, but usually decreases with increasing temperature. These insights underscore the importance of considering climatic factors—particularly temperature, humidity, and rainfall—in malaria prevention and control efforts in Bamako, Mali. Public health professionals and medical service providers should pay more attention to malaria prevention and control when the weather conditions of both high temperature and high humidity occur.

**Keywords** Effects modification · Generalized linear model · Malaria and meteorological factors

## Introduction

Malaria poses a significant global health challenge, with nearly half of the world's population at risk of contracting the disease. In 2022 alone, an estimated 249 million cases occurred across 85 countries (World Health Organization 2023; Gaither et al. 2024), with Africa bearing the highest burden, comprising 94% of global cases and 95% of deaths—estimated at 580,000 deaths in 2022 (Ministers of Health of High Burden High Impact Countries in Africa 2024) (www.who.int/news/item/06-03-2024-african-health-ministers-commit-to-end-malaria-deaths). Mali is among the top ten countries with the highest malaria cases and fatalities, accounting for 3.1% of cases and 3.3% of deaths worldwide in 2021 (World Malaria Report—WHO 2023). Bamako, the capital of Mali, experiences a notable prevalence of this disease.

Understanding the environmental factors influencing malaria transmission is pivotal for effective prevention and control strategies. Bamako, situated in the Sudanese zone, undergoes distinct seasonal climate variations, with a dry period from November to April and a rainy season from May to October. These fluctuations closely correlate with changes in malaria transmission, with the rainy season typically witnessing heightened cases due to increased mosquito vector activity (Lassana et al. 2016).

The relationship between climate and malaria has been extensively studied. Thomson et al. (2017) utilised climate data to assess its impact on malaria evaluations and found potential overestimation in three countries (Tanzania, Zanzibar, Uganda) and underestimation in three others (Mali, Senegal, Ethiopia). Adeola et al. (2019) investigated rainfall's influence on malaria transmission in South Africa's Limpopo Province and noted the variations across study areas and seasonality dependence.

While rainfall's association with malaria transmission is well-established, the role of ambient temperature as an effect modifier remains underexplored. Temperature influences various aspects of the malaria parasite's life cycle, which impacts its development rates and survival probabilities (Blanford et al. 2013; Mohammadkhani et al. 2019a). Additionally, temperature interacts with other environmental and socio-economic factors to shape transmission dynamics.

This study aimed to investigate ambient temperature's effects on malaria transmission in Bamako, Mali and explore potential effect modifiers. By analysing longitudinal malaria and meteorological data including temperature, humidity, and rainfall, we sought to bring out the complex interplay between climatic factors and malaria cases. Our study contributes to advancing the understanding of the climate-malaria relationship, which will aid global efforts in combatting this preventable and treatable disease.

The next parts of this article are structured as follows: Section "Study Area" provides an overview of the study area, section "Patients, Materials, and Methods" outlines patient demographics, materials, and methods, section "Results and Discussion" presents the results, section "Discussion" discusses them, and section "Conclusion" concludes the article.

## Study Area

Bamako is the capital and largest city of Mali. It is between 12°29′57″ and 12°42′17″ North Latitude and 7°54′22″ and 8°4′6″ West Longitude, with an area of 245 $km^2$. Bamako occupies the fringe corresponding to the zone of the Western Sudanese Savannah; therefore, it enjoys a rather humid tropical climate. The mean temperature is 27.8°. It has an important river port on the Niger and a commercial centre radiating throughout the sub-region. The city is also the main administrative centre of the country and has 3,050,570 inhabitants as of 2024. Figure 1.1 shows the geographical location of Bamako in Mali. The district of Bamako is divided into urban areas, periurban areas and peripheral villages. Malaria is hyperendemic in the peripheral villages, mesoendemic in the periurban area, and hypoendemic in the centre of Bamako, with the transmission occurring between June and November, a period that corresponds to the rainy season (Safeukui-Noubissi et al. 2004).

## Patients, Materials, and Methods

### *Data Collection*

Data collection was conducted in two reference health centres (CSRef) located on both the left and right riverbanks in Bamako. The selection process involved choosing the reference health centre of Commune III from the left riverbank and the reference health centre of Commune V from the right riverbank. These specific health centres were strategically chosen due to the expedited approval received from their respective heads to facilitate the data collection within their facilities.

Moreover, these two health centres accommodate the highest number of patients and are often referral points for several patients from surrounding neighbourhoods.

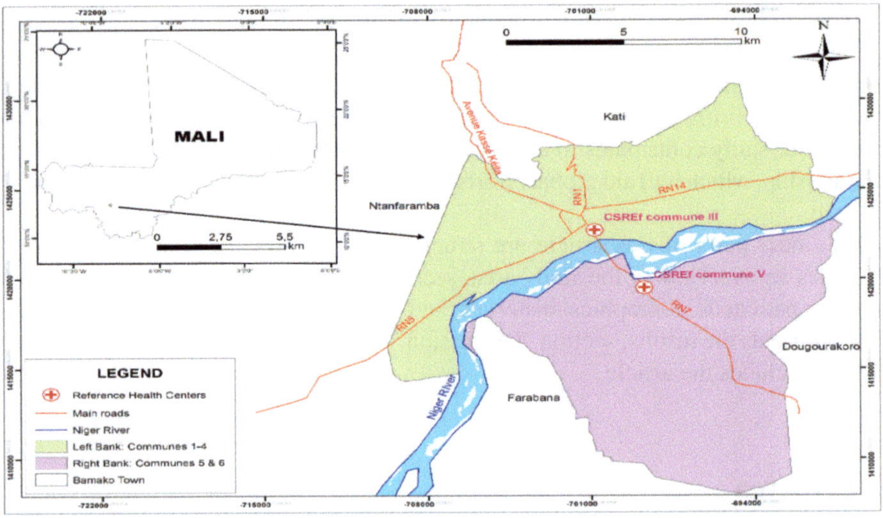


**Fig. 1.1** Location map. Map of Bamako presenting the sites where data on malaria was collected. These are CSRef Commune III coordinates: 12.64195, − 8.00663 and CSRef Commune V coordinates 12.61275, − 7.99441

As such, the collected data provided a significant representation of the population affected by malaria within the region.

Monthly data spanning from 2012 to 2022 was meticulously extracted from the archives of the two reference health centres in Bamako. This extended timeframe allowed for a thorough analysis of long-term trends and patterns in malaria cases and associated meteorological variables, thereby contributing to a comprehensive understanding of the complex interplay between climate and malaria transmission dynamics in Bamako, Mali. It is important to state that the data collection process adhered to strict anonymity protocols to protect patient privacy and confidentiality. Each data point was collected anonymously while safeguarding individuals' personal health information.

A total of 26,702 malaria cases spread over the eleven years were collected for this study, shown in Fig. 1.2.

## *Materials*

The data processing used sophisticated statistical tools, such as R, Python, and Microsoft Excel. These tools were used to run the statistical tests and construct the graphs.

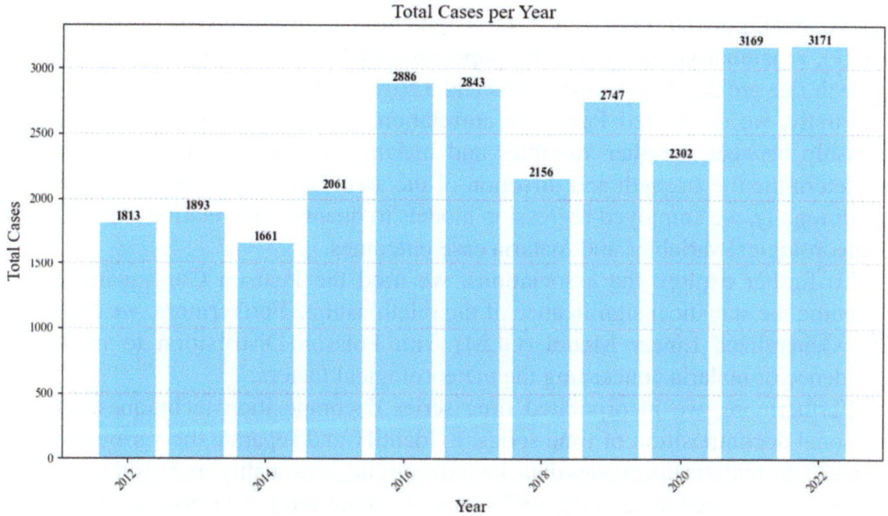


**Fig. 1.2** Distribution of the total number of malaria cases collected over the study period (from 2012 to 2022) in the two health centres

## *Method: Statistical Analyses*

This article aimed to evaluate the effect of ambient temperature and its effect modifiers on the cases of malaria in Bamako. To achieve this, we employed a comprehensive methodology involving multiple statistical tests and analyses. To ensure consistency between the malaria data (monthly data) and the weather variables (daily data), we performed monthly averages over weather data when the proportion of missing data did not exceed 10 days. If more, the entire data of the corresponding month were discarded.

Incidence is a standardised measure that takes into account the differences in population size and duration of risk. It allows comparisons to be made between different populations or periods, even if the size of the population or the duration of observation differs. It was suitable for the case of this paper, as the size of cases varied over the years and the population of Bamako might have differed for study. The incidence is expressed as follows:

$$\text{Monthly Incidence} = \frac{\text{Monthly Malaria cases}}{\text{Population at Risk (Bamako)}} \times 100{,}000$$

The population at risk is obtained by the following expression:

$$P(t) = P_0(1 + r)^{t-t_0}$$

where, $P(t)$: population of year $t$, $P_0$: population at reference year $t_0$ (2018 in this study), $r$: annual growth rate. The population of Bamako in 2018 and the annual growth rate are 2,488,641 inhabitants and 3.6, respectively (Diallo et al. 2020).

Firstly, we conducted Pearson's correlation analysis to assess the linear relationship between weather variables and malaria incidence. This test allowed us to determine the strength and direction of the association between these variables. Additionally, we employed regression models to quantify the relationships between meteorological variables and malaria case outcomes.

To further explore the associations, we used the Pearson Chi-Square Test to examine the statistical significance of the relationships. Furthermore, we employed the Generalized Linear Model (GLM) with Poisson Distribution to model the incidence of malaria concerning the meteorological factors.

Furthermore, we incorporated time series decomposition techniques, such as seasonal decomposition of time series, to identify and separate the various components of the malaria time series data, including trend, seasonality, and residual components. This enabled us to better understand the underlying patterns and fluctuations in malaria cases over time.

## Results and Discussion

### *Gender Distribution and Age Groups*

The bar chart in Fig. 1.3 shows the proportions of women and men by age group. We considered the age groups as specified in Maillard et al. (2017). The graph shows that people aged between 15 and 24 are the most affected, with a high proportion of women (20% of the total cases studied in this paper). It is important to mention that children aged 0 to 4 account for a significant proportion (around 17% of the cases studied, sexes combined).

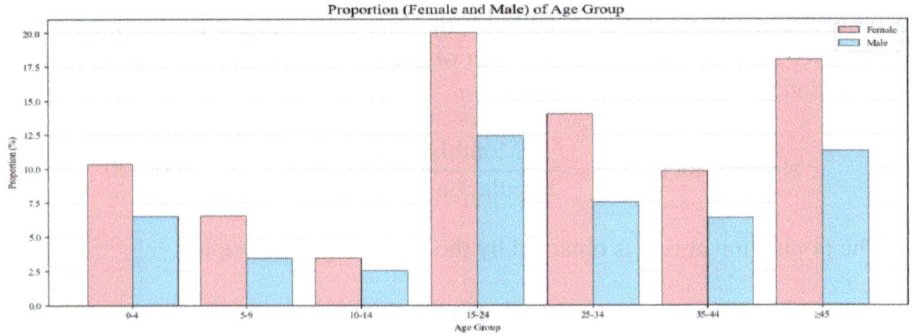


**Fig. 1.3** Gender distribution in different age groups over the period of study (from 2012 to 2022)

**Table 1.1** Monthly average of weather variables and malaria incidence over the study period (1st January 2012 to 31st December 2022)

| | Temperature (°C) | Humidity (%) | Rainfall (mm) | Incidence |
|---|---|---|---|---|
| January | 32.87 | 23.78 | 0.62 | 7.88 |
| February | 35.64 | 18.60 | 0.30 | 6.91 |
| March | 38.37 | 19.79 | 11.49 | 6.04 |
| April | **39.82** | 30.16 | 29.73 | 5.60 |
| May | 38.02 | 49.04 | 73.41 | 6.30 |
| June | 34.70 | 65.81 | 172.37 | 7.50 |
| July | 31.71 | 79.42 | 247.58 | 9.45 |
| August | 30.63 | **85.67** | **342.14** | **14.02** |
| September | 32.01 | 82.82 | 213.17 | 8.23 |
| October | 34.58 | 71.57 | 78.64 | 8.53 |
| November | 35.54 | 45.59 | 3.44 | 9.57 |
| December | 33.29 | 30.91 | 0.15 | 9.72 |

Simple bold values represent minima while starred bold values represent maxima. Incidence (number of cases per 100,000 people)

## *Descriptive Statistical*

Table 1.1 and Fig. 1.4 show the trends of malaria incidence and weather variables over the study period. As shown in Table 1.1, humidity, rainfall, and malaria incidence reach their maximum in August, while temperature registers at its lowest value. These trends are more visible in Fig. 1.4.

Figure 1.4 describes the malaria incidence curve and the bar charts of three parameters, namely, rainfall, humidity and temperature.

The curve of Fig. 1.4 indicates that all the peaks in malaria cases over the eleven years studied occurred during the rainy season and that most of them coincided with the peak in rainfall in August, the rainiest month in Bamako. We also noted that the peaks occurred in December, the coldest month of the year, which is similar to the entire study period; however, the peak of malaria decreased in April, the hottest month of the year. Figure 1.5 shows similar data, where all of the peaks in malaria incidence occurred with humidity. However, the malaria incidence negatively occurred with the maximum temperature in Fig. 1.5. The peak mostly coincided with April, which is the hottest month in Bamako.

## *Correlation Analysis*

Correlations between malaria and weather conditions included temperature, humidity, and rainfall. The weather conditions and malaria curves show that malaria

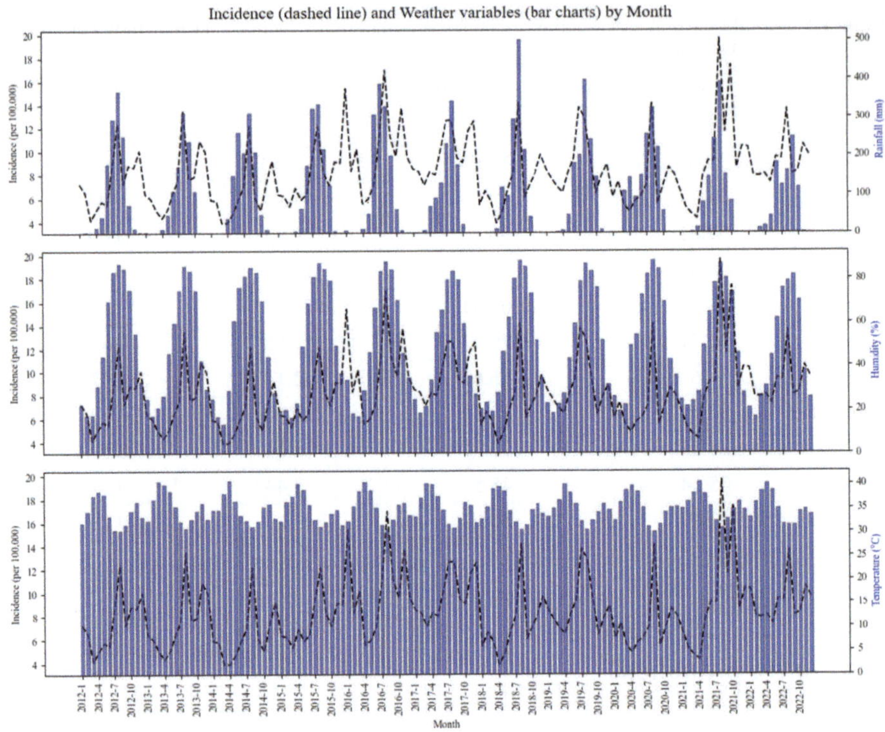


**Fig. 1.4** Evolution of malaria incidence and weather variables throughout study (from 1st January 2012 to 31st December 2022)

is present all year round in the District of Bamako. It is very frequent in July, August, and September, which correspond to the third trimester and the rainiest months of the year. It is frequent in October, November, and December, corresponding to the fourth trimester and coldest months, and less frequent from January to June, which correspond respectively to the first and second trimesters and the hottest months of the year.

In terms of indicators, malaria incidence is very high in the third and fourth trimesters, and average in the first and second trimesters of the year, respectively linked to rainfall and humidity. The peak of malaria corresponds with August, with a humidity of 85.67%, an average temperature of 30.63 °C, and rainfall of 342.14 mm. Therefore, malaria cases increase with increasing rainfall and humidity but decrease with increasing temperature. This interplay between weather variables and malaria transmission is highlighted in Fig. 1.6. Rainfall and humidity are positively associated with malaria incidence, with a Pearson correlation of 0.43 and 0.42, respectively, while temperature shows a negative association with malaria incidence (Pearson coefficient $= -0.56$). As rainfall and humidity increase, malaria incidence tends to increase as well. Increased rainfall leads to higher humidity, creating favourable conditions for mosquito survival and reproduction.

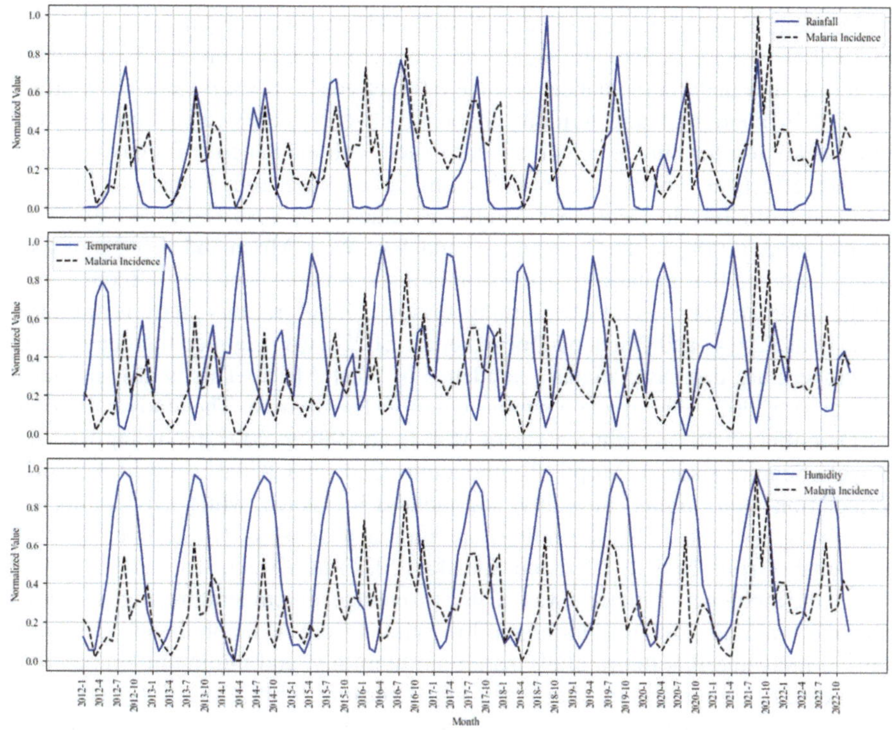


**Fig. 1.5** Normalised correlation curves between malaria incidence and weather variables over the period of study (from 1st January 2012 to 31st December 2022)

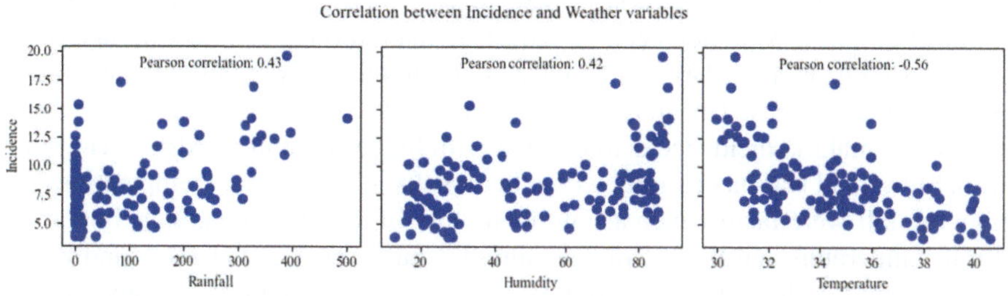


**Fig. 1.6** Correlation between malaria incidence and weather variables throughout the study (from 1st January 2012 to 31st December 2022)

## Discussion

Malaria remains a serious public health concern in Sub-Saharan Africa despite tremendous progress in lowering the global malaria burden over the previous two decades. We investigated the effects of temperature, rainfall, and humidity on malaria incidence in Bamako and explored potential modifiers over eleven years from 2012 to 2022. There are statistically significant, moderate associations between rainfall, humidity, and malaria incidence ($p$-value 0.00, effects = 0.13; $p$-value 0.00, effects = 0.59). While rainfall and humidity are positively correlated with malaria incidence in Bamako, temperature is negatively correlated with incidence in Bamako ($p$-value = 0.00, effects = −6.94). We found that malaria is rare from January to June and frequent from July to December. Studies conducted by Jean Delmont (1982) and Munier et al. (2008) in the Sudanese savannah of West Africa and Senegal, respectively, indicate that malaria transmission is discontinuous since it is frequent only in the heart and at the end of the rainy season. Therefore, this study is similar to Arab et al. (2014) and Oheneba-Dornyo et al. (2022), who claimed that the increase in temperature decreases malaria incidence in West Africa. This is also similar to Hay et al. (2002), who made a similar claim in East Africa. The study carried out by Doumbia (2010) in eight regions of Mali from 1998 to 2007, including Bamako, Kayes, Segou, Sikasso, Mopti, Tombouctou, Gao, and Kidal, showed a positive association with rainfall, humidity, and malaria incidence, while temperature was negatively associated. These findings are similar to those of (Lassana et al. 2016) in Bamako from 2001 to 2010 and Samaké (2021) in Sikasso from 2017 to 2020, where the incidence of malaria was related to average air temperature, high humidity, abundant rainfall, low insolation, and low wind speed. Akinbobola and Omotosho (2013) indicated that the rainfall, maximum humidity, and minimum temperatures are predictors of malaria incidence in tropical rainforests and Guinea Savanna areas. Moreover, studies carried out in Botswana (Chirebvu et al. 2016), Ethiopia (Sena et al. 2015), and Nigeria (Segun et al. 2020) found that rainfall, relative humidity, and minimum temperature were contributing factors to malaria incidence. The finding of Filho et al. (2023) stated that increased rainfall and humidity are positively associated with malaria transmission in new areas. Other studies conducted in Eritrea (Kifle et al. 2019; Mihreteab et al. 2020), Panama (Hurtado et al. 2018), and South Africa (Adeola et al. 2019) significantly associated higher rainfall with an increase in malaria incidences. The above findings are consistent with Liu et al. (2024), who found that malaria incidence is associated with higher rainfall in Africa. The study from Swaziland was inconsistent with other results, where warmer temperatures and high rainfall were factors increasing the number of malaria cases, including above 30° in temperature and above 5 mm in rainfall (Chuang et al. 2017). We found similar results from Wardrop et al. (2013) in China, where the rainfall was related to some areas. However, this finding accorded with the findings of Mafwele and Lee (2022) from 1901 to 2015 in Africa, which showed a positive relationship between temperature and rainfall with malaria until 1980, while malaria was negatively correlated with rainfall in the 1981–2015 period. Additionally, our findings are in agreement with the studies carried out in Bénin (AHOSSIN et al. 2024; Mehinto et al. 2020), Côte d'Ivoire (Hilaire 2017) and Sénégal (I Diouf et al. 2015), where temperature and rainfall are significantly associated with malaria incidence. Moreover, Ibrahima Diouf et al. (2022), while exploring the regional impact of future climate change on malaria in West Africa, found that malaria incidence is associated with temperature and rainfall. Similarly, Koudjom et al. (2024) found that a change in temperature and rainfall results in a significant increase in malaria prevalence in Sub-Saharan Africa. Accordingly, the average temperature has a positive association with malaria incidence in Uganda and Zimbabwe (Filho et al. 2023). The study by Liu et al. (2024) claimed that malaria incidence in Asian countries is characterised by temperature rising. According to Nili et al. (2022), the future climate in the Zahedan district in Iran will be warmer, more humid, and with higher precipitation; therefore, malaria will remain a health issue in this locality. In Indonesia, temperature and humidity are effective weather factors in malaria incidence (Mohammadkhani et al. 2019b). Furthermore, the study by Li et al. (2013) carried out in China indicated that every 1 °C increase in temperature corresponds to a 0.90% increase in malaria cases per month. Similarly, a 1% increase in relative humidity led to a malaria case increase of 3.99%. All in all, we found that meteorological conditions had a significant impact on the number of malaria cases in Bamako, the capital of Mali. A rise in rainfall, minimum temperature, and relative humidity may increase the risk of malaria transmission. Our findings provide fundamental information that can help develop a simple, precise, and economic malaria outbreak early warning system.

Whereas a rigorous methodology was applied to data collection and analysis in this paper, several limitations are worth mentioning. The first one is that data was collected from only two health centres in Bamako. Although these centres were chosen strategically, they may not fully represent malaria incidence trends across Bamako. The second limitation is that the selected health centres are located in urban areas, potentially overlooking rural regions where malaria transmission dynamics might differ significantly. Furthermore, this paper did not consider other factors influencing malaria incidences, such as socioeconomic conditions, public health interventions, migration patterns, and urbanisation. Therefore, including these variables may enhance the understanding of malaria transmission in Bamako and may be the subject of future studies.

## Conclusion

Our objective in this study was to investigate the effect of ambient temperature and its effect modifiers on malaria in Bamako. Firstly, the demographic analysis revealed that individuals aged between 15 and 24, particularly women, are the most affected, comprising approximately 20% of the total cases studied. Additionally, children aged 0 to 4 represented a significant proportion, around 17% of the cases studied, emphasising the vulnerability of young children to malaria infection.

Secondly, the trends observed in malaria incidence and weather variables over the study period highlight distinct seasonal patterns. Humidity, rainfall, and malaria incidence peak in August, coinciding with the lowest temperature values. These trends underscore the influence of climatic factors, particularly rainfall and humidity, on malaria transmission dynamics.

Furthermore, the correlations between malaria and weather conditions reveal that malaria is present throughout the year in Bamako, with peaks occurring during the third trimester (July, August, September) and the fourth trimester (October, November, and December). These periods coincide with the rainiest and coldest months of the year, indicating a strong association between rainfall, humidity, and malaria incidence.

In terms of risk indicators, the study identifies the third trimester as having a very high risk of malaria, followed by a high risk in the fourth trimester. Conversely, the risk is average during the first and second trimesters, corresponding to the hottest months of the year. The peak of malaria incidence occurs in August, characterised by high humidity (85.67%), an average temperature of 30.63 °C, and significant rainfall of 342.14 mm.

The contributions of this study to theory and practice are two-fold. First, this study provides a deeper insight into the epidemiological patterns of malaria in Bamako, Mali. By identifying the specific environmental and climatic factors that contribute to the transmission and prevalence of malaria, the research enriches the existing body of knowledge on malaria epidemiology. These findings can be used to refine theoretical models of disease transmission, helping to predict outbreaks and understand the dynamics of malaria in urban African settings. Secondly, the study introduces innovative analytical methods or frameworks that can be applied to other regions and diseases. By integrating advanced statistical analysis, the research advances the theoretical tools available for studying infectious diseases, with an emphasis on malaria.

Furthermore, the study's findings may directly inform public health strategies in Bamako, Mali. By pinpointing high-risk areas and times for malaria transmission, health authorities can better target their interventions, such as distributing insecticide-treated bed nets, conducting indoor residual spraying, and organising community health education campaigns. This targeted approach ensures that resources are used more efficiently and effectively, leading to a reduction in malaria incidence and improved health outcomes for the local population.

The findings suggest that malaria incidence increases with rising levels of rainfall and humidity but decreases with increasing temperature. These bring to light the importance of considering climatic factors, particularly temperature, humidity, and rainfall, in malaria prevention and control efforts in Bamako, Mali. Implementing targeted interventions that address seasonal variations in malaria transmission dynamics could help mitigate the impact of the disease and reduce its burden on vulnerable populations, particularly women and children.

**Acknowledgments** This document was produced with the financial support of the Prince Albert II of Monaco Foundation. The contents of this documents are solely the liability of Adama Boua Doumbia and under no circumstances may be considered as a reflection of the Prince Albert II of Monaco Foundation and/or the IPCC. The authors acknowledge the West African Science Service Centre on Climate Change and Adapted Land-Use (WASCAL) with its fully funded doctoral programme of Climate Change and Education at the University of the Gambia. The authors of this study would like to express their sincere thanks to the authorities of the two Health Centres: CSRef de Commune III from Rive Droite and CSRef de Commune V from Rive Gauche for giving them access to the data and for their contribution. They would also like to thank the authorities of the Malian Meteorological Agency "Mali-météo."

**Author's contribution** **ABD**: Conceptualisation, Methodology, Data Collection, Visualisation, Writing original draft, **KP-CK**: Conceptualisation, Methodology, Visualisation, Writing original draft, **WLF, HS & AC**: Supervision, Review, & Editing. All authours read and approved the final manuscript.

**Funding** This research received no external funding.

## References

Adeola A, Ncongwane K, Abiodun G, Makgoale T, Rautenbach H, Botai J, Adisa O, Botai C (2019) Rainfall trends and malaria occurrences in Limpopo province, South Africa. Int J Environ Res Publ Health 16(24). https://doi.org/10.3390/ijerph16245156

Ahossin R, Atchade G, Wokou G, Yabi I (2024) Rythme climatique et évolution des pathologies dans la commune de Zogbodomey au Sud-Bénin. Afrique Sci 24(3):1–16

Akinbobola A, Omotosho JB (2013) Predicting malaria occurrence in Southwest and North central Nigeria using meteorological parameters. Int J Biometeorol 57(5):721–728. https://doi.org/10.1007/s00484-012-0599-6

Arab A, Jackson MC, Kongoli C (2014) Modelling the effects of weather and climate on malaria distributions in West Africa. Malar J 13(1):1–9. https://doi.org/10.1186/1475-2875-13-126

Blanford JI, Blanford S, Crane RG, Mann ME, Paaijmans KP, Schreiber KV, Thomas MB (2013) Implications of temperature variation for malaria parasite development across Africa. Sci Rep 3:1–11. https://doi.org/10.1038/srep01300

Chirebvu E, Chimbari MJ, Ngwenya BN, Sartorius B (2016) Clinical malaria transmission trends and its association with climatic variables in Tubu village, Botswana: a retrospective analysis. PLoS ONE 11(3):1–16. https://doi.org/10.1371/journal.pone.0139843

Chuang TW, Soble A, Ntshalintshali N, Mkhonta N, Seyama E, Mthethwa S, Pindolia D, Kunene S (2017) Assessment of climate-driven variations in malaria incidence in Swaziland: toward malaria elimination. Malar J 16(1):1–10. https://doi.org/10.1186/s12936-017-1874-0

Diallo BA, Toure M, Dembele NDJ (2020) Urbanisation des zones inondables: le cas du District de Bamako. Revue Espace Geographique Et Societe Marocaine 7(2):809–820

Diouf I, Deme A, Cisse M, Gaye AT (2015) Détermination des Paramètres du Paludisme au Sénégal à partir de Données Météorologiques de Stations et de Réanalyses. XXVIIIe Colloque de l'Association Internationale de Climatologie, Liège 1:133–139

Diouf I, Adeola AM, Abiodun GJ, Lennard C, Shirinde JM, Yaka P, Ndione JA, Gbobaniyi EO (2022) Impact of future climate change on malaria in West Africa. Theoret Appl Climatol 147(3–4):853–865. https://doi.org/10.1007/s00704-021-03807-6

Doumbia S (2010) Impact du Changement Climatique sur l' Incidence du Paludisme au Mali de 1998 a 2007. In Malaria Research and Training Centre. Université de Bamako

Filho WL, May J, May M, Nagy GJ (2023) Climate change and malaria: some recent trends of malaria incidence rates and average annual temperature in selected sub-Saharan African countries from 2000 to 2018. Malar J 22:1–14. https://doi.org/10.1186/s12936-023-04682-4

Gaither C, Morgan C, Kirby R, Karema C, Gashema P, White SJ, Topazian HM, Geibrecht D, Thwai K, Boyter K, Munyaneza T, Muvunyi CM, De Dieu Butera J, Bailey JA, Mazarati J-B, Juliano JJ (2024) Prevalence of Falciparum and non-Falciparum Malaria in the 2014–15 Rwanda Demographic Health Survey. MedRxiv Preprint Server Health Sci 1–21. https://doi.org/10.1101/2024.01.09.24301054

Hay SI, Cox J, Rogers DJ, Randolph SE, Stern DI, Shanks GD, Myers MF, Snow RW (2002) Climate change and the resurgence of malaria in the East African highlands. Nature 415(6874):905–909. https://doi.org/10.1038/415905a

Hilaire MG (2017) Changement Climatique et Paludisme en Côte d'Ivoire: Représentations Sociales et Connaissance des Populations d'Adjéyaokro (Bouaké). Eur Sci J ESJ 13(26):110. https://doi.org/10.19044/esj.2017.v13n26p110

Hurtado LA, Calzada JE, Rigg CA, Castillo M, Chaves LF (2018) Climatic fluctuations and malaria transmission dynamics, prior to elimination, in Guna Yala, República de Panamá. Malaria J 17(1):1–12. https://doi.org/10.1186/s12936-018-2235-3

Delmont J (1982) Paludisme et variations climatiques saisonnieres en savane soudanienne d'Afrique de l'Ouest. Ecole des Hautes Études En Sciences Sociales 22:117–133

Kifle MM, Teklemariam TT, Teweldeberhan AM, Tesfamariam EH, Andegiorgish AK, Azaria Kidane E (2019) Malaria risk stratification and modeling the effect of rainfall on Malaria incidence in Eritrea. J Environ Public Health. https://doi.org/10.1155/2019/7314129

Koudjom E, Lokonon BOK, Egbendewe AYG (2024) Climate change, malaria prevalence and cereal yields in sub-Saharan Africa. Eur J Dev Res. https://doi.org/10.1057/s41287-024-00630-3

Lassana D et al (2016) Utilisation des facteurs climatiques pour la surveillance de la fréquence des occurrences de méningite/paludisme à Bamako. Revue Malienne d'Infectiologie et de Microbiologie 8:33–42

Li T, Yang Z, Wang M (2013) Temperature, relative humidity and sunshine may be the effective predictors for occurrence of malaria in Guangzhou, southern China, 2006–2012. Parasit Vectors 6(1):2–5. https://doi.org/10.1186/1756-3305-6-155

Liu Q, Wang Y, Deng J, Yan W, Qin C, Du M, Liu M, Liu J (2024) Association of temperature and precipitation with malaria incidence in 57 countries and territories from 2000 to 2019: a worldwide observational study. J Glob Health 14(38):13. https://doi.org/10.7189/jogh.14.04021

Mafwele BJ, Lee JW (2022) Relationships between transmission of malaria in Africa and climate factors. Sci Rep 12(1):1–8. https://doi.org/10.1038/s41598-022-18782-9

Maillard O, Pagès F, Achiraf A, Lepère J-F, Rabarison P, Filleu L (2017) Surveillance du Paludisme à Mayotte entre 2007 et 2016/Malaria Surveillance in Mayotte (France) between 2007 and 2016. Bulletin Epidemiologique Hebdomadaire 512–520

Mehinto FD, Boko M, Sedjame RA, Houssou CS (2020) Impacts du Climat sur la Santé des Enfants de 0 à 5 ans dans le Département de l ' Ouémé au Sud-Est du Bénin (Afrique de l' Ouest). J Water Environ Sci 4(1):536–553

Mihreteab S, Lubinda J, Zhao B, Rodriguez-Morales AJ, Karamehic-Muratovic A, Goitom A, Shad MY, Haque U (2020) Retrospective data analyses of social and environmental determinants of malaria control for elimination prospects in Eritrea. Parasit Vectors 13(1):1–11. https://doi.org/10.1186/s13071-020-3974-x

Ministers of Health of High Burden High Impact Countries in Africa (2024). Declaration for accelerated malaria mortality reduction in Africa: commitment that "No one shall die from malaria"

Mohammadkhani M, Khanjani N, Bakhtiari B, Tabatabai SM, Sheikhzadeh K (2019a) The relation between climatic factors and malaria incidence in Sistan and Baluchestan, Iran. SAGE Open 9(3):1–10. https://doi.org/10.1177/2158244019864205

Mohammadkhani M, Khanjani N, Bakhtiari B, Tabatabai SM, Sheikhzadeh K (2019b) The relation between climatic factors and Malaria incidence in Sistan and Baluchestan, Iran. SAGE Open 9(3). https://doi.org/10.1177/2158244019864205

Munier A, Diallo A, Sokhna C, Senghor P, Ba F, Chippaux JP (2008) Évaluation de la fiabilité du diagnostic clinique de paludisme chez les enfants consultant dans trois postes de santé périphériques, dans la zone rurale de Niakhar au Sénégal. Bull Soc Pathol Exot 102(2):113–118. https://doi.org/10.3185/pathexo3331

Nili S, Asadgol Z, Dalaei H, Khanjani N, Bakhtiari B, Jahani Y (2022) The effect of climate change on malaria transmission in the southeast of Iran. Int J Biometeorol 66(8):1613–1626. https://doi.org/10.1007/s00484-022-02305-2

Oheneba-Dornyo TV, Amuzu S, Maccagnan A, Taylor T (2022) Estimating the impact of temperature and rainfall on malaria incidence in Ghana from 2012 to 2017. Environ Model Assess 27(3):473–489. https://doi.org/10.1007/s10666-022-09817-6

Safeukui-Noubissi I, Ranque S, Poudiougou B, Keita M, Traoré A, Traoré D, Diakité M, Cissé MB, Keita MM, Dessein A, Doumbo OK (2004) Risk factors for severe malaria in Bamako, Mali: a matched case-control study. Microbes Infect 6(6):572–578. https://doi.org/10.1016/j.micinf.2004.02.007

Samaké DS (2021) Facteurs météorologiques et variations temporo-spatiales du paludisme dans le district sanitaire de Kolondiéba, Sikasso, Mali de 2017 à 2020. Université des Sciences, des Techniques et des Technologies de Bamako

Segun OE, Shohaimi S, Nallapan M, Lamidi-Sarumoh AA, Salari N (2020) Statistical modelling of the effects of weather factors on malaria occurrence in Abuja, Nigeria. Int J Environ Res Public Health 17(10):1–12. https://doi.org/10.3390/ijerph17103474

Sena L, Deressa W, Ali A (2015) Correlation of climate variability and malaria: a retrospective comparative study, Southwest Ethiopia. Ethiop J Health Sci 25(2):129–138. https://doi.org/10.4314/ejhs.v25i2.5

Thomson MC, Ukawuba I, Hershey CL, Bennett A, Ceccato P (2017) Using rainfall and temperature data in the evaluation of national malaria control programs in Africa. Am Soc Trop Med Hygiene 97(Suppl 3):32–45. https://doi.org/10.4269/ajtmh.16-0696

Wardrop NA, Barnett AG, Atkinson JA, Clements AC (2013) Plasmodium vivax malaria incidence over time and its association with temperature and rainfall in four counties of Yunnan Province, China. Malaria J 12(1):1–11. https://doi.org/10.1186/1475-2875-12-452

World Health Organization (2023) World Malaria Report. In: World Health Organization. WHO/HTM/GM. http://www.who.int/malaria/world_malaria_report_2010/en/index.html

# Chapter 2
# Climate Change Impacts on Children's Health in Nigeria

**Bolanle Kayode**

**Abstract** *Background*. The growing bodies of children are highly susceptible to the adverse effects of the changing climate, yet less priority, if any, is given to adaptive responses that mitigate their exposure to the climate change-induced risks. Despite this vulnerability, the focus of successive governments on the issue of climate change has been largely confined to economic benefits/gains with scant consideration for the impacts on human health. It is a gap that needs to be addressed. *Methodology*. The qualitative realist approach was adopted, involving review and information analysis of environmental hazards associated with the changing climate, which enabled an outline of how these impact on children's health, including the identification of mitigation measures. *Outcomes*. This study calls attention to how the changing climate fueled hazards impacts the health of children. Recommendations on adaptive responses are provided and implications of study for government and/or policy makers are highlighted. *Conclusion*. Mitigation strategies and measures that minimise the impacts of climate change hazards on children's health need to be prioritised, given their vulnerability. Recommendations are offered on steps that could be taken to reduce the impacts of climate change harmful environmental hazards on children's health.

**Keywords** Climate change · Children · Health · Mitigation · Adaptation · Resilience

## Introduction/Background

The changing climate and its associated risks that include excessive heat, flooding, erosion, desertification, drought, wildfire, air pollution etc.—undoubtedly have negative impacts on human health, especially children. Nigeria ranks seventh in the list of

B. Kayode (✉)
Faculty of Social Sciences, Department of Sociology, Bayero University, Kano, Nigeria
e-mail: bolanle.kayode@yahoo.co.uk



W. Leal Filho and F. Wolf (eds.), *Climate Change and Health: Perspectives from Developing Countries*, Climate Change Management,
https://doi.org/10.1007/978-3-031-93177-2_2

the world's top countries associated with gas flaring activities that involve the emission of carbon dioxide and release of methane (World Bank 2021), both of which currently account for about 80% of global warming. In their analysis of children's exposure to climate change hazards, the Children's Climate Risk Index (CCRI) ranks Nigeria as the second most vulnerable country in the world (Leadership Newspaper 2023). The United Nations International Children's Emergency Fund (UNICEF) also found that not less than 113, 730,000 children in Nigeria are affected by climate change, being a statistic that represents 51% of the 223 million people in the country (Punch Newspaper 2024). Largely due to global warming, climate change has not developed without negative consequences for human health. In Nigeria, though there may not have been reported severe climate change-fueled disasters such as volcanos, tornados, tsunami, earthquakes, or hurricanes, the country nevertheless has been experiencing adverse impacts of environmental hazards associated with the changing climate (Ovuakporaye et al. 2012). Excessive heat, flooding, storming weather, drought, water borne diseases, air pollution, wildfire smoke, etc. that are commonly associated with climate change are not only uncommon in nigeria, but have negative impacts on human health, especially the children (Ezezue et al. 2018). The impacts are not limited to disruption of shelters, access to medical care, clean water, and increased food insecurity but also include direct exposure to pollutants, exacerbation of heat-related illnesses and sometimes, chronic illnesses. Flooding, for instance, is known to trigger water supply disruptions that could make children vulnerable to displacement, homelessness, drowning, loss of close family members, and exposure to infectious diseases, while hot temperature/excessive heat is known to impact the ability of children to learn, affect sleep patterns, and lessen their concentration in the classroom or in completing their homework. Research (Killgore 2010) has evidenced a connection not only between a lack of sleep and cognitive disruption but also learning difficulties.

For millions of children living in communities where oil extraction and other gas flaring are common activities, the adverse effects of pollution for their health can be worse due to emissions into their physical environment (Ayoola 2011). Extreme weather conditions such as hot temperature, flooding, and pollution are also not unusually connected with the intensity and frequency of greenhouse gas emission in the oil producing regions (Ajugwo 2013). In addition to the raging flame and loud noise that are associated with oil and gas extraction, the black carbon and methane, both of which are climate forcers, including volatile organic compounds known to be dangerous air pollutants are being emitted into the living environment with negative impacts for human health. The failure to protect children from climate-induced hazards is not unlikely to neglect their exposure to dangers that hinder their ability to thrive/flourish, therefore underscoring the need for mitigation strategies. The significant public health implications of this study are threefold:

(1) To create awareness on the impacts of climate change-induced hazards on children's health
(2) To identify potential responses that mitigate risks associated with climate change induced hazards on children's health

(3) To reveal the implications for government and/or policy makers.

## Method

This study delves into the impacts of climate change on children's health in Nigeria with the aim of identifying adaptive responses that mitigate the adverse impacts of climate change hazards on their health. The study adopted a realist qualitative approach, underpinned by the assumption that the 'same intervention will not work everywhere or for everyone' (Wong et al. 2016). In its striking difference from other research methodologies, the realist qualitative approach allows for the conceptualisation and conduct of qualitative studies using 'constructs'. As much as society may wish to construct a 'one-cap-fits all' approach in addressing impacts of climate change for different subpopulations, the possibility of attaining such a singular approach is rare or perhaps impossible. In particular, the unique vulnerability of children's health, due to their less ability to remove themselves from harm reinforces the need for adaptive responses to mitigate their exposure to climate change-induced hazards. In adopting the realist qualitative methodology, this study: (a) uses electronic and non-electronic data bases to gather information on the research topic; (b) reviews and evaluates what is already known about climate change-fueled hazards, especially the impacts for children's health; and lastly (c) recommends adaptive responses to mitigate the exposure of children's health to the environmental hazards associated with climate change.

## Conceptual Issues

The impact of the changing climate on health is widespread, and children are most susceptible given their physiological uniqueness, in addition to their limited understanding about the adverse effects of climate change hazards on health in comparison to adults. A child may be unable to remove him- or herself from harm arising from climate-induced hazards and unaware of the potential lifelong consequences they often bear on health and wellbeing. Oftentimes, children may not understand the dangers of unpleasant weather conditions such as dusty or smoky environments and are unlikely to notice when the quality of air is poor, such that they breathe in pollutants. Different regions of oil-rich Nigeria may arguably be disproportionately exposed to climate change stressors, such that whilst in oil producing areas such as the Niger Delta, there could be higher levels of emissions from gas flaring and oil extraction activities, while climate change stressors in the northern region could include extreme temperature, drought, and flooding associated with seasonal changes that make the physical environment unfriendly (Information Portal 2019). Notwithstanding these disproportionalities, children can be more greatly exposed to these hazards due to their vulnerability.

Meanwhile, the focus of successive governments on the issue of climate change has been largely confined to economic benefits/gains with scant consideration for the impacts on human health. The national statement of the Nigerian government during the 28th United Nations Climate Change Conference (aka COP28) in Dubai in December 2023 mainly generalised initiatives to reduce the carbon footprint in the country (e.g., deploying fleets of electric buses, banning gas-flaring, etc.). How the changing climate adversely affects human health and wellbeing, especially children, and mitigation initiatives that may reduce their vulnerabilities to these adversities are neither being discussed nor prioritised for solutions.

In the sixty-page National Action Plan on Gender and Climate Change for Nigeria (2020), one of the stated objectives is to increase the understanding of climate change impact and promote mitigation initiatives that minimise risks associated with climate change among youths and other vulnerable groups (4.1.2, sub-section i & v). However, adaptive responses that may help reduce the effects of climate stressors for this vulnerable group are not available. There are no clear action plans to mitigate the susceptibility of children to the adverse impacts of climate change therefore a gap that needs to be addressed. Against this background, this paper asks the question: Do climate change hazards impact the health of children in Nigeria, and if yes, how? The paper recommends adaptive responses to mitigate the effects of climate change fueled hazards on children's health.

## Review of Studies on the Changing Climate and Impacts on Human Health

In their study on resilience planning for climate hazards (i.e. heat waves, heavy downpours, flooding, pollution etc.) and health impacts with two tribal governments in Virginia, Hutton et al. (2023) identified that the impacts of environmental hazards on human health have been increasing and will continue to increase due to the changing climate (Hutton et al. 2023). Identifying the adverse effects on human, the authors indicated that the hazards induced by the changing climate impact health directly (i.e., heat-related illnesses, exposure to pollutants, exacerbation of chronic illnesses, etc.) and indirectly (i.e. food insecurity, limited access to medical care, and loss of clean water, shelter, and sanitation, etc.), in addition to disruptions caused to infrastructures. The authors emphasised that there is no 'one-size-fits-all' approach to mitigating the adverse effects of climate-induced hazards given the non-uniqueness in communities or groups reinforces the significance of distinctive mitigation and resilient initiatives to reduce the impacts of climate change-induced hazards on the health of children.

The Environmental Protection Agency in the United States provided a report (EPA 2023) that highlighted five climate-related environmental hazards associated with children's health and wellbeing: extreme heat, poor air quality, changes in seasonality, flooding, and infectious diseases. Spolighting the uniqueness of children's vulnerability to climate change-induced hazards the significance of having mitigation strategies that addresses these impacts on their health was emphasized by the report (EPA 2023).

Lal et al. (2019) had earlier argued that conditions such as changes in temperature that promote bacterial growth, including poor management of wastewater, are commonly associated with illnesses that can lead to death in children, if left unattended.

Ezezue et al. (2018), in their research on the 'Effective Management of Flood in Nigeria', likewise acknowledged the increasing change in climate that may be connected with flooding incidents. In their results analysis, the authors found that despite the vulnerability to flooding in the communities studied, there was no proper flood management hence, consequences that included submergence (of buildings, farmlands and livestock), death, and increased poverty due to flooding.

These strands of research demonstrate an emerging consensus that climate change-induced hazards have impacts on human health. Children in particular are susceptible to the hazards, therefore underscoring the need to have practicable action plans that include mitigation initiatives with immediate and long-term benefits to reduce exposure to climate change-fueled hazards. This paper will now focus on climate change-induced hazards and the impacts these could have for children's health in Nigeria. Mitigation initiatives to reduce the impacts of climate change-induced hazards on children's health, as well as implications for government and policy makers are discussed.

## Climate Change Induced Hazards and Impacts for Children's Health

Among the many harmful environmental hazards associated with the changing climate and having negative impacts on the health of children are desertification, drought, flooding, hot temperature, pollution and food insecurity. These climate change-associated hazards and their impacts for children's health are discussed as follows:

**Desertification** The changing climate, coupled with human actions such as tree felling and burning to make charcoal have been contributing to desertification in Nigeria, especially in the northern part of the country and no doubt with unfavourable impacts. Desertification is not unusually associated with the loss of biological diversity, alteration in the geo-chemical composition of the soil, water scarcity, and reduction in agricultural yields, all of which are contributory factors to food insecurity, reduced economic growth, and disease burdens. Wildfire activity, commonly connected to desertification releases smoke which produce a complex mixture of particulates and organic compounds that threaten human health and can make children prone to illnesses that include asthma, lung infection, airborne diseases, and other respiratory conditions with lifelong impacts. Gladwell et al. (2013) found that hot temperatures often associated with desertification may result in low school attendance for children and affect their participation in outdoor or recreational activities, thereby hampering their interaction with nature that has the potential to make them less vulnerable to childhood stress and improve their mental health.

**Drought** Exacerbation of water stress by climate change whereby there is extreme limited water resources, leading to increased competition and, even conflict is not uncommon in many households with children in Nigeria. Due to water shortages, drought increases the use of contaminated water that poses a significant threat to the lives of children with the likely increase in sanitation-related diseases, considered to be a leading cause of death in young children. According to research (UNICEF 2023), inadequate water, sanitation, and hygiene-related diseases continue to be responsible for the deaths of around 400,000 children under the age of five each year or 1000 every day. Disrupting weather patterns especially in the northern part of the country, are often related to the exacerbation of water scarcity, period of low or limited rainfall, hot temperatures can including contamination of water supplies, all of which may impact the quantity and quality of the water that children need to survive. Shortages of water, poor quality drinking water, dry and dusty conditions, and poor air quality are all associated with drought. Drought can additionally lead to a shortage of land to grow crops thereby negatively affecting food production and nutrition in children. A national emergency of food security was declared by the federal government of Nigeria in July 2023 due to an apparent inconsistent and unrealistic access to enough food for active healthy living (Kayode 2023). Yobe and some other states especially in the northeast and central parts of Nigeria have experienced drought as well as an invasion of quella birds and locust that destroy plants/crops. Drought can often lead to adverse plant responses, which leads to reduced crop, forage yields, or total forage failure but may also disrupt food production and distribution, limit food availability, and increase the cost of food items, all of which can make it harder for children to get healthy food and adequate nutrition for their physical and mental growth. An additional threat to human health is the common use of grass pollen by farmers during drought, whereby a variety of grasses are used to fertilize plants with similar species. According to Meltzer et al. (2009), some grasses and trees are known to produce aeroallergens that can potentially trigger health concerns such as allergies and respiratory diseases such as asthma with poorer health outcomes in children.

**Flooding** Though floods are considered a natural disaster with far reaching effects on people, the changing climate may be exacerbating its occurrence. In Nigeria, tornados and flooding due to effects of climate change such as heavy rainfall are becoming increasingly common, with associated risks that include pluvial flooding, riverine flooding, overrunning of water banks, and damages to homes, infrastructures, etc. More than thirteen flood-prone states in Nigeria (Adamawa, Zamfara, Abia, Benue, Akwa-Ibom, etc.) were affected by flood, and in some cases, storm surge incidents that caused damages to properties and injuries to people when the Cameroonian government opened the flood gate of Lagdo dam, following excessive rainfall in the last quarter of 2023. As a result, family homes were damaged, farmlands were destroyed, and many lost their accommodations or belongings due to submergence (National Emergency Management Agency, Zamfara 2023). The experience left children temporarily displaced, and reported increase in the level of injuries resulting from falls, especially children. Mould also built up in homes, which not only increased the vulnerability of children to respiratory diseases such as asthma but also exacerbation of unsanitary conditions that led to diarrhoea (National Emergency Management Agency, Zamfara 2023). According to Chakraborty (2019), anxiety or worries over housing uncertainty among children and/or their exposure to family members' stress over damage to property and belongings are likely to affect their mental health.

The overflow of riverbanks due to extreme weather conditions, such as heavy rainfall during the wet season, is an additional hazard in some flood-prone states and poses significant risks to human health, especially children. In Adamawa state, the heavy downpour that precipitated a large release of water from Lagdo dam caused a large displacement of over 8594 households with more than 51,043 individuals, the majority of whom were women, children, and the elderly (National Emergency Management Agency, Adamawa 2023). When flooding occurs, it hinders people's access to their homes, cause huge losses in the agricultural sector, jeopardize the growth of newly planted and sprouting crops with consequences for food availability, while damages are also caused to significant infrastructures such as roads, shelter facilities, health facilities, IDP sites, and schools. For children living close to areas with a high level of oil production activities, flooding increases their exposure to chemical and biological pollutants and poses additional risks such as ingesting or drinking contaminated water, with poorer outcomes for their safety, health and general wellbeing. Runoff from flooding can oftentimes introduce disease-carrying organisms into water bodies and contaminate drinking water supplies in the communities, with children having a higher risk of illnesses. Additionally, flooding can bring about sewage spills that can be hazardous to health, especially children, and cause allergy reactions due to bacterial mould and viruses (National Oceanic and Atmospheric Administration).

**Hot Temperature** The changing climate has been leading to extended warm seasons and shorter cool seasons (U.S Environmental Protection Agency). According to a report (EPA 2023), climate change impacts are generally expected to become worse as the Earth continues to warm. Nigeria largely contributes to the changing climate through the emission of carbon dioxide from gas flaring and methane releases from gas venting, both of which make up a significant percentage of global warming (World Bank 2021). As opined by Gobler (2020), increased growth of bacteria as a result of warming water temperature is not unconnected to warm temperatures, with the potential for vulnerability to waterborne hazards. While different regions in Nigeria are disproportionally affected by excessive heat, studies (Limaye et al. 2018) show that warm temperatures can and do have a wide range of health impacts on human health regardless of the locations/regions and are particularly pronounced in children, further worse for those with pre-existing conditions. Xu et al. (2014) evidenced the connections between children with pre-existing conditions and their vulnerability to the effects of heat. Heat illnesses from extreme temperature have reportedly led to death in children, with infants being the most vulnerable as their bodies are less able to tolerate heat. Severe heat can have multiple effects on children's wellbeing, as for instance, it can be uncomfortable for their learning, lead to poor school attendance, decrease motivation to complete school work, affect their concentration in the classroom, lead to poor quality of sleep, cause skin infection such as eczema, increase risk of heat stroke among young athletes, and cause anxiety, all of which may in turn make children vulnerable to poor mental health conditions. Burke et al. (2018) found that children may experience psychological or cognitive effects from exposure to stress following or preceding severe weather that includes extreme temperature. Extreme temperature is also commonly being associated with increasing water temperatures that encourage the growth of harmful bacteria that can make people especially, children sick. Warmer temperatures, for instance do not only increase mosquito breeding but also other pest outbreaks and the biting rates, which may cause illness and fatal sickness, if not treated promptly. In Nigeria, over 250,000 children reportedly die every year from malaria disease (Westfield Development 2023), while the World Health Organization also made a declaration in 2023 that African regions carry a higher proportion of the global malaria burden, and children under five years account for about 80% of all malaria deaths in the region (WHO 2023).

**Pollution** As a result of oil flaring activities that emit black carbon, which is a component of fine particulate matter that has negative effects on human health, Nigeria is considered to be contributing to over 7 million deaths a year from air pollution all over the world (UNEP 2023). Oil flaring is associated with oil exploration processes and though this is considered a dangerous practice with severe human health and environmental effects, it is practiced in oil producing communities involved in oil refinery activities, oil rigs, oil wells, landfills, etc. According to research (Burke et al. 2021), gas flare is associated with the release of toxic chemicals that not only include black carbon with impact for the climate, but in addition a dangerous pollutant with adverse effects for human health. Children are particularly at risk for respiratory health concerns from air pollution, that includes inhaling poor air quality, choked up feelings, oxygen starvation, inadequate air in the lungs, respiratory infection, asthma, oxidative stress etc., all of which may lead to panic attacks, lung injury or even death in extreme cases (Mathiarasan and Hüls 2021). The exposure of children to pollutants from wildfires or fire disasters, not uncommon in Nigeria due to hot temperatures and deforestation, may also have consequences for children's health, including respiratory illnesses, reduced lung functioning, poor sleep patterns, etc. According to Garcia et al. (2021) wildfires are known to increase particulate matter and toxic gas concentrations, both of which have the potential to trigger or worsen asthma, most especially for the vulnerable population.

**Food insecurity** Agriculture is the major source of food for Nigeria's teeming population and, therefore, a significant factor in human development and sustenance, especially for the physical and mental growth of children. However, the changing climate has been negatively affecting the sector. Poor growth in the agricultural sector has led to the current food insecurity in Nigeria and is not unconnected to climate change environmental hazards, with implications being constraints in the production of livestock, increase in crop failure, rising cost of food items due to scarcity, etc. In other words, the lack of purchasing power by many Nigerians due to current economic hardship may not be the only reason for being dood insecure, but also as a result of climate induced hazards that impacts the level of food production. Adverse or irregular weather conditions leading to desertification, drought, and flooding are linked to the changing climate and impact how much food is available for consumption. Hutton et al. (2022) linked food insecurity to chronic diseases such as diabetes, hypertension, and heart disease. It is common in Nigeria for soil to get eroded by large amounts of flowing water during heavy rainfall in the wet season, ruining crops, destroying agricultural products, and threatening wildlife. In some regions of the country such as the northwest and northeast, the non-availability of forest and water as a result of drought associated with extended periods of low or limited rainfall and high temperature prevents the cultivation of enough food, with adverse impacts for the most vulnerable that includes children. Additionally, the atmospheric contaminants that are popularly associated with gas flaring, especially in the oil producing regions of the country, can acidify the soil, with adverse effects that include depletion of soil nutrients, and leading to stunted growth of crops, withering away of young crops, and decreased nutritional value. Research (Keim 2008) found an association between reduced accessibility to food and an exacerbation of negative impacts on population health.

**Illness/Infectious Diseases** Methane releases from the gas venting that is practised in Nigeria, especially in the oil producing communities, is associated with pollution, and so also are other oil exploration processes. Children, due to their growing bodies and limited ability to remove themselves from harm, are most susceptible to health impacts from induced hazards, be it physical, social, or psychological. In particular, the exposure of children to methane can lead to a reduction in the amount of oxygen breathed from the air and may cause slurred speech, vision problems, nausea, vomiting, headache, or even memory loss (Public Health England 2019). New babies born into an environment with poor air quality may also likely develop asthma that makes them prone to frequent illnesses, some of which may have lifelong effects on their behaviour and/or learning. Research (Ovuakporaye et al. 2012) has found an association between pollutants and negative health impacts that included deformities in children, skin problems, and lung damage. Furthermore, the rapid breeding of insects, water contamination, unsanitary conditions from flooded latrines, etc., are associated with severe weather conditions, and may trigger airborne and/or waterborne diarrhoeal diseases but also, generally hazardous to human health with long term ramifications that include death if not properly treated, with children being most susceptible. Lal et al. (2019) likewise opined that conditions that promote bacterial growth and poor management of waste water, such as changes in temperature are connected to poor health.

## Adaptation Responses to Mitigate Impacts of Climate Hazards on Children's Health

As countries all over the world are pledging commitments to limit the emissions of greenhouse gases contributing to climate change, proactive steps also need be taken to reduce the impacts of its associated hazards on human health. This is especially for children given they are more susceptible to these hazards than the adult population. The neglect of children's health exposure to climate-induced hazards can lead to poorer health outcomes that hinder their ability to grow properly and thrive. The below highlighted adaptation responses, though not an exhaustive list, may mitigate the impacts of climate induced hazards on children's health:

(a) During extreme heat, parents and guardians should encourage children to drink enough water, wear light clothes to prevent overheating, plan outdoor play for when the temperature is cooler, avoid leaving children in packed cars, and teach children the importance of of using shade areas such as under trees or canopies to rest or cool off.
(b) Keep accommodations and learning spaces well ventilated. This strategy can protect children from hot temperature and mitigate impacts of this climate change hazard on their health. Strategies that include the building of walls to prevent flooding, temporary sites to serve as shelter when there are flood incidents, attention to advance warnings, etc. will not only promote the safety of children during climate change adversities but help to reduce their exposure to hazards.
(c) For families living in areas prone to climate change-induced hazards such as flood, consider having a contingency plan for safety. The contingency plan may, for instance, focus on where the family may find temporary accommodation in case of home submergence, what the family may do to protect children from wading in the floodwaters, how the family will continue to have access to clean drinking water, etc. Help children to understand what this means, such that if or when the hazard occurs, they know what the family will do. Adopting these strategies can reduce anxiety, trauma, or stress associated with the exposure to or experience of climate change hazards and is, therefore, a way to promote their safety and strengthen their resilience.
(d) Discourage children from playing outside when the air quality is poor, especially when there has been wildfire or in places with heavy smoke. When a child has been affected by severe weather conditions, such as contracting diarrhoea from drinking contaminated water or developing respiratory ill-health due to poor air quality, parents or caregivers should ensure that they promptly access medical attention.
(e) Hospital visits increase when the climate gets warmer, therefore adequate planning by parents/caregivers can help ensure that children receive support and attention at the hospital, when required. This is more so that hospitals may be prioritising acute cases given the current shortages in medical personnel, as many are quitting for better opportunities in the western countries.

(f) At the state level, strengthening the preparedness and response capacities of at-risk communities is likely to mitigate the adversities associated with climate-induced hazards, especially for children who are considered the most vulnerable. For instance, the various branches of the State Emergency Management Agency (SEMA) could have a contingency plan for the evacuation and sheltering of people during climate-induced disasters. The plan may also include devising strategies for mobilising non-food items such as mattresses, blankets, and hygiene kits to reduce health risks.

(g) Collaboration with stakeholders such as the media and local organisations including religious bodies can enhance social engineering in the protection of children's health from climate-induced hazards. The collaboration may help sensitise the general public on the need to have clear waterways/channels, promote understanding on the benefits that flood-proof structures could offer, provide enlightenment on how interior barrier constructions may prevent recurrent incidents in flood prone areas etc.

## Implications for Government and/or Policy Makers

1. Given the provisions within the National Action Plan on Gender and Climate Change for Nigeria (paragraph 4.1.2, sub-section i & v), it becomes paramount to promote mitigation initiatives in minimizing risks associated with climate change-fueled hazards that impact the health of children. Devising ways to educate children and others such as teachers, parents, and guardians on the importance of protecting children from climate change-induced hazards is crucially important. England, for instance, recently set a target for all educational settings to have a climate action plan in place by 2025, part of which commits students (including children) to learning about the potential effects of climate-change induced hazards. It is a good policy that can be emulated.
2. Consider embarking on sensitisation campaigns to ensure unobstructed drainage systems as well as having contingency plans (aka emergency response plan) for the immediate relocation of residents, as this may mitigate the effect that flooding can have on people, especially the more vulnerable children.
3. The current economic hardship in Nigeria makes it impossible for citizens to access resources such as tax credits from the government to install air conditioning systems in homes or schools (as is the practice in the developed countries) that reduce the impacts of harsh weather conditions. Although Nigeria lacks this capacity, there are other practicable mitigation strategies that can help to reduce the impacts of climate change on human health. The tree planting programme that government has already embarked on needs to be intensified, as it can improve afforestation, help to off-set heat, and provide cool-off points for children during extreme temperatures. Additionally, the programme may help to intercept and/or reduce surface runoff, therefore lessening the likelihood of flooding, for which children are most endangered.

4. Collaboration with the healthcare sector is essential to address the current syndrome of healthcare professionals abandoning their jobs for offers in the western countries. Impacts of hot temperatures and other extreme weather conditions on children's health are likely to increase their visits to emergency departments and/or to pediatricians for ill health concerns therefore, the need for a sustainable healthcare system that can meet their health needs.
5. Sensitising citizens to build their homes with materials that do not absorb heat, enlightening those living in flood-prone areas on the importance of elevating properties during construction as well as encouraging them to build sea/water walls with concrete materials that are resistant to erosion will likely mitigate some of the climate change-induced hazards that impact the health of children. The same is true for the promotion of strategies that reduce greenhouse gases, the implementation of anti-flaring policies for natural gas waste from oil production, including having an effective monitoring process that enhances compliance with regulations on emission. Taking these steps may help to protect children's health from environmental pollution and keep them safe.
6. Promotion of strategies that discourage human practices that contribute to the emission of greenhouse gases which warms the atmosphere can reduce the hazards associated with climate change and, most importantly, limit their adverse impacts on children's health. The decision taken by the government in December 2023 to exempt liquefied petroleum gas (LPG) from customs duty and VAT payment is a welcome development. This will likely increase the supply of liquefied petroleum gas (LPG) to meet local demands, reduce market prices for the product, discourage the popular practice of cooking with firewood, charcoal, or sawdust, and prevent the practice of burning trees for charcoal production, which is considered to be a major driver of deforestation with adverse impacts on children' health.
7. A further consideration to reduce the impact of climate-fueled hazards on the health of children could be improving town planning procedures to enhance good drainage systems that facilitate adequate outflow capacity, reduce the incidence of submergence, home loss, or even drowning incidents. For instance, dredging the waterways could prevent water overflow and provide mitigation for flooding. Therefore, enacting and enforcing laws and/or strict regulations that prevent people from building houses on water ways, curtailing the common practices of carrying out economic activities in unauthorised areas, preventing the mounting of illegal construction on lands along the river banks, etc. may all need to be prioritised.

## Discussions, Outcomes and Conclusion

This study delved into climate change-fueled hazards and their impacts on children's health in Nigeria. In answering the research question posed by study, the different ways that climate change-induced hazards impact children's health were discussed while recommendations were offered on adaptation responses to mitigate the hazards. In conclusion, children are more susceptible to adverse effects of the changing climate given their growing bodies, but little or no attention is being accorded at the moment to minimise the exposure of children's health to climate change-induced risks in Nigeria. This paper highlighted how the health of children is impacted by climate change-associated hazards and recommended adaptation responses to mitigate these negative impacts. If steps must be taken to protect children's health from the negative impacts of climate change, adaptation responses that reduce their exposure to the hazards associated with climate change must be prioritised for implementation. Whilst it is positive that Nigerian government is remaining optimistic that the Country would become carbon–neutral in the near future as predicted during the 'United Nations Climate Change Conference' (aka COP28) held in December 2023, it would be useful nonetheless that top priority be given to the implementation of adaptive responses to mitigate climate change hazards that impact on children's health, given their vulnerability. As the country is being financially supported by the Green Climate Fund (GCF) and Adaptation Fund to boost the implementation of climate plan and adaptation efforts, a sticking consideration should be that a portion is committed to resources that provide short and long- term mitigations to minimise the effects of climate change hazards on children's health.

## References

Ajugwo AO (2013) Negative effects of gas flaring: the Nigerian experience. J Environ Pollut Hum Health 1(1):6–8

Ayoola TJ (2011) Gas flaring and its implication for environmental accounting in Nigeria. J Sustain Dev 4(5):244–250

Burke SE, Sanson AV, Van Hoorn J (2018) The psychological effects of climate change on children. Curr Psychiatry Rep 20(5):1–8

Burke M, Driscoll A, Heft-Neal S, Xue J, Burney J, Wara M (2021) The changing risk and burden of wildfire in the United States. Proc Natl Acad Sci 118(2)

Chakraborty J, Collins TW, Grineski SE (2019) Exploring the environmental justice implications of Hurricane Harvey flooding in Greater Houston, Texas. Am J Publ Health 109(2):244–250

EPA (2023) Climate change and children's health and well-being in the United States. U.S. Environmental Protection Agency, EPA 430-R-23-001

Ezezue BO, Agha NC, Ndieze J, Dickson RS (2018) Effective management of flooding in Nigeria: a study of selected communities in Anambra State. https://www.semanticscholar.org. Corpus ID: 234363036

Garcia E, Rice MB, Gold DR (2021) Air pollution and lung function in children. J Allergy Clin Immunol 148(1):1–14

Gladwell VF, Brown DK, Wood C, Sandercock GR, Barton JL (2013) The great outdoors: how a green exercise environment can benefit all. Extreme Physiol Med 2(1):1–7

Gobler CJ (2020) Climate change and harmful algal blooms: insights and perspective. Harmful Algae 91:101731

Hutton NS, McLeod G, Allen TR, Davis C, Garnand A, Richter H, Romero C (2022) Participatory mapping to address neighborhood level data deficiencies for food security assessment in Southeastern Virginia, USA. Int J Health Geogr 21(1). https://doi.org/10.1186/s12942-022-00314-3

Hutton NS, Yusuf W, Palma J (2023) Resilience planning with two tribal governments in Virginia for climate hazards and health impacts. Natural Hazards Center Public Health Disaster Research Report Series, 30. University of Colorado Boulder. https://hazards.colorado.edu/public-health-disaster-research/resilience-planning-with-two-tribal-governments-in-virginia-for-climate-hazards-and-health-impacts

Kayode BO (2023) Food insecurity in Nigeria—call for palliative inclusiveness. Open access. https://www.researchgate.net/profile/Bolanle-Kayode

Keim ME (2008) Building human resilience: the role of public health preparedness and response as an adaptation to climate change. Am J Prev Med 35(5):508–516

Killgore WD (2010) Effects of sleep deprivation on cognition. Prog Brain Res 185:105–129

Lal A, Fearnley E, Wilford E (2019) Local weather, flooding history and childhood diarrhoea caused by the parasite *Cryptosporidium* spp.: a systematic review and meta-analysis. Sci Total Environ 674:300–306

Leadership Newspaper (2023) https://leadership.ng/unicef-warns-of-climate-risk-on-nigerian-children

Limaye VS, Vargo J, Harkey M, Holloway T, Patz JA (2018) Climate change and heat-related excess mortality in the Eastern USA. EcoHealth 15(3):485–496

Mathiarasan S, Hüls A (2021) Impact of environmental injustice on children's health—interaction between air pollution and socioeconomic status. Int J Environ Res Public Health 18(2):795

Meltzer EO, Blaiss MS, Derebery MJ, Mahr TA, Gordon BR, Sheth KK, Simmons AL, Wingertzahn MA, Boyle JM (2009) Burden of allergic rhinitis: results from the Pediatric Allergies in America survey. J Allergy Clin Immunol 124(3):S43–S70

National Action Plan on Gender and Climate Change for Nigeria (2020) https://www.climatechange.gov.org

National Emergency Management Agency, Zamfara (2023) https://www.msn.com/en-xl/news/other/flood-wreaks-havocdestroys-residential-buildings-farmlands-in-zamfara/ar-AA1pLgys

National Emergency Management Agency, Adamawa (2023) https://www.nema.gov.ng/nema-allay-fears-of-nigerians-overrelease-of-excess-water-from-lagdo-dam-in-cameroon/

National Oceanic and Atmospheric Administration. National Weather Service (NWS) Preliminary US Flood Fatality Statistics. https://www.weather.gov/arx/usflood. Accessed on: 24 Jan 2024

Nigerian Government, Information Portal (2019) http://www.nigeria.gov.ng/index.php/. Accessed 18 Dec 2023

Ovuakporaye SI, Aloamaka CP, Ojieh AE, Ejebe DE, Mordi JC (2012) Effects of gas flaring on lung function among residents of ib Gas flaring community in Delta State, Nigeria. Res J Env Earth Sci 4(5):525–528

Public Health England Centre for Radiation (2019) Chemical and Environmental Hazards

Punch (2024) http://punchng.com/110m-nigeria-children-face-climate-change-challenge-unicef/

UNICEF (May, 2023) Triple threat: how disease, climate risks, and unsafe water, sanitation and hygiene create a deadly combination for children. https://www.unicef.org

UNEP (September 2023) Pollution action note—data you need to know. https://www.unep.org

U.S. Environmental Protection Agency (2024) Environmental justice. Available at https://www.epa.gov/environmentaljustice. Accessed: 26 Mar 2024

Westfield Development (2023) https://www.westfielddevelopment.org/malaria-in-nigeria/

WHO (2023) https://www.who.int/news-room/fact-sheets/detail/malaria

Wong G, Greenhalgh T, Westhorp G, Manzano A, Greenhalgh J, Jagosh J (2016) RAMESES II reporting standards for realist evaluations. BMC Med 14(1)

World Bank (2021) Global gas flaring tracker report. https://thedocs.worldbank.org/en/doc/1f7221545bf1b7c89b850dd85cb409b0-0400072021/original/WB-GGFR-Report-Design-05a.pdf

Xu Z, Sheffield PE, Su H, Wang X, Bi Y, Tong S (2014) The impact of heat waves on children's health: a systematic review. Int J Biometeorol 58(2):239–247

# Chapter 3
# Navigating Healthcare Systems Challenges Amidst Climate Change Impacts in Coastal Bangladesh

**Byomkesh Talukder, Krishna Prosad Mondal, Md. Mehedi Hasan, Monishankar Sarkar, Reza Salim, and Tawhidul Islam**

**Abstract** Climate change is having many impacts on Bangladesh's coastal regions, including sea level rise, salinity intrusion, floods, cyclones, and coastal erosion. These changes are seriously disrupting local ecosystems and putting a lot of strain on the area's healthcare systems. As people in these coastal areas try to deal with these problems, their access to basic health services is becoming increasingly difficult. This issue affects not just individuals, but the whole healthcare system, including public and private sectors, disease management, government efforts, and support from various organisations. Even though these risks are urgent, there has not been enough research on how climate change specifically affects healthcare in coastal Bangladesh. This study uses a variety of methods, including secondary data, field observations, interviews with experts, and GIS techniques to identify four main healthcare challenges: (1) weaknesses in infrastructure; (2) limited access to health services; (3) issues with water, sanitation, and hygiene (WASH) programmes; and (4) challenges in preventive healthcare. The study's findings provide useful insights into the situation in coastal Bangladesh and can help guide the development of effective

B. Talukder (✉)
Global Health, Florida International University, Miami, USA
e-mail: btalukde@fiu.edu

K. P. Mondal · T. Islam
Institute of Remote Sensing and GIS, Jahangirnagar University, Dhaka, Bangladesh
e-mail: s.t.islam@juniv.edu

Md. M. Hasan
Department of Geography, Dhaka College, Dhaka, Bangladesh

M. Sarkar
Institute of Disaster Management and Vulnerability Studies, University of Dhaka, Dhaka, Bangladesh
e-mail: ms@du.ac.bd

R. Salim
Amader Gram Cancer Care & Research Center, Bagerhat, Bangladesh
e-mail: rezasalim@bfes.net



W. Leal Filho and F. Wolf (eds.), *Climate Change and Health: Perspectives from Developing Countries*, Climate Change Management,
https://doi.org/10.1007/978-3-031-93177-2_3

policies and interventions. These recommendations are also relevant for other coastal areas facing similar vulnerabilities worldwide.

**Keywords** Climate change · Disaster risks · Healthcare vulnerabilities · Coastal Bangladesh

## Introduction

The dynamic coast of Bangladesh is highly vulnerable to a range of rapid and slow-onset events, including sea level rise, coastal flooding, erosion, and cyclones, all of which disrupt the socio-ecological systems in the region (GOB 2018; Ashrafuzzaman et al. 2022; Rahman et al. 2022; Talukder et al. 2023; Adshead et al. 2024). While Bangladesh was ranked seventh by the Global Climate Risk Index as one of the countries most affected by climate change between 1999 and 2018 (Eckstein et al. 2021), its coastal region is the most vulnerable. This vulnerable region covers approximately 47,211 $km^2$ (Moslehuddin et al. 2015). From an administrative perspective, 19 of Bangladesh's 64 districts are classified as coastal districts (GOB 2018). According to the 2011 Population Census, this zone is home to around 35 million people in 6.85 million households (BBS 2012).

Climate change is making cyclones more intense and more common in Bangladesh. This is a major concern for its low-lying coastal areas, which are particularly vulnerable to storm surges (Dasgupta et al. 2010). For example, in 2024, Cyclone Remal brought heavy rain and strong storm surges, causing severe 5- to 8-foot flooding in several coastal districts. Khulna, Satkhira, and Pirojpur were among the hardest-hit areas (IFRC 2024).

Coastal Bangladesh is also grappling with serious issues caused by rising sea levels, a problem mainly driven by climate change. Experts predict that sea levels could climb nearly a meter by 2100 (Ashrafuzzaman et al. 2022). This rise poses major risks to the environment, the economy, and millions of people in low-lying coastal areas. The increase in sea levels will greatly affect coastal communities, damage infrastructure, and impact public health. As the sea encroaches, it will also make the water saltier, harming health and weakening infrastructure resilience in these areas.

Coastal flooding, driven by rising sea levels and frequent cyclones, has serious implications in damaging infrastructure, displacing communities, and impacting critical services such as health, education, electrification, and economic stability (Smith et al. 2021; Adshead et al. 2024). For example, major cyclones such as Sidr and Aila have had severe impacts on vulnerable groups like women, children, and the elderly population and have destroyed infrastructure and socio-ecological networks (Kabir et al. 2016; Roy 2020). Cyclone Remal in 2024 poses health risks, including potential outbreaks of diseases due to standing water and damaged sanitation facilities. Vulnerable populations, including pregnant women, face additional challenges in accessing healthcare (ReliefWeb 2024).

Despite the many ways climate change affects health and healthcare systems in coastal Bangladesh, these challenges are still not fully understood. This chapter examines the different issues coastal healthcare systems face due to climate change impacts in the region.

## Methodology

This study is based on both primary and secondary data. The primary data comes from the author's extensive observations in the southwest coastal region of Bangladesh. The authors have been engaged in research and NGO activities in this area for over 25 years. For example, Reza Salim, the fourth author, has run the Amader Gram Cancer Care & Research Center for over 25 years, focusing on innovative, transformative, and community-oriented healthcare services addressing cancer and other non-communicable diseases (NCDs) in southern Bangladesh. The second and third authors were born and raised in the southwest coastal region, providing them with firsthand experience of the impacts of cyclones and salinity on human health. They are currently involved in developing studies and research in coastal Bangladesh. The last author has led numerous research projects funded by international organisations such as the UNDP and the World Bank, focusing on the impacts of climate change. The first author has published an article on collective intelligence for addressing community and planetary health issues resulting from salinity prompted by sea level rise (Talukder et al. 2023), based on Bangladesh's southwest coast. All the authors have shared their experiences of the impacts of climate change on southwest coastal health systems. Their observations provide context-rich insights that are difficult to capture through other methods like surveys or experiments. In addition to onsite observations, the study aimed to visually assess the current health infrastructure and environmental conditions of coastal regions. Field visits were conducted to hospitals, clinics, and community health centres to evaluate the condition of the infrastructure, the availability of resources, and environmental challenges such as erosion and salinity.

Key informant interviews were conducted to validate the authors' insights on the impacts of climate change on health systems. The interviewees included 2 Upazila Health Officers, 3 community health workers, 3 pharmacists, an informal village health care provider, and members of communities affected by climate change: 5 elderly farmers who have faced multiple cyclones while engaging in agricultural activities, 3 college-level educators, 5 female college students, and 3 reporters from 1 local newspaper and 2 national newspapers. Before the interviews, informed consent was obtained (Fig. 3.1).

Secondary data were gathered through a narrative literature review of academic articles, government reports, and NGO publications related to the impacts of climate change on healthcare systems in coastal Bangladesh.

Geographic Information System (GIS) techniques were employed to identify health centres in areas vulnerable to sea level rise in coastal Bangladesh. The analysis

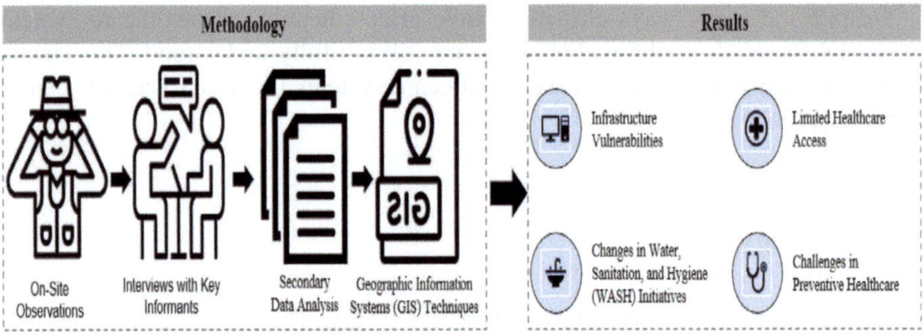


**Fig. 3.1** Research methodology and result

utilised three key datasets: administrative shapefiles, gridded population density data, and digitised healthcare data. The administrative shapefiles, representing divisional country boundaries, and river shapefiles were obtained from the Bangladesh Agricultural Research Council (BARC). The population density data for Bangladesh in 2020 was derived from WorldPop (2018) and adjusted to align with the corresponding United Nations Population Division (UNPD) estimates. Healthcare data were gathered from Google Earth and Google Maps by identifying healthcare facilities through map tags and high-resolution satellite imagery. Data on potential sea level rise were sourced from Rekacewicz (2006). All datasets were compiled and analysed using ArcGIS 10.8 software.

## Results and Discussion

The results and discussion sections are organised into (i) an overview of the current coastal healthcare systems and (ii) healthcare systems challenges amidst climate change. The study's findings and associated discussion are presented below

### (i) An Overview of the Present Coastal Health Care Systems

The coastal regions of Bangladesh face significant challenges in providing quality health services to the poor and marginalised populations living in these areas (Iqbal 2019; Huda et al. 2020; Zafri et al. 2021). Poverty, remoteness, and vulnerability to climate change impacts such as cyclones, sea-level rise, and salinity intrusion make it difficult for the coastal populations to access health services (Rahman and Ahmad 2018; Iqbal 2019; Huda et al. 2020; Zafri et al. 2021) for climate-related common health issues such as scarcity of drinking water, hypertension, skin diseases, gastrointestinal issues, reproductive health problems, waterborne diseases, kidney diseases, malnutrition, and impact on mental health (Rahman and Ahmad 2018; Huda et al. 2020; Zafri et al. 2021; Talukder et al. 2023).

Vulnerable groups in coastal Bangladesh face additional health challenges due to climate change. For instance, women in these areas contend with limited access to resources and services, gender inequalities, illiteracy, traditional norms, and lack of participation in decision-making processes, all of which increase their vulnerability (Kathinka et al. 2023). Climate change events such as cyclones, storm surges, floods, and waterlogging adversely affect agriculture and water sectors, impacting women's livelihoods (Ashrafuzzaman et al. 2022). These climate-induced changes significantly affect women's mental health as they deal with increased workloads, water scarcity, food insecurity, and social insecurity (Goudet et al. 2023). Limited access to safe drinking water forces people in coastal Bangladesh to use contaminated sources like ponds or canals, increasing the risk of waterborne diseases such as diarrhoea and cholera (Talukder et al. 2016; Ashrafuzzaman et al. 2022).

Coastal healthcare encompasses the health infrastructure and health services available within these regions (Frieden et al. 2023), which face unique challenges due to their geographic location and vulnerability to climate change impacts. The health structure of healthcare in coastal Bangladesh comprises primary, secondary, and tertiary services and infrastructure, including hospitals, clinics, and community health centres (see Fig. 3.2). However, these systems are not always efficient. Maintaining the health infrastructure is challenging due to a lack of coordinated and effective health services. Our field observations, key informants, and literature review confirm that the coastal health system struggles to organise and deliver services effectively in these hard-to-reach areas due to inadequate health facilities and a shortage of skilled medical human resources such as doctors, pharmacists, and nurses (Huda et al. 2020; Zafri et al. 2021). Disabled and disadvantaged groups such as children, marginalised people, and women, in particular, face even greater barriers to accessing healthcare in these regions (Huda et al. 2020). Additionally, high poverty rates in many coastal districts hinder access to proper healthcare facilities, especially for marginalised groups (Huda et al. 2020).

(ii) **Healthcare Systems Challenges Amidst Climate Change**

## *Weaknesses in Infrastructure*

Climate change impacts create spatial and temporal vulnerabilities (Ahmed and Eklund 2021; Adshead et al. 2024). These vulnerabilities affect health infrastructure (Nahian 2023) in various ways, directly and indirectly, as shown in Table 3.1. Spatial vulnerability refers to the different degrees of susceptibility to hazards or risks in the coastal region of Bangladesh (Del Pinto et al. 2024). The spatial vulnerabilities are: (i) flooding and storm surges, (ii) sea level rise, (iii) erosion, (iv) salinity intrusion, and (v) cyclones. Temporal vulnerabilities in coastal Bangladesh refer to the susceptibility to risks and hazards that change over time (IPCC 2021). The temporal vulnerabilities in coastal Bangladesh are: (i) excessive rainfall and flooding, (ii) tidal surges, (iii) seasonal variability in water availability, and (iv) seasonal shifts in weather patterns.

Sea levels are rising in coastal Bangladesh, with projections indicating a one-meter increase over the next 50 years (see Fig. 3.3). Many healthcare facilities in these

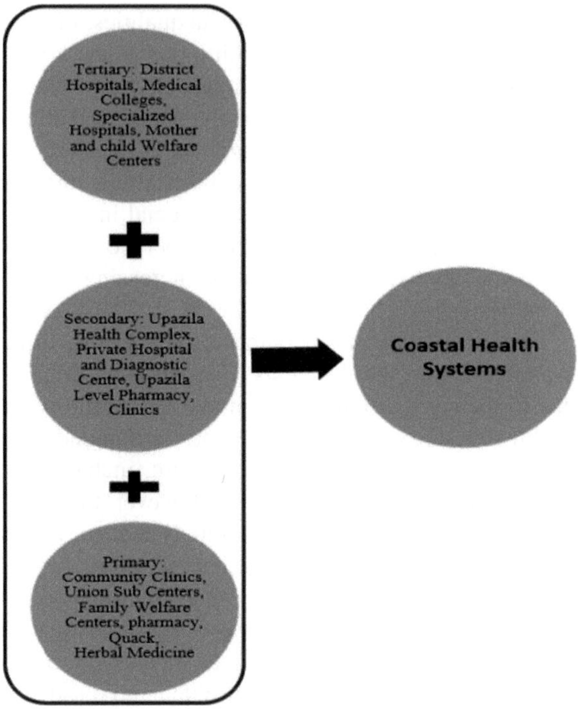


**Fig. 3.2** Bangladesh coastal health structure (based on field observation)

coastal regions are located in areas where this rise will occur. A one-meter increase in sea level will affect all these facilities, leading to saline intrusion, flooding, and damage to healthcare infrastructure, including buildings. This will seriously threaten the overall healthcare system in coastal Bangladesh. Besides, healthcare facilities in low-lying areas, as shown in Fig. 3.4, are susceptible to inundation due to spatial vulnerabilities such as cyclones and storm surges. These facilities are also at risk of flooding due to excessive rainfall.

Coastal erosion is a frequent and significant destructive force in Bangladesh, affecting socio-ecological systems (Islam et al. 2023). Healthcare facilities in highly eroded coastal areas will decline, making it harder for vulnerable communities to access healthcare. Sea level rise is causing significant surface, groundwater, and soil salinity in coastal Bangladesh (Sarker et al. 2021). Salinity intrusion is causing freshwater scarcity in coastal Bangladesh (Rahman et al. 2017). For instance, in many saline areas, maintaining sterile conditions for medical procedures is compromised due to the scarcity of freshwater, making it challenging to provide effective medical care.

All temporal vulnerabilities (see Table 3.1) significantly impact health service demand (UNDP 2022). For instance, during the monsoon season, there is a notable increase in waterborne diseases and injuries due to heavy rainfall and flooding. Intense rainfall and resulting flooding can also cause sudden and severe damage

**Table 3.1** Spatial and temporal vulnerability of health infrastructures in coastal Bangladesh and impacts on health service delivery

| Vulnerabilities | | Impacts on health service delivery | | | | | |
|---|---|---|---|---|---|---|---|
| | | Healthcare service delivery | Affecting access to essential medical care | Damage to healthcare facilities | Hamper medical supplies, equipment, etc. | Impact personnel during emergencies | Temporary closures and disruptions of services |
| Spatial | Flooding and storm surges | ✔ | ✔ | ✔ | ✔ | ✔ | ✔ |
| | Sea level rise | ✔ | ✔ | ✔ | ✔ | ✔ | ✔ |
| | Erosion | | | ✔ | | | ✔ |
| | Salinity intrusion | | | ✔ | ✔ | | ✔ |
| | Cyclones | ✔ | ✔ | ✔ | ✔ | ✔ | ✔ |
| Temporal | Excessive rainfall and flooding | ✔ | ✔ | ✔ | ✔ | ✔ | ✔ |
| | Tidal Surges | | | ✔ | | | ✔ |
| | Seasonal variability in water availability | | | ✔ | ✔ | | ✔ |
| | Seasonal shifts in weather patterns | ✔ | | | | | |

*Note* "✔" indicates the impacts of spatial and temporal vulnerabilities on health service delivery. This table has been generated based on field observations and support from key informants

to health infrastructure, disrupting services and the medical supply chain and exacerbating emergency and general healthcare disparities (Rezwana 2017).

The impacts of intense flooding, cyclones, and increasing salinity cause coastal populations to be temporarily or permanently displaced, straining existing health infrastructure and overwhelming resources in host communities (WHO 2015; Chowdhury et al. 2020; Adshead et al. 2024). Both spatial and temporal vulnerabilities worsen coastal areas' waterborne and vector-borne disease conditions and trigger seasonal outbreaks. This situation seriously threatens coastal communities, which already struggle with inadequate health infrastructure and limited human resources

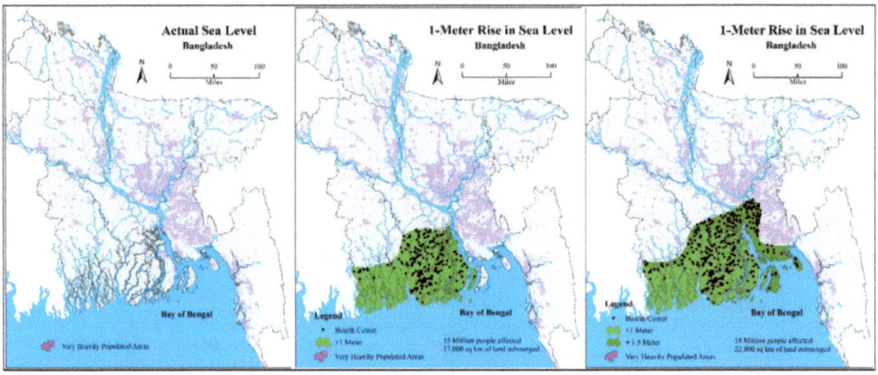


**Fig. 3.3** Maps based on Rekacewicz (2006) and WorldPop (2018) illustrate the probable consequences of rising sea levels in Bangladesh, depicting current conditions and predicting changes with 1 and 1.5-m increments. These forecasts show that a 1-m rise could flood 17,000 km$^2$ and affect 15 million people, while a 1.5-m rise could inundate 22,000 km$^2$ and impact 18 million people. The rising sea levels pose growing dangers to densely populated areas and critical infrastructure, including health centres. A 1-to-1.5-m rise in sea level would significantly impact coastal health infrastructure, potentially inundating or affecting most facilities with salinity in both water and soil, making essential health care inaccessible for millions

**Fig. 3.4** Healthcare facilities surrounded by saline water in the southwestern coastal region of Bangladesh (*Source* Picture taken by the authors at Shyamnagar Upazila on the southwest coast of Bangladesh 2024)

(Iqbal 2019; Zafri et al. 2021; Adshead et al. 2024). Both spatial and temporal vulnerabilities hinder agricultural productivity in coastal areas, leading to food and nutrition insecurity. This results in malnutrition, which increases the burden on existing medical interventions to address nutrition-related diseases (Thomson et al. 2018).

## *Limited Access to Health Services*

Coastal embankments, which are part of the coastal road network, often serve as vital access routes to healthcare facilities. Cyclones frequently cause significant damage to these roads (see Fig. 3.5). Additionally, coastal flooding from storm surges, heavy rainfall, and prolonged waterlogging further deteriorates these crucial routes. This damage makes it difficult for coastal communities to reach healthcare services. Geographic remoteness, inadequate transportation, and socioeconomic disparities further compound the challenges faced by marginalised and low-income coastal populations. Vulnerable groups, including pregnant women, the elderly, and people with disabilities, face especially significant hurdles in accessing care, particularly during emergencies. The damaged road network creates both physical and logistical barriers for coastal residents seeking healthcare (WHO 2015; Afroz et al. 2018; Islam et al. 2021; Bernard et al. 2022).

**Fig. 3.5** Destruction of the road network ([A], [B], and [C]) and coastal embankment ([D]) during the recent cyclone Remal in 2024. These roads and the embankment serve as access routes to healthcare facilities (*Source* Pictures taken by the authors during Remal 2024)

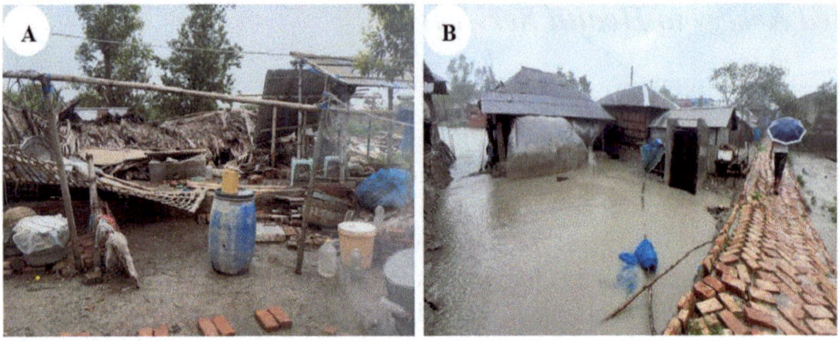

**Fig. 3.6** Vulnerable communities collected drinking water ([A]) in plastic containers after cyclone Remal in 2024, as freshwater became scarce. The submersion of latrines ([B]) contaminated the area with human waste, exposing the vulnerable community to waterborne diseases (*Source* Pictures taken by the authors during Remal 2024)

## *Issues with Water, Sanitation, and Hygiene (WASH) Programmes*

Access to WASH is essential for addressing the health, socio-economic, and environmental challenges faced by coastal communities in Bangladesh (World Bank 2021). Climate change significantly hampers WASH programmes in coastal Bangladesh due to vulnerabilities that affect freshwater resources both spatially and temporally. Climate change is expected to increase the salinity of surface and groundwater, leading to saltwater intrusion into freshwater aquifers and surface water sources, thereby reducing access to safe drinking water (World Bank 2021). Additionally, floods and cyclones damage WASH infrastructure, such as tubewells and latrines (see Fig. 3.6), limiting access to clean water and sanitation and increasing the risk of waterborne diseases such as cholera and diarrhoea (Sanitation and Water for All 2022; ACAPS 2024)). Disruptions in water and sanitation services hinder proper hygiene practices, making communities more vulnerable to infections and diseases (Jerin et al. 2023; WaterAid, n.d.a, b).

## *Challenges in Preventive Healthcare*

Preventive healthcare and improving access to these services are key goals in public health initiatives (Cruden et al. 2016). Like other parts of Bangladesh, preventive healthcare in coastal Bangladesh consists of various programmes and initiatives that aim to reduce disease incidence and promote overall health and well-being. The key preventive healthcare measures are: (i) immunisation programmes, (ii) maternal and child health, (iii) noncommunicable disease (NCD) control, (iv) infectious disease prevention, (v) awareness: health education and promotion, (vi) vitamin distribution and (vii) screening and early detection (Iqbal 2019; Rawal et al. 2019; WHO 2019; ICDDRB 2020; Kabir et al. 2023; WB 2023).

Cyclones, floods, and prolonged waterlogging in coastal Bangladesh hamper vaccination coverage by damaging road networks and health facilities. During cyclones, flooding, and excessive rainfall, women and children in geographically isolated areas struggle to access emergency health care, which hinders maternal and child health services (Kabir et al. 2014; WHO et al. 2023). Due to spatial and temporal vulnerability, saline water affects the effort to prevent noncommunicable diseases (NCDs) such as hypertension (Hoque et al. 2016; Tsai et al. 2024) and communicable diseases, including waterborne illnesses such as diarrhoea (Asma and Kotani 2021). In areas affected by cyclones and floods, health promotion, the distribution of Vitamin A and other supplements, and health screening face challenges in reaching vulnerable people due to logistical disruptions.

## Conclusion and Recommendations

Climate change profoundly affects Bangladesh's coastal regions through rising sea levels, increasing salinity, frequent cyclones, and severe flooding. These changes disrupt socio-ecological systems and threaten vulnerable communities and public health. According to research conducted by NASA, projections suggest that sea levels could rise significantly by the end of the century. This rise will have major consequences for coastal populations and their healthcare systems.

This chapter addresses the specific challenges facing healthcare in coastal Bangladesh, emphasising the urgent need for policy interventions and adaptive measures. The already damaged healthcare infrastructure in these regions is under growing pressure due to the escalating impacts of climate change. Coastal communities, especially those in low-lying and remote areas, face significant barriers to accessing essential health services, further complicated by damage to infrastructure such as roads and medical facilities.

To effectively address these challenges, immediate and strategic action is required to develop healthcare systems that can endure and adapt to climate impacts. Strengthening preventive healthcare and improving access to essential services are critical steps in mitigating the adverse effects on vulnerable populations and fostering a sustainable and healthy future.

From the study, the key recommendations are:

- Detailed mapping and future projections of healthcare system challenges caused by climate change in coastal Bangladesh.
- Incorporate climate change adaptation into healthcare policies and planning.
- Allocate resources to build climate-resilient infrastructure and healthcare services.
- Enhance coordination and collaboration among government agencies, NGOs, and other stakeholders.
- Advocate for climate justice and equity in healthcare policy decisions.

- Support interdisciplinary research to create evidence-based decision-making.

## References


ACAPS (2024) Impact of tropical cyclone Remal in Bangladesh. ACAPS. https://www.acaps.org/fileadmin/Data_Product/Main_media/20240612_ACAPS_Bangladesh_-_Impact_of_Tropical_Cyclone_Remal.pdf

Adshead D, Paszkowski A, Gall SS, Peard AM, Adnan MSG, Verschuur J, Hall JW (2024) Climate threats to coastal infrastructure and sustainable development outcomes. Nat Clim Change 1–9

Ahmed S, Eklund E (2021) Climate change impacts in coastal Bangladesh: migration, gender and environmental injustice. Asian Affairs 52(1):155–174

Afroz S, Cramb R, Grünbühel C (2018) Vulnerability and response to cyclones in coastal Bangladesh: a political ecology perspective. Asian J Soc Sci 46(6):601–637

Ashrafuzzaman M, Gomes C, Guerra J (2022) Climate justice for the southwestern coastal region of Bangladesh. Front Clim 4:881709

Asma KM, Kotani K (2021) Salinity and water-related disease risk in coastal Bangladesh. EcoHealth 18(1):61–75

BBS (2012) Bangladesh Population and Housing Census 2011: National Report (Volume 1: Analytical Report). Bangladesh Bureau of Statistics. Ministry of Planning, Government of the People's Republic of Bangladesh

Bernard A, Long N, Becker M, Khan J, Fanchette S (2022) Bangladesh's vulnerability to cyclonic coastal flooding. Nat Hazard 22(3):729–751

Chowdhury MA, Hasan MK, Hasan MR, Younos TB (2020) Climate change impacts and adaptations on health of Internally Displaced People (IDP): an exploratory study on coastal areas of Bangladesh. Heliyon 6(9)

Cruden G, Kelleher K, Kellam S, Brown CH (2016) Increasing the delivery of preventive health services in public education. Am J Prev Med 51(4):S158–S167

Dasgupta S, Huq M, Khan ZH, Murshed M, Ahmed Z, Mukherjee N, Khan MF, Pandey K (2010) Vulnerability of Bangladesh to cyclones in a changing climate: potential damages and adaptation cost. The World Bank, Development Research Group, Environment and Energy Team

Del Pinto M, Chmutina K, Palaiologou F, Bosher L (2024) The role of the spatial network in urban disaster risk variations: reimagining the notion of spatial vulnerability at the urban scale. Int J Disas Risk Sci 1–14

Eckstein D, Künzel V, Schäfer L (2021) Global climate risk index 2021: who suffers most from extreme weather events? Weather-related loss events in 2019 and 2000–2019. Germanwatch e.V., Bonn. ISBN: 978-3-943704-84-6

Evertsen KF (2023) Women as subjects of risk in Bangladesh's coastal riskscapes. Sociol Rev 003802612311752–003802612311752. https://doi.org/10.1177/00380261231175223

Frieden TR, Lee CT, Lamorde M, Nielsen M, McClelland A, Tangcharoensathien V (2023) The road to achieving epidemic-ready primary health care. The Lancet Public Health 8(5):e383–e390

GOB (2018) Bangladesh delta plan 2100. General Economics Division (GED) Bangladesh Planning Commission Government of the People's Republic of Bangladesh. Print (plandiv.gov.bd)

Goudet J-M, Arif FB, Owais H, Ahmed HU, Ridde V (2024) Climate change and women's mental health in two vulnerable communities of Bangladesh: an ethnographic study. PLOS Global Public Health 4(6):e0002080. https://doi.org/10.1371/journal.pgph.0002080

Hoque MA, Scheelbeek PFD, Vineis P, Khan AE, Ahmed KM, Butler AP (2016) Drinking water vulnerability to climate change and alternatives for adaptation in coastal South and South East Asia. Clim Change 136:247–263

Huda FA, Mahmood HR, Hossain AT, Khan J, Faruk O, Razzak ZS, El Arifeen S et al (2020) Analysis of health needs and health system response in the coastal districts of Bangladesh

ICDDRB (2020, July 23) Analysis of health needs and health system response in the coastal districts of Bangladesh. Final study report. UK Government. https://assets.publishing.service.gov.uk/media/5f34f2478fa8f5174ac78f24/icddrb_Final_Study_Report_23_July_2020_DFID.pdf

IFRC (2024) Bangladesh: cyclone Remal. International Federation of Red Cross and Red Crescent Societies. Retrieved 24 July 2024, from https://www.ifrc.org/emergency/bangladesh-cyclone-remal

IPCC (2021) Climate change 2021: the physical science basis. Contribution of Working Group I to the sixth assessment report of the intergovernmental panel on climate change. IPCC report

Iqbal MH (2019) Disparities of health service for the poor in the coastal area: does Universal health coverage reduce disparities? J Market Access Health Policy 7(1):1575683

Islam MN, van Amstel A, Islam MM, Shahin M, Miraj M, Ghosh S, Islam I et al (2021) Climate change impacts and mitigation approach: coastal landscape, transport, and health aspects. In: Bangladesh II: climate change impacts, mitigation and adaptation in developing countries, pp 41–101

Islam MS, Crawford TW, Shao Y (2023) Evaluation of predicted loss of different land use and land cover (LULC) due to coastal erosion in Bangladesh. Front Environ Sci 11:1144686

Jerin T, Chowdhury MA, Azad MAK, Zaman S, Mahmood S, Islam SLU, Jobayer HM (2023) Extreme weather events (EWEs)-Related health complications in Bangladesh: a gender-based analysis on the 2017 catastrophic floods. Nat Hazard Res

Kabir R, Khan HT, Ball E, Caldwell K (2014) Climate change and public health situations in the coastal areas of Bangladesh. Int'l J Soc Sci Stud 2:109

Kabir R, Khan HT, Ball E, Caldwell K (2016) Climate change impact: the experience of the coastal areas of Bangladesh affected by cyclones Sidr and Aila. J Environ Publ Health

Kabir A, Karim MN, Billah B (2023) The capacity of primary healthcare facilities in Bangladesh to prevent and control non-communicable diseases. BMC Primary Care 24(1):60

Moslehuddin AZM, Abedin MA, Hossain MA, Habiba U (2015) Soil health and food security: perspective from southwestern coastal region of Bangladesh. In: Food security and risk reduction in Bangladesh, pp 187–212

Nahian MA (2023) Public health impact and health system preparedness within a changing climate in Bangladesh: a scoping review. Challenges 14(1):4

Rahman MM, Ahmad S (2018) Health, livelihood and well-being in the coastal delta of Bangladesh. In: Ecosystem services for well-being in deltas: integrated assessment for policy analysis, pp 131–145

Rahman MA, Dawes L, Donehue P, Rahman MR (2022) Transformation of the coastal social-ecological system in southwest Bangladesh due to empolderment. Water Hist 14(2):147–167

Rahman MTU, Rasheduzzaman M, Habib MA, Ahmed A, Tareq SM, Muniruzzaman SM (2017) Assessment of fresh water security in coastal Bangladesh: an insight from salinity, community perception and adaptation. Ocean Coast Manag 137:68–81

Rawal LB, Kanda K, Biswas T, Tanim MI, Poudel P, Renzaho AM, Ahmed SM et al (2019) Non-communicable disease (NCD) corners in public sector health facilities in Bangladesh: a qualitative study assessing challenges and opportunities for improving NCD services at the primary healthcare level. BMJ Open 9(10):e029562

Rekacewicz P (2006) Impact of sea level rise in Bangladesh. GRID. https://www.grida.no/resources/5648

ReliefWeb (2024, May 29) Bangladesh: cyclone Remal 2024 situation report no. 03. https://reliefweb.int/report/bangladesh/bangladesh-cyclone-remal-2024-situation-report-no-03-29-may-2024

Rezwana N (2017) Disasters, gender and access to healthcare: women in coastal Bangladesh. Routledge

Roy S (2020) Transforming gendered lives and livelihoods in post-disaster settings in the Bangladesh Sundarbans Forest. In: Climate change, hazards and adaptation options: handling the impacts of a changing climate, pp 463–479

Sanitation and Water for All (2022) Climate task team briefing paper. https://www.sanitationandwaterforall.org/sites/default/files/2022-04/2022SMM-SWA%20Climate%20Task%20Team-Briefing%20paper_0.pdf

Sarker MMR, Van Camp M, Hossain D, Islam M, Ahmed N, Karim MM, Walraevens K et al (2021) Groundwater salinization and freshening processes in coastal aquifers from southwest Bangladesh. Sci Total Environ 779:146339

Smith AC, Tasnim T, Irfanullah HM, Turner B, Chausson A, Seddon N (2021) Nature-based solutions in Bangladesh: evidence of effectiveness for addressing climate change and other sustainable development goals. Front Environ Sci 9:737659

Talukder B, Saifuzzaman M, van Loon GW (2016) Sustainability of agricultural systems in the coastal zone of Bangladesh. Renewable Agric Food Syst 31(2):148–165

Talukder B, Salim R, Islam ST, Mondal KP, Hipel KW, Orbinski J (2023) Collective intelligence for addressing community planetary health resulting from salinity prompted by sea level rise. J Clim Change Health 10:100203

Thomson MC, Grace D, De Fries RS, Metcalf CJE, Nissan H, Giannini A (2018) Climate impacts on disasters, infectious diseases and nutrition

Tsai C, Hoque MA, Vineis P, Ahmed KM, Butler AP (2024) Salinisation of drinking water ponds and groundwater in coastal Bangladesh linked to tropical cyclones. Sci Rep 14(1):5211

UNDP (2022) Vulnerability of health care facilities to climate change. United Nations Development Programme. https://www.undp.org/sites/g/files/zskgke326/files/2022-11/Vulnerability%20of%20Health%20Care%20Facilities%20to%20Climate%20Change%202022%20Report.pdf

WaterAid (n.d.a) WASH and climate: policy and financing disconnects in Bangladesh. Retrieved from https://www.wateraid.org/bd/sites/g/files/jkxoof236/files/wash-and-climate---policy-and-financing-disconnects-in-bangladesh.pdf

WaterAid (n.d.b) Climate resilience WASH programming in coastal areas of Bangladesh: an end-line study. Retrieved from https://www.wateraid.org/bd/sites/g/files/jkxoof236/files/climate-resilience-wash-programming-in-coastal-areas-of-bangladesh-an-end-line-study_0.pdf

WB (2021, April 26) Building coastal resilience to protect lives and livelihoods in Bangladesh. https://www.worldbank.org/en/results/2021/04/26/building-coastal-resilience-to-protect-lives-and-livelihoods-in-bangladesh

WB (2023, August 30) World Bank helps Bangladesh improve primary healthcare in cities. World Bank. https://www.worldbank.org/en/news/press-release/2023/08/30/world-bank-helps-bangladesh-improve-primary-healthcare-in-cities

WHO (World Health Organization) (2015) Vulnerability and adaptation to climate change in coastal and drought-prone areas of Bangladesh: health and WASH. Retrieved from https://research.fit.edu/media/site-specific/researchfitedu/coast-climate-adaptation-library/asia-amp-indian-ocean/bangladesh/WHO.--2015.--Vulnerability-and-Adaptation-to-Climate-Change-in-Coastal-and-Drought-Prone-Areas-of-Bangladesh-Health-and-WASH.pdf

WHO, UNICEF and UNFPA (2023) Protecting maternal, newborn and child health from the impacts of climate change: a call for action. World Health Organization. https://www.unfpa.org/sites/default/files/resource-pdf/HRP%20CLIMATE%20CHANGE%20IMPACT_WEB_V18_SPREADS_final.pdf

WHO (2019, February 5) Bangladesh adopts multisectoral action plan for noncommunicable disease control and prevention. World Health Organization. https://www.who.int/bangladesh/news/detail/05-02-2019-bangladesh-adopts-multisectoral-action-plan-for-noncommunicable-disease-control-and-prevention

WorldPop, School of Geography and Environmental Science, University of Southampton; Department of Geography and Geosciences, University of Louisville; Département de Géographie, Université de Namur; & Center for International Earth Science Information Network (CIESIN), Columbia University (2018) Global high resolution population denominators project—funded by The Bill and Melinda Gates Foundation (OPP1134076). https://doi.org/10.5258/SOTON/WP00675

Zafri NM, Nurullah M, Neema MN, Waliullah M (2021) Spatial accessibility to healthcare facilities in coastal region of Bangladesh. Int J Health Plann Manage 36(3):643–655

# Chapter 4
# On the Frontline of Climate Change and Health: A Health Worker Eyewitness Report from the Global South

**Charlotte Njua Mbuh**

**Abstract** Countries in the Global South are highly vulnerable to climate change's health impacts. While extensively researched, quantitative evidence has provided minimal insights into daily realities. To address this gap, in 2023 The Geneva Learning Foundation (https://www.learning.foundation/climate) organised an experience-sharing event dedicated to climate impacts on health for its 50,000+ international network of health and other workers in the Global South. An online survey collected data on perceptions and observed impacts. A total of 4700 health practitioners from 68 countries registered, with 1,260 submitting observations. Analysis revealed high levels of concern across multiple domains, with food security, malnutrition, vector-borne diseases, and water-borne diseases being the most frequently cited impacts. Qualitative analysis provided additional insights, including the tendency to contextualise climate change within broader environmental degradation, highlight complex interrelated impact pathways on community health, and identify vulnerabilities of specific groups such as women, pastoralists, etc. The study also uncovered examples of health workers collaborating with communities to build climate resilience. These cases emphasise the potential to leverage health workers' close community relationships, trusted positions, and expert knowledge to support community-focused efforts in limiting climate change health impacts in the Global South. This approach could significantly contribute to adaptation strategies and resilience-building.

**Keywords** Climate change and health · Global South · Health workers · Communities · Qualitative research · Malnutrition · Infectious disease · Non-communicable disease · Mental health · Equity · Health services

C. N. Mbuh (✉)
The Geneva Learning Foundation, Geneva, Switzerland
e-mail: climate@learning.foundation

 
W. Leal Filho and F. Wolf (eds.), *Climate Change and Health: Perspectives from Developing Countries*, Climate Change Management,
https://doi.org/10.1007/978-3-031-93177-2_4

## Introduction

The Intergovernmental Panel on Climate Change (IPCC) concluded with very high confidence that climate-related illnesses, premature deaths, malnutrition and threats to mental health and wellbeing are increasing (Intergovernmental Panel on Climate Change 2022). A similar conclusion has been drawn by the World Health Organization (WHO), which estimates that, between 2030 and 2050, climate change is likely to cause an additional 250,000 deaths a year just from malnutrition, malaria, diarrhoea, and heat stress (World Health Organization 2019). Similarly, the World Bank estimates that a warmer climate could lead to at least 21 million additional deaths by 2050 from just five health risks: extreme heat, stunting, diarrhoea, malaria, and dengue (New Program to Protect Millions from Rising Climate-Related Deaths and Illness n.d.).

These impacts will not be felt equally. According to the World Bank, more than half the world's population (4.5 billion people) are at high risk of experiencing an extreme weather event; 2.3 million of these vulnerable people are living in poverty and almost 400 million are living in extreme poverty (Doan et al. 2023). Climate change is magnifying existing health inequalities, both between and within countries (Romanello et al. 2023). Those most at risk include groups such as women (Desai and Zhang 2021), children (Weeda et al. 2024), older people, and people with pre-existing health conditions (New Program to Protect Millions from Rising Climate-Related Deaths and Illness, n.d.). Climate change impacts are also most likely to affect countries in the Global South, particularly in Africa (Wright et al. 2024). Within countries, disadvantaged populations are most vulnerable and least able to adapt to a changing environment (Climate Inequality Report 2023, n.d.).

The Geneva Learning Foundation, a Swiss non-profit with a mission to research, develop, and scale up new ways to learn and lead against critical threats to our societies, facilitates peer learning activities for health and humanitarian workers based in the Global South or other disadvantaged settings. Through a digital platform, it aims to connect health workers in different countries and facilitate the exchange of information through international networks—helping them to learn from each other and from participating global experts who act as "guides on the side".

A major focus of TGLF activities has been immunization. By joining a "Movement for Immunization Agenda 2030 (IA2030)", immunization professionals pledge to work toward global IA2030 targets and to support each other in their efforts to improve vaccine coverage. Regular "Teach to Reach: Connect" events provide an opportunity for health workers to share their experiences of particular areas of immunization practice or other health topics (Jones et al. 2023).

Relatively few studies have provided an integrated, first-hand account of the health impacts of climate change at the community level. This study set out to leverage the extensive TGLF health worker network in the Global South to generate quantitative and qualitative insights into local climate impacts on communities at an unprecedented scale.

## Methodology

In 2023, TGLF organised a special event dedicated to climate change and health. It was intended to provide an opportunity for members of the TGLF global community to share their experiences of a changing climate and how it is affecting the health of their local communities. Separate events were held for English- and French-speaking participants.

In advance of the event, participants were invited to complete an online survey that captured information about themselves, their local areas, their experience of a local changing climate, and the impacts on health they have observed. The survey combined a series of multiple-choice questions as well as free-text fields.

The survey asked respondents to say whether they had directly observed any of 20 climatic and environmental impacts locally, derived from an analysis of the existing literature. They were also asked whether they had observed any of 17 health impacts, also derived from the existing literature.

English and French versions of the survey were produced, with French language responses translated by DeepL translator for qualitative analysis. Thematic analysis was used to draw out insights relating to individual health impacts and cumulative effects on communities.

## Results: Quantitative Responses

A total of 4700 health practitioners from 68 countries registered for this special event, and 1206 of them submitted observations—generating a uniquely large **collection of data on climate-related health impacts in the Global South** (Foundation et al. 2023). Almost three-quarters of survey respondents were from the subnational level.

Respondents reported direct local experience of a **wide range of climatic and environmental impacts** (Fig. 4.1). From a list of 20 potential impacts, the three most selected were:

- Impacts on **farming** (land- or water-based)—49.2% of respondents.
- Impacts on the distribution of **disease-carrying insects**—47.6% of respondents.
- **Urban areas becoming hotter**—46% of respondents

Across this group, **13 out of 20 impacts were reported by at least 25% of respondents**, illustrating the wide range of effects linked to climate change being noticed at the community level.

**Health impacts** were very commonly reported (Fig. 4.2). Presented with a list of 17 potential health impacts, respondents most reported:

- Increased **malnutrition** or undernutrition—59.3% of respondents.
- Increased **water-borne diseases**—59.2% of respondents.

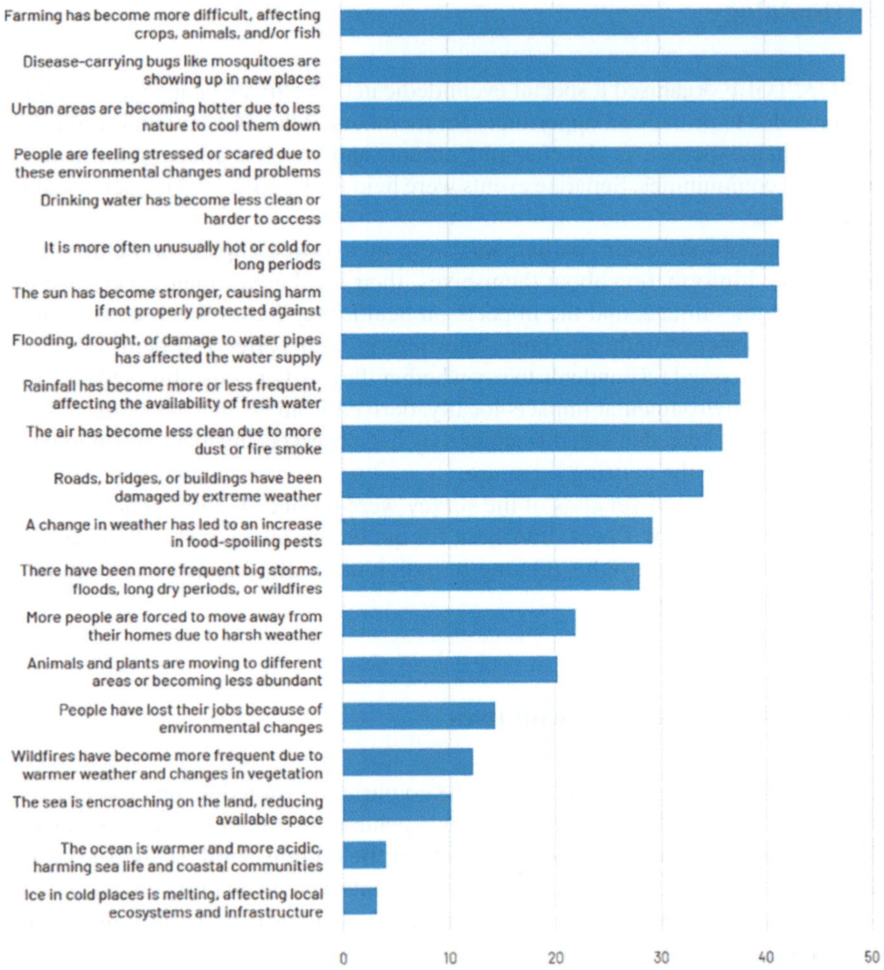


**Fig. 4.1** Climate and environmental impacts noted by survey respondents (% of total respondents, n = 1206)

- Changes to the incidence or distribution of **vector-borne diseases**—51.1% of respondents.

Again, a wide variety of impacts were reported, with **13 out of 17 impacts being reported by at least 25% of respondents**.

Among respondents, **93% believed there was a link between climate change and health**. They were also **fairly confident in this connection** (average score of 4.1 out of 5). Notably, women were more likely than men to report climate and environmental changes and health impacts. Responses differed between countries, with respondents from Kenya being particularly likely to report climate impacts.

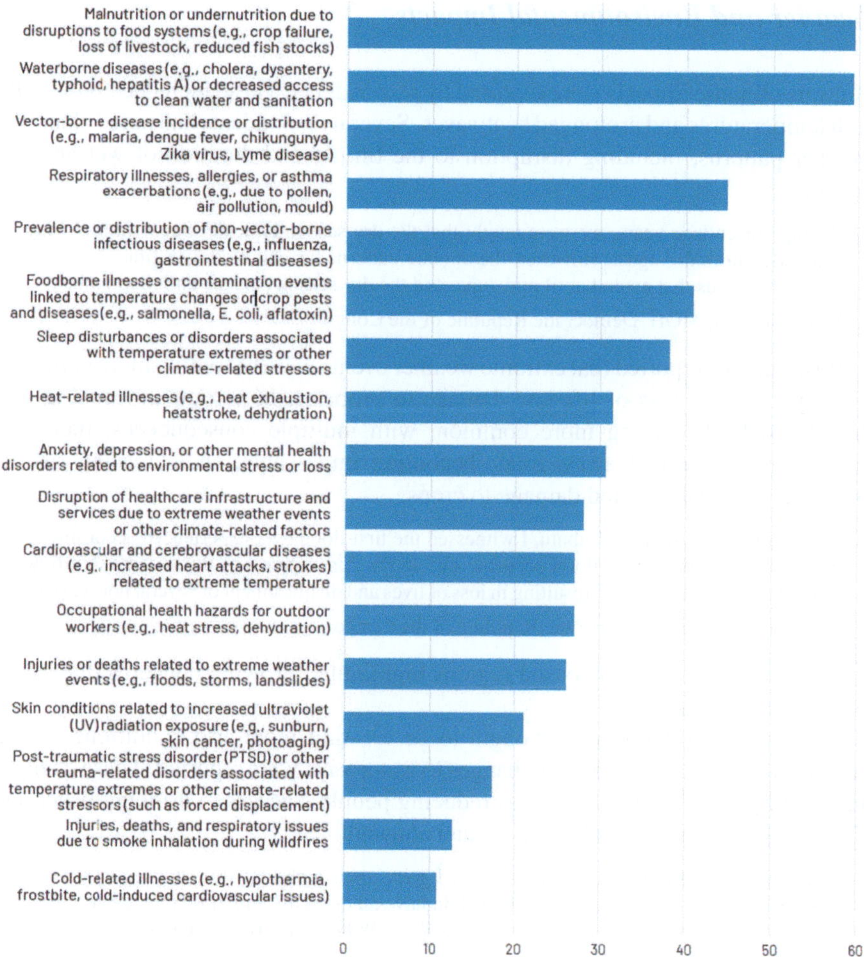


**Fig. 4.2** Health impacts noted by survey respondents (% of total respondents, n = 1206)

## Results: Qualitative Findings

Survey respondents were asked to provide examples of the climatic changes they had noticed locally (survey prompt: "Tell us more" about the changes in health that you have observed) and to describe the local climate-related health impacts they had witnessed (survey prompt: Would you like to "share your experience" in relation to climate and health? "We would like to understand your story."). Responses were analysed thematically according to the climate and environmental impacts and the nature of the health impacts.

## Weather and Environmental Impacts

Multiple climatic impacts were reported by respondents. These included abnormally high temperatures and prolonged heatwaves. Several respondents had noted changing weather patterns, including disruption to the timing and duration of wet and dry seasons.

> During our college years, we were taught that the dry season typically began in mid-May and ended in mid-August. However, the climate patterns have shifted over time. Now, we experience rains that extend until mid-June, and the dry season persists until September.
>
> Man, National, MOH, Democratic Republic of the Congo

Respondents reported that extreme weather events were becoming more common, leading to injury, loss of life and damage to property. In particular, flooding was thought to be becoming more common, with multiple consequences, including housing being washed away, roads becoming impassable, an increasing risk of erosion and landslides, and damage to crops.

> During the year I spent in Lubutu, I witnessed the first flood ever experienced according to the local population. This flood occurred due to the convergence of three rivers that flow through the town of Lubutu, resulting in loss of lives and displacement of several households.
>
> Man, Region, MOH, Democratic Republic of the Congo

Several respondents in coastal regions highlighted growing concerns due to sea level rise and storms.

Higher temperatures were thought to be creating more arid conditions and an increased risk of drought. Major changes to waterways were reported. These changes have multiple implications, such as reducing people's access to drinking water and disrupting both agricultural practices and aquaculture.

> Going back home to the community where I grew up as a child, I was shocked to see that most of the rivers we used to swim and fish in have all dried up, and those that are still there have become very shallow so that you can easily walk through a river you required a boat to cross in years past.
>
> Man, National, MOH, Nigeria

Drier conditions were also seen to be leading to growing numbers of wildfires. These have both short-term impacts, such as injury, as well as longer-terms effects due to declining air quality (responsible for an estimated 6.7 million premature deaths a year, 89% in the Global South; World Health Organization 2022) and loss of agricultural land or productivity.

## Direct Impacts on Health

Respondents noted a range of ways in which climate change is directly affecting the health of local populations, either because of **higher temperatures and heatwaves** (Gibb et al. 2024) or as a consequence of **extreme weather events**.

Examples reported included an increased risk of heatstroke, sunstroke and dehydration. Flooding was widely reported to have led to injury and loss of life because of damage to properties or transport.

> Where I live in Buea, the flood from Mount Cameroon took away all belongings of people in my neighborhood and killed a secondary school student who was playing football with his friends.
>
> Woman, Region, Education or Research Organization, Cameroon

## *Impacts on Food Security*

One of the most concerning impacts of climate change was seen in **food production**, undermining the ability of households to feed themselves and to support themselves economically and compromising wider food security (Iannotti et al. 2024). Growing levels of **malnutrition**, including among children, are the inevitable result.

Changing weather patterns were thought to be having a major impact on farming. The growing unpredictability of weather, for example, is affecting farmers' ability to manage their crops effectively.

> Because of climate change, farmers are not able to study the weather as before in order to know the rainy and dry seasons to prepare and produce a good yield.
>
> Woman, Health Facility, MOH, Ghana

Extreme weather is also disrupting agricultural productivity. Both drought and excessive rainfall, including floods, can have devastating effects on crops. Additional challenges include the increasing difficulty of transporting food when roads are damaged or impassable. In some cases, the availability of fresh food has been reduced or prices increased, affecting wider food security.

> The rainy season is short-lived, resulting in meager harvests by its end. In terms of health, the area has a high rate of chronic malnutrition among young children, as well as cases of anemia. Nouna is categorized as part of the regions known as the "Granary" or Agro-Pastoral Zone. These areas traditionally provide food during the lean season to other regions and even neighboring countries. In recent years, nutritional surveys have highlighted a serious issue of malnutrition among young children.
>
> Women, National, MOH, Burkina Faso

Healthcare facilities can be among those affected by reduced food security.

> My facility is a referral center that receives patients from five neighboring states. Hospitals are always full to capacity and the shortage of food and fruits increases hardship and suffering.
>
> Man, Health Facility, MOH, Nigeria

As well as impacts on crops, **livestock** can be affected by reduced access to water and food or increased exposure to infectious diseases. In many settings, livestock make a critical contribution to economic wellbeing as well as foodstuff, so livestock losses can have a devastating impact.

> The weather pattern has changed so much in my community of Kajiado, affecting pastoralists. Many animals have died and as animals were the main source of financial stability, many homes are now starving due to lack of rainfall.
>
> Woman, Kajiado, Kenya

Multiple examples were reported of an increase in the prevalence of **malnutrition**. Groups such as young children and pregnant women were felt to be at particular risk, the latter at especial risk of anaemia.

> Climate change has had a significant impact on agricultural production in terms of food products. The traditional patterns of dry and rainy seasons are no longer followed due to factors like uncontrolled forest fires. This disruption has led to a decrease in the production of food items. Additionally, there's a concerning trend of inadequate utilisation of staple foods like corn and cassava, which are often used in the production or distillation of indigenous alcohol. As a result, the community is facing a surge in cases of malnutrition, highlighting the urgency to address the challenges posed by climate change on both agricultural practices and food security.
>
> Man, District, MOH, Democratic Republic of the Congo

## *Infectious Disease*

Much of the discourse relating to climate change impacts on health has focused on infectious disease (Mora et al. 2022), particularly the changing distribution of disease vectors. Respondents identified multiple ways in which the prevalence of infectious diseases is changing, due to biological and/or social drivers.

Many respondents reported that **vector-borne diseases**, particularly **malaria**, were on the rise locally, often linked to larger mosquito populations. One key driver was thought to be environmental changes that favour mosquito breeding, such as more pools of water following storms or floods.

> In my community in Mbaagwa Clan of Ikurav-Ya District of Kwande LGA, there was heavy rainfall one certain night. The community members woke up the following day to see water flooding their environment. Without any help, they have to wait for nature to take its course. After a few days, the water disappeared. Not long ago that small children and adults started getting sick. Their major complaint was fever, loss of appetite, vomiting, and general weakness of the body. A careful microscopic and laboratory analysis showed malaria parasites.
>
> Man, District, Ministry of Environment, Nigeria

**Behavioural changes** can also lead to increased exposure to mosquitoes or other insect vectors, such as sleeping outside on hot nights. In some cases, mosquitoes' biting habits were thought to be changing.

> I have observed sleep disturbance due to heat in a community without an electricity supply. They are forced to sleep outside their houses leading to mosquito bites which have increased the incidence of malaria fever in the community.
>
> Woman, National, MOH, Nigeria

Many concerns were also raised about an increasing prevalence of **water-borne diseases**, such as diarrhoeal disease. Flooding can overwhelm sanitation systems, leading to environmental contamination with gastrointestinal pathogens. With water scarce, use of contaminated water sources has also been seen to increase.

> Because of poor drainage, when it rains, sewage overflows and you find most of the places are in very bad condition. When children go out to play, pick things up, and put them into their mouths, they get sick stomach aches.
>
> Woman, Health Facility, MOH, Kenya

As well as these acute effects, more intense rainy seasons can lead to displacement and adoption of more **unsanitary practices**.

> In Lake Chad, during the rainy season, the various arms of the lake expand, causing flooding that affects the villages situated along its shores. This flooding often forces the inhabitants to relocate to higher ground. However, a significant portion of the population around the lake lacks proper sanitation facilities, leading to open defecation near the water's edge. As the water levels rise, this practice contributes to the spread of diseases, particularly when access to healthcare facilities becomes challenging.
>
> Man, Health Facility, MOH, Chad

A further issue highlighted was the increased risk of **zoonotic spillovers**, the transmission of animal pathogens to humans because of increased contact between humans and wild animals, migration of people whereby they encounter new pathogens, or changes in the distribution of disease vectors.

> The region's architecture is outdated, and the city has transformed into a migratory hub where diverse behaviours converge. New diseases emerge, and the indigenous population is grappling with illnesses that were previously unknown to them.
>
> Man, Region, NGO, Niger

## *Asthma, Allergies and Other Non-communicable Diseases*

Among the indirect effects of climate change are those relating to **air quality**, which is affected by dust storms and particulate matter released during wildfires (Johnston et al. 2024), and by **behavioural changes** driven by climate effects. These can have significant implications for a range of non-communicable diseases, particularly respiratory diseases.

Although air quality was frequently raised as an important health issue, triggering or exacerbating **respiratory conditions**, a wide range of factors were seen to be affecting air quality, not necessarily related to climate. This illustrates how climate effects are not experienced independently, but interact with other aspects of environmental change, such as deforestation, industrial activity and vehicle use, particularly in urban areas.

> There is a huge dumping site in Korle-Bu which always burns tyres and rubbish, making the air around the community very polluted. Lots of people living in the community go to

> health facilities with respiratory tract infections, asthma attacks, and many other respiratory infections. The burning site is posing a lot of health implications for the indigenous people living in the community.
>
> Man, National, MOH, Ghana

Other reported impacts included **unhealthier diets** due to shortages of fresh produce. Those with pre-existing chronic conditions, such as respiratory disorders or cardiometabolic conditions, were thought to be at particular risk of adverse effects due to greater heat stress and other reasons.

## *Mental Health Impacts*

A potentially under-appreciated aspect of climate change is the impact it can have on **mental health** (White et al. 2023). Although a hotter world is likely to be more stressful for many, it is likely that the indirect social effects of climate change will have the greatest impact on mental wellbeing.

One commonly reported issue was alterations to **sleep patterns**, known to be associated with mental health (Scott et al. 2021). Sleep disturbance may be a result of higher temperatures, making sleeping indoors impossible or due to loss of property or displacement.

> Residents of Lubumbashi, for example, experience shorter and less restful sleep due to the difficulty in cooling the body down, exacerbated by a decrease in tap water availability.
>
> Man, Health Facility, NGO, Democratic Republic of the Congo

The sleep of parents of young children may be particularly affected.

> The amount of sun has increased, making it too hot inside the bedroom at night. This is affecting newborn babies who cry at night. Because there is no light, no fans, the poor people are really suffering as a result of the impact of climate change.
>
> Man, Health Facility, Private Industry, Nigeria

Many respondents noted that climate change was making everyday life more difficult, leading to increased levels of **stress, worry, and anxiety**. Sometimes, this was thought to reflect the difficulties of living in a hotter, harsher world.

> People who could afford air-conditioned houses enjoyed life while others who could not did continue to adjust to endure the heat. People were not looking healthy. Poverty and stress were written on people's faces.
>
> Woman, District, NGO, Nigeria

More commonly, however, stress and anxiety were linked to climate-related impacts on **lives and livelihoods**, particularly food insecurity and growing economic hardship.

> There is despair and hopelessness as hunger bites harder due to crop failure and people lose their sources of livelihood without any foreseeable solution in sight.
>
> Woman, National, MOH, Nigeria

> When it comes to the impact of climate change on mental health, we can observe a significant disturbance in the well-being of farmers. Even just a couple of days without rainfall can trigger a sense of sadness among them. Instances of minor depression have been noted among household heads who helplessly witness their crops withering due to inadequate moisture.
>
> Man, District, MOH, Burkina Faso

Again, these impacts have social as well as individual consequences. Respondents provided examples where people affected were being driven to adopt risky behaviours, leading to an increase in substance abuse and alcohol misuse due to adversity.

> Due to climate and weather changes, food production has declined. There has been job loss, low income, and depression. Also, men became alcoholics, which is now a national menace the government is trying to fight.
>
> Man, Health Facility, MOH, Kenya

> Notably, cases of nervousness and increased consumption of local alcoholic drinks during periods of intense heat have been observed, leading to physical violence, accidents, and incidents on public highways.
>
> Man, Health Institution, MOH, Togo

The experiences shared indicate widespread concern about the mental health impacts of climate change, particularly driven by social disruption. They illustrate the need to consider a broad concept of health when considering the health impacts of climate change and when developing appropriate responses. This is of particular importance given that, in many settings in the Global South, mental health services are limited or non-existent (Wainberg et al. 2019).

## *Social Impacts*

Respondents identified how climate change is already having a profound effect on lives and livelihoods, with multiple consequences for health and wellbeing.

When agricultural land is lost, areas become uninhabitable or when making a living from farming is no longer economically viable, households may be displaced or forced to migrate, typically to urban centres.

> In the lowlands of Kasese district where I come from, people have become homeless due to melting ice causing floods even in areas where floods previously did not happen.
>
> Woman, Health Facility, MOH, Uganda

Displaced people may find themselves in informal settlements where conditions favour the spread of infectious diseases, owing to high population densities and poor sanitation. Displacement can also trigger conflict between new arrivals and those already present.

> The imbalanced climate also strains human relationships within cities. People find it challenging to accommodate others in their homes due to the discomfort caused by the climatic conditions. This can lead to tension and strained interactions among residents.
>
> Man, District, NGO, Côte D'Ivoire

Competition for natural resources such as water or land can also be a source of tension in rural areas.

> The effects of climate change are evident in the region through frequent and intense winds accompanied by dust from November to May, spanning about seven months. The vegetation cover, including trees, firewood, and green grasses, has become scarce, leaving only sand and stunted thorny shrubs. The region's fauna has also been affected. Ongoing conflicts between herders and farmers over grazing grounds persist, exacerbating food insecurity.
>
> Man, Region, MOH, Chad

Those living in coastal areas may be particularly vulnerable to sea-level rise and storm surges.

> In recent years, tidal waves and coastal windstorms have pounded the shores of Fuveme and its environment with very little protection … the once vibrant fishing community is also gradually being lost due to increased coastal erosion and flooding.
>
> Woman, District, Education or Research Organisation, Ghana

Pastoral/nomadic populations may also be at risk, with climate change threatening traditional ways of life.

> The weather pattern has changed so much in my community of Kajiado, affecting pastoralists. Many animals have died, and as animals were the main source of financial stability, many homes are now starving due to lack of rainfall.
>
> Woman, International, Private Industry, Kenya

The social consequences include unemployment, poverty, and displacement.

Women were also seen to be a group at particular risk of climate-related health impacts, illustrating the finding that climate change impacts differ by gender (van Daalen et al. 2020). As primary caregivers in many settings, the impact of additional childhood disease will fall primarily on women. Domestic duties, mostly carried out by women, are likely to become more difficult. With water less available, it was suggested that menstrual hygiene may become more challenging.

> Prolonged drought dries up the dirty community stream that serves both livestock and residents. This makes it difficult for community members to access water and much harder for menstrual hygiene management for teenage girls leading to an increase in infections in the unbearable heat. Due to the difficulty in managing the monthly menstrual cycle due to limited access to water sanitation hygiene and period poverty, many teenage girls prefer to get pregnant to save them the worry of menstruating monthly for 9 months.
>
> Woman, District, NGO, Nigeria

Other gender-related issues were highlighted. These included an increased risk of **gender-based violence** or of exiting the education system**. Women also typcially have less economic independence or reserves to draw upon during times of financial hardship**. As one respondent noted, loss of livelihoods can force women into desperate measures, such as sex work, in order to acquire the resources to survive.

In the process of carrying out my research on gendered responses to climate change in the Sunyani municipality, I just learned a young lady had to enter into sleeping with men for her needs due to the fact that farming within the area has been affected by climate change. This led to her contracting a sexually transmitted disease that she is now suffering to cure.

Woman, Region, Ministry of Gender, Ghana

## *Access to and Quality of Healthcare*

The examples described above illustrate how climate change can trigger or exacerbate multiple health conditions. In addition, respondents noted how climate-related factors can significantly affect access to health services.

Extreme weather events can damage or lead to the closure of healthcare facilities—temporarily or permanently.

In the facility where I work, floods always enter in when it rains. This affects services for two to three days to allow for the facility to be cleaned up. People are sick and suffer very much.

Woman, Health Facility, MOH, Kenya

Kohistan is a remote district in the northeast of Khyber Pakhtunkhwa; last year flash floods destroyed many health facilities and markets and people had to travel a long distance to reach medical facilities.

Man, Region, Other Global Health Partner, Pakistan

In some cases, access can be disrupted by extreme weather events, such as storms or heavy rains, and the damage they can do to transport infrastructure. This can disrupt access to routine or emergency healthcare.

There was also a pregnant woman in labour. Unfortunately, they couldn't get a strong boat or canoe that could stand the high current waves coming from the seaside. In the process of searching for a better means of taking her to the nearest health centre, she got exhausted and died.

Woman, District, NGO, Nigeria

During the rainy season, it is very difficult for people to seek care for their health needs. They wait for the condition to get worse before coming to the facility.

Man, Health Facility, MOH, Ghana

Transport difficulties can also disrupt medical logistics, leaving facilities short of essential medicines.

Access to health services can also be disrupted by financial barriers, with people unable to afford health services.

> As a result of the disruption in the seasonal shifts, a modest family reliant solely on agriculture experienced the tragic death of their young son within their community. The critical factors involved were as follows: their crop yield plummeted to zero due to their inability to manage the erratic changes in the seasons, and malnutrition, likely compounded by other illnesses, afflicted the family. Faced with financial constraints stemming from the complete failure of their agricultural efforts, they resorted to providing home-based care for their family. Tragically, their youngest son paid the ultimate price with his life. In summary, the ever-changing climate dynamics have left us disoriented and uncertain about the future.
>
> Man, Health Facility, MOH, Democratic Republic of the Congo

## *Responding to the Crisis*

The responses give a strong sense that climate effects on health are already being felt. The view from frontline health workers is that climate change is being experienced at local levels and is interacting with many other factors, often in complex interlinked ways, to affect the health of local populations. As was also noted, this suggests that adaptive solutions need to be multifactorial if they are to address the full range of impacts communities are likely to face.

> Public health is also severely impacted. Diseases linked to poor sanitation, contaminated water, and unhygienic practices, such as those related to dirty hands, have become more prevalent. The scarcity of safe drinking water further exacerbates health issues. Unfortunately, effective waste management policies are lacking, contributing to environmental degradation and health risks.
>
> A crucial factor intertwined with these challenges is overpopulation. The rapid growth of our population adds additional pressure on already strained resources and infrastructure.
>
> **It is evident that these issues require a comprehensive and multifaceted approach to address the complex interactions between climate change, infrastructure, public health, and demography.**
>
> Man, District, NGO, Democratic Republic of the Congo

Although it was not explicitly requested in the survey, some respondents described responses to the climate crisis being taken at the country or more local levels. Multiple countries have developed climate change adaptation strategies or specific initiatives. Examples cited include the Green Ghana strategy (Green Ghana Project—Operation 5000000 Trees 2021) and a One-Health community early warning system in Niger.

> In response to these challenges, a comprehensive One-Health project has been implemented. As part of this project, a Community Early Warning and Emergency Response System has been established to enhance the monitoring of climate-related impacts and facilitate a swift response to emerging health issues. The implementation of this system has proven effective in addressing various alerts and crises in a timely manner.
>
> Man, District, NGO, Niger

Respondents also described the actions they were taking locally, such as planting trees and increasing the amount of fresh produce they grow. Others are working collectively with communities to build resilience to climate change and protect public health.

> In the fight against global warming, we work as a team by forming environmental circles. Regarding the changes we encounter, we highlight disasters, erosions, and fires that occur from time to time, air pollution caused by dust, the overwhelming sun, floods, and many other environmental problems mainly related to deforestation and poor waste management in our environment. These issues expose us to various health problems. With a total of 50 people, which forms 10 circles of 5 people each, we have organised ourselves to raise awareness in the community and explain to them the dangers we face due to the behaviours of environmental incivility that we engage in through our daily actions. Our goal is to raise awareness and foster a sense of climate justice. Among the health problems encountered, we can observe psychosomatic issues. We point out diseases that arise from psychological stress related to disasters, and others fall ill due to an unhealthy environment. Additionally, we highlight infectious and parasitic diseases, allergies, and general systemic diseases, all with their multiple complications. This serves as a lesson in environmental citizenship.
>
> Man, Region, NGO, Democratic Republic of the Congo

## Discussion

This study aimed to generate insights into the experience of climate change and its impacts on health at the community level in the Global South, drawing on the intimate knowledge of community health held by local health practitioners.

The respondents' contributions highlight the reality of climate change as it is being experienced in countries in the Global South, including those designated as climate vulnerable. They emphasise several notable aspects of climate impacts on health.

First, health workers are already noticing climate impacts on health. In general, these are not new or qualitatively different impacts. Rather, climate change is exacerbating existing health challenges or unwinding progress made in the past. Previously controlled infectious diseases are returning, malnutrition is surging, and people's capacity to access health services is being disrupted. For these respondents at least, the health impacts of climate change are not a distant threat but are being experienced now. This is consistent with World Meteorological Estimates that extreme weather, climate, and water-related events were responsible for more than 2 million deaths between 1971 and 2021 (Economic Costs of Weather-Related Disasters Soars but Early Warnings Save Lives 2023).

Second, the contributions highlight how health workers on the ground see climate effects within a broader, more holistic context of environmental and ecosystem disruption. Climate change is affecting the natural environment, but so too is changing land use such as deforestation, urban sprawl, and industrialisation. These factors interact with one another to affect the environment in which people live, and it is often typically difficult to ascribe a specific cause to emerging health problems. Indeed, climate change can be described as a "threat multiplier", acting in concert with multiple other factors to affect living conditions. It may be relatively rare for people to be displaced solely because of climate change, but climate change is likely to be a factor in the majority of displacements (Siegfried 2023).

A third theme emerging from the contribution is the complexity of the health impacts being driven by climate change. Although hotter temperatures and heatwaves can create new health risks, both impacts are indirect and follow complex causal pathways. Climate impacts, interacting with other environmental disruptors, are having a major impact on social determinants of health, the socio-environmental factors that are the key drivers of health status (Ziegler and Muchira 2023).

A fourth theme apparent in many contributions is that the most disadvantaged groups and populations are the ones most affected. In some cases, this is because of biological vulnerabilities, such as the presence of pre-existing health conditions or old age, or the specific nutritional and other needs of children and pregnant women. Mostly, however, it reflects the lack of capacity to respond to a changing and more challenging world. Subsistence farmers or pastoralists often already lead a precarious existence, where even small shifts can have catastrophic consequences on the ability to derive a living. The poorest in society are least able to afford the adaptations required because of a changing climate.

Further research could examine how health workers could provide additional insights into the experience of climate change at the community level and its impacts on health. This could provide a more holistic view of climate impacts, in terms of changing disease burdens (infectious and non-infectious) as well as changes in health-seeking behaviour and access to health services. This more granular and integrated perspective could be used to underpin targeted and tailored responses at national and subnational levels.

There are also opportunities to explore how health workers can work in partnership with communities to prioritise challenges and co-design local risk-mitigation and adaptation strategies.

## Limitations

The major strengths of this study are its large sample size and specific focus on the Global South (particularly Africa). Key limitations include the fact that the sample is self-selected and non-representative of health workers as a whole. The impacts reported are the personal perspectives of health workers, and collection of data to verify the statements made was beyond the scope of the project.

## Conclusions

Although this analysis is not based on a representative sample, it is based on inputs from a large number of respondents primarily based in countries in the Global South. The contributions are not intended to prove that climate change is affecting health: many other rigorous academic studies have demonstrated that beyond reasonable doubt. Rather, they bring to life the reality of climate change and its health impacts as it is being experienced by health professionals and the communities with whom they work.

Previous TGLF studies have demonstrated that members of the TGLF network are highly motivated and have a strong connection to their local communities, seeing it as their role to safeguard the health of local people (Jones et al. 2022). Participation is entirely voluntary, and participants receive no financial compensation for taking part in training activities. Indeed, many use their own resources to participate or implement projects.

Health workers also hold a trusted and influential position within communities (IPSOS Global Trustworthiness Index 2023). In this study, more than three-quarters of respondents described themselves as "part of the family" or "a trusted friend". Some are already demonstrating their commitment to communities by working with them to identify ways to enhance resilience to climate change and safeguard public health.

There is an opportunity to build on this commitment to communities, this position of trust, and the specialists' knowledge of health to better protect communities from the potential health impacts of climate change (Campbell-Lendrum et al. 2023). As previous TGLF peer-learning activities have demonstrated, there is a great appetite among health workers in the Global South to connect to their peers, to share experiences and learn from each other, and to gain input from global experts. As well as empowering health workers with new knowledge, these connections are reported to be inspiring and strengthening motivation (The Geneva Learning Foundation 2023).

Establishing a global network of "climate-concerned" health workers could facilitate the rapid dissemination of knowledge among groups closely connected to communities, provide the inspiration to make a difference, and create a pathway for the adoption and adaption of evidence-based practices from the global level. Such activities could contribute to a more resilient health system (Mosadeghrad et al. 2023) and empower local community-led action to drive adaptation to climate change to mitigate risks to health (Pisor et al. 2022).

## References

Campbell-Lendrum D, Neville T, Schweizer C, Neira M (2023) Climate change and health: three grand challenges. Nat Med 29(7):1631–1638. https://doi.org/10.1038/s41591-023-02438-w

Climate Inequality Report 2023 (n.d.) Fair Taxes for a Sustainable Future in the Global South. WID—World Inequality Database. https://wid.world/news-article/climate-inequality-report-2023-fair-taxes-for-a-sustainable-future-in-the-global-south/

Desai Z, Zhang Y (2021) Climate change and women's health: a scoping review. GeoHealth 5(9). https://doi.org/10.1029/2021gh000386

Doan M, Hill R, Hallegatte S, Corral P, Brunckhorst B, Nguyen M, Freije-Rodriguez S, Naikal E (2023) Counting people exposed to, vulnerable to, or at high risk from climate shocks a methodology. https://documents1.worldbank.org/curated/en/099602511292336760/pdf/IDU07639ca570f3cb048db09bf60fc2cc82df22d.pdf

Economic Costs of Weather-Related Disasters Soars but Early Warnings Save Lives (2023, May 23). World Meteorological Organization. https://wmo.int/news/media-centre/economic-costs-of-weather-related-disasters-soars-early-warnings-save-lives

Foundation TGL, Jones I, Mbuh C, Sadki R, Eller K, Rhoda D (2023, December 1) On the frontline of climate change and health: a health worker eyewitness report. Zenodo. https://zenodo.org/records/10204660

Gibb K, Beckman S, Vergara XP, Heinzerling A, Harrison R (2024) Extreme heat and occupational health risks. Ann Rev Publ Health 45(1). https://doi.org/10.1146/annurev-publhealth-060222-034715

Green Ghana Project—Operation 5000000 Trees (2021) Mlnr.gov.gh. https://greenghana.mlnr.gov.gh

Iannotti L, Kleban E, Fracassi P, Oenema S, Lutter C (2024) Evidence for policies and practices to address global food insecurity. Ann Rev Publ Health 45(1). https://doi.org/10.1146/annurev-publhealth-060922-041451

Intergovernmental Panel on Climate Change (2022) Climate Change 2022: impacts, adaptation and vulnerability. In: IPCC sixth assessment report. IPCC. https://www.ipcc.ch/report/ar6/wg2/

IPSOS Global Trustworthiness Index (2023) Who does the world trust? https://www.ipsos.com/sites/default/files/ct/news/documents/2023-10/Ipsos-global-trustworthiness-index-2023.pdf

Johnston FH, Williamson G, Borchers-Arriagada N, Henderson SB, David MJS, Bowman (2024) Climate change, landscape fires, and human health: a global perspective. Ann Rev Publ Health 45(1). https://doi.org/10.1146/annurev-publhealth-060222-034131

Jones I, Watkins KE, Sadki R, Brooks A, Gasse F, Yagnik A, Mbuh C, Zha M, Steed I, Sequeira J, Churchill S, Kovanovic V (2022, August 22). IA2030 Case Study 7. Motivation, learning culture and programme performance. Zenodo. https://zenodo.org/records/7004304

Jones I, Mbuh C, Sadki R (2023) Listening and learning at teach to reach 7 (IA2030 listening and learning report 4). Zenodo (CERN European Organization for Nuclear Research). https://doi.org/10.5281/zenodo.7766585

Mora C, McKenzie T, Gaw IM, Dean JM, von Hammerstein H, Knudson TA, Setter RO, Smith CZ, Webster KM, Patz JA, Franklin EC (2022) Over half of known human pathogenic diseases can be aggravated by climate change. Nat Clim Change 12(12). https://doi.org/10.1038/s41558-022-01426-1

Mosadeghrad A, Parvaneh Isfahani M, Eslambolchi L, Zahmatkesh M, Afshari M (2023) Strategies to strengthen a climate-resilient health system: a scoping review. Global Health 19(1). https://doi.org/10.1186/s12992-023-00965-2

New Program to Protect Millions from Rising Climate-Related Deaths and Illness (n.d.) World Bank. https://www.worldbank.org/en/news/press-release/2023/12/03/health-program-protect-millions-from-climate-related-deaths-illness

Pisor AC, Basurto X, Douglass KG, Mach KJ, Ready E, Tylianakis JM, Hazel A, Kline MA, Kramer KL, Lansing JS, Moritz M, Smaldino PE, Thornton TF, Jones JH (2022) Effective climate change adaptation means supporting community autonomy. Nat Clim Chang 12(3):213–215. https://doi.org/10.1038/s41558-022-01303-x

Romanello M, Di Napoli C, Green C, Kennard H, Lampard P, Scamman D, Walawender M, Ali Z, Ameli N, Ayeb-Karlsson S, Beggs PJ, Belesova K, Ford LB, Bowen K, Cai W, Callaghan M, Campbell-Lendrum D, Chambers J, Cross TJ, Robin K (2023) The 2023 report of the lancet countdown on health and climate change: the imperative for a health-centred response in a world facing irreversible harms. Lancet 402(10419). https://doi.org/10.1016/s0140-6736(23)01859-7

Scott AJ, Webb TL, Martyn-St James M, Rowse G, Weich S (2021) Improving sleep quality leads to better mental health: a meta-analysis of randomised controlled trials. Sleep Med Rev 60(60):101556. https://doi.org/10.1016/j.smrv.2021.101556

Siegfried K (2023, November 15) Climate change and displacement: the myths and the facts. UNHCR. https://www.unhcr.org/news/stories/climate-change-and-displacement-myths-and-facts

The Geneva Learning Foundation (TGLF). (2023) Teach to reach 9: shared experiences (1.0). Teach to reach: connect 9, Online. The Geneva Learning Foundation. https://doi.org/10.5281/zenodo.10062521

van Daalen K, Jung L, Dhatt R, Phelan AL (2020) Climate change and gender-based health disparities. Lanc Planet Health 4(2):e44–e45. https://doi.org/10.1016/s2542-5196(20)30001-2

Wainberg ML, Scorza P, Shultz JM, Helpman L, Mootz JJ, Johnson KA, Neria Y, Bradford J-ME, Oquendo MA, Arbuckle MR (2019) Challenges and opportunities in global mental health: a research-to-practice perspective. Curr Psychiat Rep 19(5). https://doi.org/10.1007/s11920-017-0780-z

White BP, Breakey S, Brown M, Smith JA, Tarbet A, Nicholas PK, Viamonte AM (2023) Mental health impacts of climate change among vulnerable populations globally: an integrative review. Ann Glob Health 89(1):66–66. https://doi.org/10.5334/aogh.4105

Weeda LJZ, Bradshaw CJA, Judge MA, Saraswati CM, Le PN (2024) How climate change degrades child health: a systematic review and meta-analysis. Sci Total Environ 170944–170944. https://doi.org/10.1016/j.scitotenv.2024.170944

World Health Organization (2019, August 9) Climate change. Who.int; World Health Organization: WHO. https://www.who.int/health-topics/climate-change#tab=tab_1

World Health Organization (2022, December 19) Ambient (outdoor) air quality and health. Who.int; World Health Organization: WHO. https://www.who.int/news-room/fact-sheets/detail/ambient-(outdoor)-air-quality-and-health

Wright CY, Kapwata T, Naidoo N, Asante KP, Arku RE, Cissé G, Simane B, Atuyambe L, Berhane K (2024) Climate change and human health in Africa in relation to opportunities to strengthen mitigating potential and adaptive capacity: strategies to inform an African "Brains Trust." Ann Glob Health 90(1). https://doi.org/10.5334/aogh.4260

Ziegler C, Muchira J (2023) Climate change. Primary Care Clin off Pract 50(4):645–655. https://doi.org/10.1016/j.pop.2023.04.010

# Chapter 5
# Epidemiological Modelling of Climate-Dependent Diseases: Case of Dengue Fever in Burkina Faso

**Domêzag Jean-Jacques Dabiré, Pascal Yaka, and Essoninam Passiké Pokona**

**Abstract** With the aim of strengthening dengue fever (DF) surveillance and management in Burkina Faso, our study was carried out in the Sahel, Centre and Hauts-Bassins regions. In addition to improving knowledge of DF-climate relationships, the study aimed to characterise the seasonal evolution of DF, highlighting the climatic parameters most strongly correlated with this disease, and develop models for forecasting DF incidence based on these climatic parameters. This work was carried out between 2016 and 2022 using monthly climate, epidemiological, and demographic data. First, linear regression models were generated. In a second stage, backward elimination calibration, followed by manual calibration, was used to eliminate the least relevant parameters to obtain the final models. Third, the variance inflation factor test, Breusch-Pagan test, Durbin-Watson test or Likelihood ratio test were used to verify statistical hypothesis. Fourth, the Taylor diagram and a graphical representation were used to validate the models. The incidence of DF can be predicted by: sunshine duration, rainfall, atmospheric pressure and vapor pressure in the Centre region; maximum temperature, minimum temperature, wind and atmospheric pressure in the Hauts-Bassins region; rainfall and wind in the Sahel region. These models showed fairly conclusive results during validation.

**Keywords** Epidemiological model · Climate-human health · Dengue fever · Burkina Faso

D. J.-J. Dabiré (✉)
AGRHYMET/ANAM, Ouagadougou, Burkina Faso
e-mail: domezagdabire@gmail.com; domezag_dabire@yahoo.fr

P. Yaka
PNUD, Ouagadougou, Burkina Faso

E. P. Pokona
GIZ, Lome, Togo



W. Leal Filho and F. Wolf (eds.), *Climate Change and Health: Perspectives from Developing Countries*, Climate Change Management,
https://doi.org/10.1007/978-3-031-93177-2_5

# Introduction

Vector-borne diseases are the most common health risk worldwide (Nakhapakorn et al. 2005). After malaria, dengue fever (DF) is the second most common infectious disease worldwide, in terms of morbidity and mortality (Wu et al. 2010). In tropical and subtropical regions where DF is present, 2500 million people are exposed to it (Gharbi et al. 2011), with 390 million infections (Wu et al. 2010) and 20,000 deaths recorded annually (Gharbi et al. 2011). In Burkina Faso, DF has been resurgent since the middle of the 2010–2019 decade with numerous cases regularly reported (Seogo et al. 2021). Historically, in Haute-Volta, today's Burkina Faso, the first cases of DF were reported in the literature in 1925 (Amarasinghe et al. 2011). Other studies have reported outbreaks, in particular in 1982 (Amarasinghe et al. 2011), 2006 (Collenberg et al. 2006), 2013 (Tarnagda et al. 2014), and 2016 (Seogo et al. 2021). As a result, DF constitutes an additional concern for the country, due to its limited income. Indeed, the country must find additional resources for staff training and the purchase and provision of diagnostic tests. Added to this are the economic costs incurred by the unavailability of sick workers and those caused by deaths. In this context of poverty, these economic losses could hamper the country's development (Seogo et al. 2021).

Several factors can influence the emergence or resurgence of DF cases. These include social factors (human behaviour); biological factors (viral, human and vector organisms, introduction of new serotypes, herd immunity of the population); environmental factors (land use, urbanisation, type of vegetation cover); population movements such as migrations and their evolution; and climatic factors (Nakhapakorn et al. 2005; Amarakoon et al. 2008; Zambrano et al. 2012). Regarding climatic factors, parameters such as temperature, humidity, evaporation, and rainfall act directly on the development conditions of mosquitoes and the virus responsible for DF. Some of these parameters can also act indirectly on mosquito development conditions, by regulating evaporation and influencing land occupation and use.

Given the above, a strategy for monitoring and managing DF seems necessary. One aspect of this strategy is the ability to predict the appearance and evolution of DF cases. Indeed, early warning of DF epidemics could improve the effectiveness of vector control campaigns and help target preventive actions. Early interventions could delay the spread of epidemics, reducing their impact on the healthcare system (Gharbi et al. 2011). Healthcare facilities could adapt their response in terms of beds availability and mobilisation of human and material resources (Gubler 1998).

To do this, our study aims to contribute to a better understanding of the relationship between climate and human health. More specifically, in each climatic zone of Burkina Faso, it aims to characterise the seasonal evolution of DF, highlight the main climatic parameters influencing the onset and evolution of DF, and propose prediction models based on these relevant climatic parameters.

## Study Zones

For the purposes of this study, we have chosen a study area in each of Burkina Faso's climatic zones. Located in the heart of West Africa, this country has a dry season from November to May and a rainy season from June to October. The dry season is generally cold from November to February and hot from March to May. The country is subdivided into three climatic zones: the Sahelian zone (rainfall below 600 mm), the Sudano-sahelian zone (rainfall between 600 and 900 mm) and the Sudanian zone (rainfall above 900 mm). The regional level attracted our attention as it is the second administrative and health level after the national level. For the Sahelian climatic zone, the Sahel region was chosen; the Centre region for the Sudano-sahelian zone and the Hauts-Bassins region for the Sudanian zone (Fig. 5.1). These regions were chosen because all or most of their surface area lies within the climatic zones concerned, but also because of their population density and the quality of the data available. Located in the extreme north of the country, the Sahel region covers an area of 36,166 km$^2$ with a population density of 31 inhabitants/km$^2$. The Centre region is located in the centre of the country, with a density of 1014 inhabitants per km$^2$ over an area of 2826.26 km$^2$. With a surface area of 25,499 km$^2$, the Hauts-Bassins region is located in the west of Burkina Faso with a density of 88 inhabitants per km$^2$ (NISD 2022).

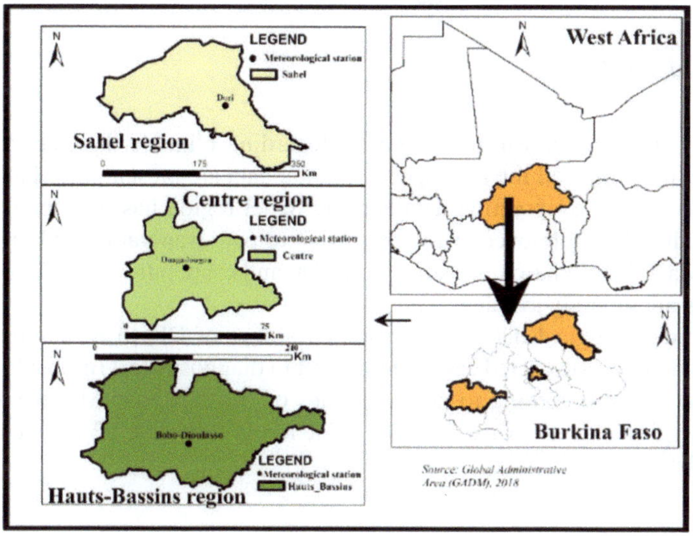


**Fig. 5.1** Study zones (Sahel, Centre and Hauts-Bassins regions)

**Table 5.1** Characteristics of the data used for each region

| Type of data | Parameters (lag 0–6 for climatic parameters) | Time step and period | Data source |
|---|---|---|---|
| Climate | Minimum temperature (Tnm), maximum temperature (Txm), mean temperature (Tmoym), Minimum humidity (Hnm), maximum humidity (Hxm), mean humidity (Hmoym), vapour pressure (Em), sunshine duration (Im), precipitation (Pm), atmospheric pressure (Pam), reference evapotranspiration (ETPm), pan evaporation (Evm), wind speed (Vm) | Monthly 2016–2022 | National Meteorology Agency (NAMA) |
| Epidemiology | Number of DF cases | Monthly 2016 to 2022 | Sectoral statistics departement of health |
| Demography | Number of persons | Monthly 2016 to 2022 | National Institute of Statistics and Demography (NISD) |

## Materials and Methodology

### *Data and Tools*

The study data are shown in Table 5.1. Regarding climatic data, apart from the Centre region where all the announced climatic data are present, the Hauts-Bassins region has no pan evaporation data and the Sahel region has no vapour pressure, sunshine duration, reference evapotranspiration, pan evaporation and atmospheric pressure data. Precipitation, sunshine duration, and evaporations are monthly sums; the others are monthly averages.

For epidemiological data, we considered a case of DF, to be any case with a clinical diagnosis (suspected case) and/or a rapid diagnostic test (probable case) and confirmed or not according to the guidelines of Burkina Faso Ministry of Health (Ministry of Health, Burkina Faso 2017). R for Windows 4.2.2 has been used to process data.

### *Methodology*

The methodology is defined in 7 steps.

## *Data Preparation*

We transformed the number of DF cases into DF incidence according to Eq. (5.1) in order to compare data from the 3 regions.

$$IDF = \frac{100,000 * NDF}{P} \tag{5.1}$$

IDF: DF incidence; NDF: number of DF cases; P: regional population.

## *Seasonal Variability of IDF*

The aim was to produce a graphical representation of IDF based on monthly averages for the period 2016–2022. This approach aimed to highlight the periods and seasons that favour the start, development, and end of DF cases.

## *Statistical Processing*

Statistical processing involved Pearson correlation analysis. This analysis was based on the correlation coefficient (Eq. 5.2) and the statistical significance (*p*-value) at the 5% threshold.

$$r = \frac{Cov(X, Y)}{\sigma_X \sigma_Y} \tag{5.2}$$

*r*: correlation coefficient; $Cov(X, Y)$: covariance of variables X and Y; $\sigma_X \sigma_Y$: standard deviation of variables X and Y.

The correlation was calculated firstly between IDF data and climatological parameters, and secondarily between climatological parameters. The aim was to retain only those climatic variables that correlated significantly with the IDF and were not strongly correlated with other climatic parameters ($r < 0.85$). Given the life cycle of the DF vector (around 1–3 months), these analyses were carried out at several lag levels of climatological data in relation to epidemiological data: 0, 1, 2, 3, 4, 5 and 6 months.

## *Modelling*

The period from January 2016 to December 2021 was used to train the models. The parameters retained from the correlation analysis were used for modelling by

**Table 5.2** Hypothesis test of models and their characteristics

| Test | Characteristics | Interpretation |
|---|---|---|
| VIF (Variance Inflation Factor) (Akaike 1974) | Linear independence of residuals | if VIF < 5, no collinearity if VIF > 5, collinearity exist |
| Breusch-Pagan test (Breusch et al. 1979) | Constancy of residuals | if *p*-value < 0.05, heteroscedasticity if *p*-value > 0.05, homoscedasticity |
| Durbin-Watson test (DW) (Durbin et al. 1971) | Autocorrelation of residuals | if 1.5 < DW value < 2, no autocorrelation otherwise there are autocorrelation |
| Likelihood ratio test | Level of model improvement | if *p*-value > 0.05, no difference between 2 models if *p*-value < 0.05, there is difference between 2 models |

multiple linear regression using ordinary least squares (OLS) or generalised least squares (GLS) (Nakhapakorn et al. 2005; Gama et al. 2013; Pokona et al. 2021).

## *Calibration*

First, calibration using the backward method in R software enabled the least relevant variables to be eliminated (Pokona et al. 2021; Jolion 2003). Then, through manual elimination, variables with high inflation and those with insignificant coefficients were removed. Finally, by analysis of variance (ANOVA), the best models regressed by GLS were retained.

## *Hypothesis Testing*

Models are stable if residuals are constant, linearly independent, and not autocorrelated (Tomassone et al. 1983) or are autocorrelated but with good allowance for this autocorrelation (Fox et al. 2018). The criteria are presented in Table 5.2.

## *Validation*

The period from January to December 2022 was used to validate the models. The Taylor diagram and graphical representation were used to compare observed and simulated values. The Taylor diagram gives the standard deviation ($\sigma$), the root mean square error (RMSE), and correlation coefficient (r). Thus, a model will be

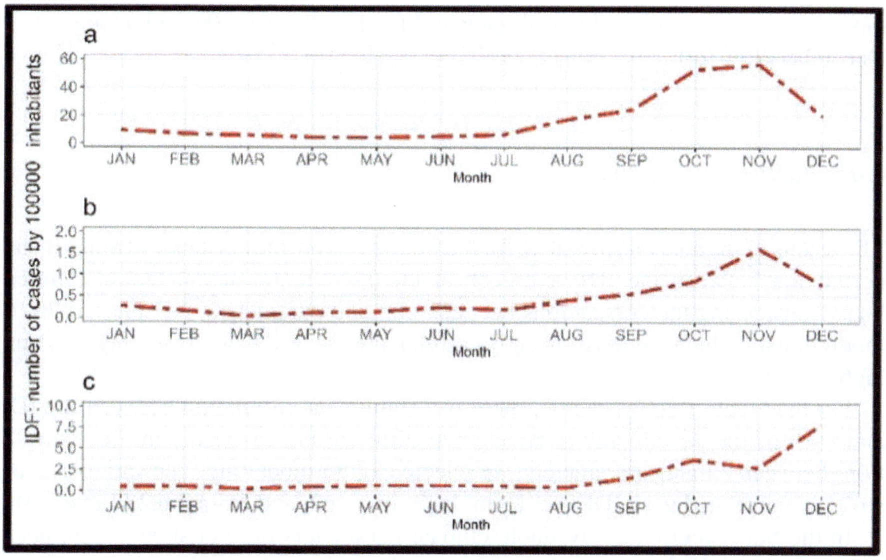


**Fig. 5.2** Seasonal evolution of IDF: **a** in Centre region; **b** in Hauts-Bassins region; **c** in Sahel region

validated if r, in absolute value, tends towards 1, RMSE is low, and $\sigma$ is close to the reference[1] (Taylor 2001). The graphical representation shows whether the simulated values can indicate the start, peak and end of DF disease.

## Analysis

### *Seasonal Variability of IDF*

**Centre region**: IDF analysis shows that the disease is present throughout the year. It varies from 4 cases/100,000 inhabitants in May to 54 cases/100,000 inhabitants in November. IDF increases progressively during the rainy season, peaking 1–2 months after the end of the rains, in November (Fig. 2a).

**Hauts-Bassins region**: IDF remains low throughout the year, with periods when it is almost zero (March–April). It varies between 0 cases/100,000 inhabitants and 2 cases/100,000 inhabitants in November. IDF increases progressively during the rainy season, peaking 1 month after the end of the rains, in November (Fig. 2b).

**Sahel region**: IDF remains low from January to August; it increases during the rainy season, peaking in December, three months after the end of the rainy season. It

[1] Observed data $\sigma$.

ranges from 0 cases/100,000 inhabitants in March to 8 cases/100,000 inhabitants in December (Fig. 2c).

## *Correlations Analysis*

The results of the analysis revealed, for the Centre region, the existence of a significant correlation between the IDF and each of the 13 climatic parameters at different lags. Considering the correlations between climatic parameters, we selected 20 for modelling, of which the most strongly correlated were: Pm_lag3, Hnm_lag2 et Pam_lag6 (Table 5.3).

For the Hauts-Bassins region, there is a significant correlation between the IDF and each of the 12 climatic parameters at different lags. Considering the correlations between climatic parameters, we selected 22 for modelling. The variables most strongly correlated with IDF are: Hnm_lag3, Im_lag3, and Hnm_lag2 (Table 5.4).

In the Sahel region, all available climate parameters did not show a significant correlation with IDF: these include average relative humidity and minimum temperature. Considering the correlations between climatic variables, only 6 parameters at different lags were retained (Table 5.5).

**Table 5.3** *r* and *p*-value of the variables retained in the Centre region

| Parameters | Vm_lag0 | Em_lag1 | Vm_lag1 | Im_lag2 | Hnm_lag2 |
|---|---|---|---|---|---|
| *r* | – 0.46 | 0.36 | – 0.37 | – 0.43 | 0.60 |
| *p*-value | $9.7 \times 10^{-6}$* | $6.4 \times 10^{-4}$* | $5.0 \times 10^{-4}$* | $3.6 \times 10^{-5}$* | $1.8 \times 10^{-9}$* |
| Parameters | Im_lag3 | Pm_lag3 | Em_lag4 | Im_lag4 | Tnm_lag4 |
| *r* | – 0.49 | 0.61 | 0.45 | – 0.37 | 0.31 |
| *p*-value | $1.8 \times 10^{-6}$* | $6.1 \times 10^{-10}$* | $1.4 \times 10^{-5}$* | $5.6 \times 10^{-4}$* | $4.4 \times 10^{-3}$* |
| Parameters | Vm_lag4 | Pm_lag4 | Em_lag5 | Tnm_lag5 | Vm_lag5 |
| *r* | 0.27 | 0.42 | 0.3 | 0.45 | 0.38 |
| *p*-value | $1.1 \times 10^{-2}$* | $6.4 \times 10^{-5}$* | $5.0 \times 10^{-3}$* | $1.7 \times 10^{-5}$* | 3.0 $10^{-4}$* |
| Parameters | Pam_lag5 | Tnm_lag6 | Vm_lag6 | Evm_lag6 | Pam_lag6 |
| *r* | – 0.31 | 0.47 | 0.32 | 0.52 | – 0.58 |
| *p*-value | $4.5 \times 10^{-3}$* | $5.1 \times 10^{-6}$* | $3.0 \times 10^{-3}$* | $3.4 \times 10^{-7}$* | $7.1 \times 10^{-9}$* |

* Significance at 5%

**Table 5.4** $r$ and $p$-value of the variables retained in the Hauts-Bassins region

| Parameters | Tnm_lag0 | Vm_lag0 | Hxm_lag1 | Vm_lag1 | Im_lag2 | Txm_lag2 |
|---|---|---|---|---|---|---|
| $r$ | – 0.25 | – 0.29 | 0.23 | – 0.39 | – 0.33 | – 0.34 |
| $p$-value | $2.0 \times 10^{-2}$* | $7.4 \times 10^{-3}$ | $2.9 \times 10^{-2}$* | $2.2 \times 10^{-4}$* | $1.7 \times 10^{-3}$* | $1.2 \times 10^{-3}$* |
| Parameters | Hnm_lag2 | Vm_lag2 | Pm_lag2 | Im_lag3 | Txm_lag3 | Hnm_lag3 |
| $r$ | 0.42 | – 0.25 | 0.33 | – 0.46 | – 0.41 | 0.47 |
| $p$-value | $5.0 \times 10^{-5}$* | $1.7 \times 10^{-2}$* | $2.0 \times 10^{-3}$* | $9.3 \times 10^{-6}$* | $1.0 \times 10^{-4}$* | $4.3 \times 10^{-6}$* |
| Parameters | Pm_lag3 | Im_lag4 | Txm_lag4 | Hnm_lag4 | Pm_lag4 | |
| $r$ | 0.37 | – 0.34 | – 0.31 | 0.39 | 0.30 | |
| $p$-value | $4.3 \times 10^{-4}$* | $1.3 \times 10^{-3}$* | $4 \times 10^{-3}$* | $1.8 \times 10^{-4}$* | $5.1 \times 10^{-3}$* | |
| Parameters | Em_lag5 | Im_lag5 | Tnm_lag6 | Vm_lag6 | Pam_lag6 | |
| $R$ | 0.24 | – 0.22 | 0.27 | 0.28 | – 0.28 | |
| $p$-value | $1.8 \times 10^{-3}$* | $3.7 \times 10^{-2}$* | $1 \times 10^{-2}$* | $7.9 \times 10^{-3}$* | $7.8 \times 10^{-3}$* | |

* Significance at 5%

**Table 5.5** $r$ and $p$-value of the variables retained in the Sahel region

| Parameters | Vm_lag1 | Hnm_lag3 | Vm_lag4 | Pm_lag4 | Tmoym_lag6 | Vm_lag6 |
|---|---|---|---|---|---|---|
| $r$ | – 0.22 | 0.26 | 0.22 | 0.51 | 0.24 | 0.24 |
| $p$-value | $3.6 \times 10^{-2}$* | $1.4 \times 10^{-2}$* | $0.4 \times 10^{-1}$* | $4.0 \times 10^{-7}$* | $2.2 \times 10^{-2}$* | $2.1 \times 10^{-2}$* |

* Significance at 5%

## Modelling and Calibration

For each study area, the selected parameters were used for modelling. After calibration, the results in Tables 5.6 and 5.7 were obtained.

In the Centre region, Im_lag2, Pm_lag3, Em_lag4, and Pam_lag6 are the variables present in the final model. In fact, the coefficient of Pam_lag6 is not significant but this variable contributes to the model's stability by containing variance inflation.

In Hauts-Bassins region, Txm_lag4, Pam_lag6, Tnm_lag0, and Vm_lag0 are explanatory variables, but the last two help to contain variance inflation.

Pm_lag4 and Vm_lag4 are the variables of the model for the Sahel region.

**Table 5.6** Final models and characteristics in the Centre and Sahel regions

| Region | Model equation | AIC | BIC | Pseudo $R^2$ (Cragg and Uhler) |
|---|---|---|---|---|
| Centre | $IDF_C = \exp(-0.005 Im\_lag2 + 0.002\,Pm\_lag3 + 0.048 Em\_lag4 - 0.096 Pam\_lag6 + 99.237) - 1$ | 140.381 | 160.871 | 0.37 |
| Sahel | $IDF_S = \exp(0.882\,Vm\_lag4 + 0.005 Pm\_lag4 - 1.043) - 0.3$ | 174.268 | 187.928 | 0.26 |

**Table 5.7** Final models and characteristics in the Hauts-Bassins region

| Region | Model equation | $R^2$ | $R^2$ adjusted | *p*-value |
|---|---|---|---|---|
| Hauts-Bassins | $IDF_{HB}$ = (0.137Txm_lag4 - 0.160Tnm_lag0 + 0.025Vm_lag0 + 0.486Pam_lag6 -496.323)^(-0.42) – 0.5 | 0.23 | 0.181 | 0.001555 |

**Table 5.8** Results of hypothesis tests

| Models | VIF | Breusch-Pagan | Durbin-Watson | Likelihood ratio test | Conclusion |
|---|---|---|---|---|---|
| $IDF_C$ | 1.07 to 1.36 no colline-arity | *p*-value = 0.793 homosc-edastic | Not required | 54.994 *p*-value < 0.0001 | Criteria of validation are ok |
| $IDF_{HB}$ | 1.10 to 2.07 no colline-arity | *p*-value = 0.688 homosc-edastic | 1.845 *p*-value = 0.359 not autocorrelated | Not required | Criteria of validation are ok |
| $IDF_S$ | = 1.09 no colline-arity | *p*-value = 0.120 homos-cedastic | Not required | 14.198 *p*-value = 0.0008 | Criteria of validation are ok |

## *Verification of Hypothesis Tests*

The verification tests showed satisfactory results (Table 5.8). In the Centre region, the likelihood ratio test between the final model retained and the final model without correlation structure, which is equal to 55 (*p*-value < 0.0001), showed a significant difference in favour of the final model, proof of good autocorrelation support. The same result was obtained for Sahel region, with 14 as the likelihood ratio test (*p*-value = 0.0008).

## *Validation of Models*

For the Centre and Sahel regions, the Taylor diagram shows fairly strong model forecasting ability. Standard deviations are fairly close and correlation coefficients fairly high. RMSE are low.

Model forecasting ability for the Hauts-Bassins region is less strong. RMSE is fairly high, with very different standard deviations. The correlation coefficient is fairly high (Table 5.9).

The analysis of the seasonal cycle of observed and forecast data for 2022 showed that the models for the Centre and Sahel regions achieved a good seasonal representation of IDF. The result is less satisfactory for the Hauts-Bassins region (Fig. 5.3).

**Table 5.9** Taylor diagram results

| Regions | Data type | σ | RMSE | r |
|---|---|---|---|---|
| Centre | Observed | 0.6 | 0.54 case/100000 inhabitants | 0.73 |
| | Simulated | 0.7 | | |
| Sahel | Observed | 0.7 | 0.43 case/100000 inhabitants | 0.76 |
| | Simulated | 0.6 | | |
| Hauts-Bassins | Observed | 1.3 | 0.9 case/100000 inhabitants | 0.67 |
| | Simulated | 0.6 | | |

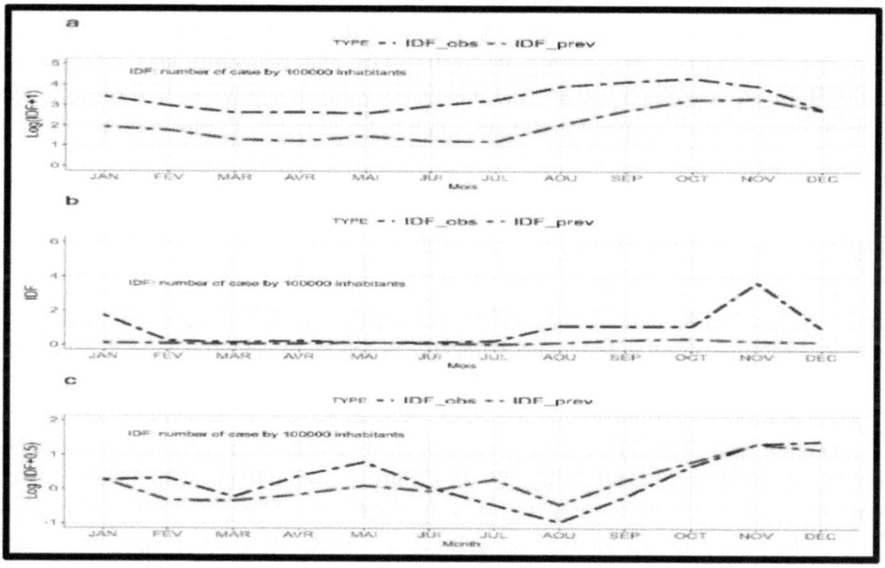


**Fig. 5.3** Observed and simulated data of IDF in 2022: **a** Centre region; **b** Hauts-Bassins region; **c** Sahel region

# Discussion

## *Seasonal Variability of IDF*

In all three regions, DF incidence can be observed almost all year round, which could indicate the existence of favourable climatic conditions, particularly humidity (availability of water), for vector development through the year. Indeed, the existence of wastewater and water storage for various uses could explain this availability during the dry season. The IDF then gradually increases with the onset of the rainy season, reaching its maximum 1–3 months after the end of the rains. The regularity of the rains increases the number of water points necessary for the formation of larval breeding sites. This situation is corroborated by the work of Cazelles et al. (2005) and Poveda et al. (2000). Seogo et al. (2021), based on the work of Lozano et al. (2012), confirmed that the period between October and November corresponds both to the end of the season and to an increase in Aedes populations. The peak IDF production period varies from region to region. This disparity is probably due to the life cycle of the mosquito vector in general and the lifespan of the adult mosquito in particular, both of which are strongly influenced by local environmental conditions. Beserra et al. (2006) and Failloux et al. (2003) have stated that the lifespan of Aedes mosquitoes that have reached the imago stage can vary from two to five weeks, or even eighty days under certain conditions. The gap between the end of the season and peak of IDF could be explained by the fact that the end of the rains marks the start of a period when water is available for vectors, with fewer hazards likely to destroy mosquito habitats (larval breeding sites, refuges for imago mosquitoes) and influence their flight activity. Such hazards include heavy rain, strong winds and run-off. The harmful action of these phenomena has been highlighted by Gama et al. (2013).

## *Influences of Explanatory Variables*

In the study areas, statistical analysis has revealed that rainfall, temperature, humidity, vapour pressure, wind, atmospheric pressure at sea level, and sunshine duration are correlated with IDF at different lags. The focus here is on the lags used in the models. Several authors like Singh et al. (2022), Ouattara et al. (2022), Cazelles et al. (2005), and Gagnon et al. (2001) have succeeded in establishing an association between DF epidemics and climatic variability. For the Centre region, IDF is negatively correlated with Im_lag2 and Pam_lag6; it is positively correlated at Pm_lag3 and at Em_lag4. In the Hauts-Bassins region, IDF is negatively correlated with Txm_lag4, Pam_lag6, Tnm_lag0, and Vm_lag0. In the Sahel region, Vm_lag4 and Pm_lag4 are positively correlated with IDF. Like the work of Islam et al. (2023) in Bangladesh, climatic parameters present various correlations with IDF, depending on the influence of these parameters on the dynamic of DF.

## *Centre Region*

High IDF in November is explained by low Im_lag2 (2 months before November), low Pam_lag6 (6 months before November), high Pm_lag3 (3 months before November) and high Em_lag4 (4 months before November). Several authors have confirmed the existence of relationships between DF and climatic parameters with time lags. This is the case of Amarakoon et al. (2008), who showed that links existed between DF and temperature with a lag of up to 6 months, and between DF and precipitation with a lag of 1–3 months. For Wang (2018), based on the work of other authors, Aedes eggs can survive in the absence of water for 4–6 months when dissected and then hatch on contact with water to give larvae. However, dissected eggs require favourable conditions of temperature and relative humidity for their preservation and to keep them ready to hatch when water is available. This is what Li et al. (2013) report when they state that temperature and relative humidity are crucial factors in the spread of the DF virus at every stage of the mosquito's life cycle. The correlation between IDF and these 4 variables in the Centre region could therefore be explained by the fact that Aedes, the vector of the disease, requires certain climatic conditions for its various aquatic and aerial development phases. Em_lag4 would represent the temperature and humidity conditions influencing dissected eggs pending the availability of water; Em_lag4 is calculated from temperature and used to calculate relative humidity. As for Pm_lag3, it not only provides the vital environment essential for dissected eggs to continue their cycle, but also extends mosquito distribution zones for new broods and supplies water to existing breeding grounds. Finally, Pm_lag3 regulates relative humidity, which can ensure the longevity of adult mosquitoes (Gama et al. 2013). Im_lag2 is thought to influence vector (larvae) behaviour, as confirmed by Christophers (1960). He found that exposing Aedes larvae to light in the laboratory tended to cause them to cluster at one point in the crop; this is detrimental to them because of the competition for nutrients that is created by clustering. Reduced sunlight duration would mean less exposure to light and therefore a greater number of vectors available for DENV transmission. Sunshine duration also indirectly affects temperature; a high sunshine duration could correspond to high temperatures that are then harmful to the vector during these development phases. Morin et al. (2013), citing other authors, asserted that high temperatures damage mosquitoes and limit virus transmission. Pam_lag6 is considered mosquito flight activity, as shown by the work of Li et al. (2013). The authors were relatively evasive, but it is thought to act as a low-frequency signal explaining mosquito flight activity 6 months later. From the moment water is available, it takes around 3 months to see cases of DF. From hatching to adulthood, it takes between 4 and 10 days. According to Beserra et al. (2006), Failloux et al. (2003), Natal (2002), and Wang (2018), the adult phase can last from 14 to 35 days on average, but up to 80 days under certain conditions. This gives a total of 18–90 days. Added to this is the incubation period of the disease, which lasts around 10 days, and the duration disease manifestation, which can last up to 7 days. Bearing in mind that most patients are generally tested at the end of the process described above, it could take between 35 and 107 days between the availability of water and the cases observed. Our results fall well within this range. In fact, from the beginning of the adult stage, the mosquito can potentially transmit the disease. However, this requires several conditions to be met, namely, that the mosquito can take a blood meal from a person infected with DF, that it can bite a host not immune to the same virus serotype, and that it can do so after the extrinsic incubation period (EIP). Encountering eligible hosts capable of generating mass transmission would be more likely following displacement of the vector or hosts. All these transmissions conditions would therefore require time, hence the 3- month delay.

## *Hauts-Bassins Region*

The higher the variables Txm_lag4, Tnm_lag0, Vm_lag0 and Pam_lag6 are, the lower the IDF. The lower these variables are, the greater the IDF. Thus, a high IDF in November is explained by low Txm_lag4, Pam_lag6, Tnm_lag0, and Vm_lag0 in July (4 months before November), May (6 months before November), and November, respectively. As in the Centre region, Aedes requires favourable climatic conditions during these different development phases. Txm_lag4 would represent the temperature conditions influencing dissected eggs while waiting for the availability of water.

Next, Pm_lag3 (not present in the model but significantly positively correlated with IDF) provides the living environment for the dissected eggs, extends the Aedes distribution area, feeds existing roosts, and regulates relative humidity. Pam_lag6 may influence the flight activity of mosquitoes. Tnm_lag0 may damage the vector and limit virus transmission at the time of infection. The references given for the Centre region are valid for the Hauts-Bassins. Wind (Vm_lag0) in the Hauts-Bassins region can be relatively high (between 1.5 and 4.3 m/s), which would result in low vector density and therefore low IDF values (Li et al. 2013). In addition, Vm_lag0 affects the vector's biting activity: bites are effective when the wind is relatively weak and difficult when the wind is relatively strong. From the moment water is available, it takes around 3 months to observe cases of DF. Given the interval of 35–107 days between water availability and observed cases, our results are in good agreement.

## *Sahel Region*

The higher Vm_lag4 and Pm_lag4 are, the higher the IDF. The lower these variables are, the lower the IDF. Thus, a high incidence in December is explained by high Pm_lag4 and relatively high Vm_lag4 in August (4 months before December). As in the previous two regions, Pm_lag4 provides the living environment for dissected eggs, extends the Aedes range, feeds existing breeding sites, and regulates relative humidity. As for the wind, its value is generally low (between 0.32 and 2.40 m/s), which provides favourable conditions for mosquito development by keeping larval breeding sites in place. The work of Gama et al. (2013) provides further details on this subject. However, an increase in wind speed improves the mobility of vectors and therefore the spread of the virus. Arsin et al. (2020) confirmed that the vector can travel on average 40 m/day but can reach 100 m/day with favourable winds. From the moment water is available, it takes approximately 4 months to observe cases of DF. Given the interval of 35–107 days between water availability and observed cases, our results are slightly above the upper limit of the interval. There may be several reasons for this, such as differences in the duration of certain phases of the mosquito's life or in the social behaviour of the hosts towards the disease. In addition, the time scale chosen for the study (monthly) could also mean that the incidence in the month of December could be due to cases observed only in the first decade, reducing the elapsed time to 100 days instead of 120 days. Finally, the quality of the collected data or the absence of real links could also explain this discrepancy. As in the Centre and Hauts-Bassins regions, the time required for conditions of transmission would explain the 4-month delay. The region's low wind speed and relatively low population density (31 inhabitants/km$^2$) reinforce the idea that the vector needs more time to spread the disease.

## *Model Quality*

Model quality depends on the local climatic context (Morin et al. 2013). In the Centre and Sahel regions, periods of low IDF as well as periods of higher incidence are fairly well reproduced in the models, with respective r values of 0.73 for the Centre region and 0.76 for the Sahel region. The model for the Hauts-Bassins region has an r value of 0.67; the different phases of IDF's evolution are moderately reproduced. For Dubé et al. (2007), comparing model predictions with observed data remains the ideal way of testing model validity, by demonstrating a certain level of agreement between the two.

Statistically, these models presented fairly satisfactory results.

## *Limitations of Study*

Despite the results obtained in this study, certain limitations can be noted.

The lack of depth of epidemiological data is a crucial problem in epidemiological modeling. The study was carried out over 7 years, from 2016 to 2022, which does not allow the consideration of many situations. In addition, the quality of the epidemiological and climatic data, the failure to consider explanatory variables other than climatic (migration of populations, social characteristics, soils, vegetation, hydrology etc.), and the lack of a detailed study of the impact of climate change on the health of the population, are some of the main problems encountered. There is also the absence of an epidemiological threshold clearly defined by the Burkina Faso health authorities, which makes it impossible to work on the basis of a reference. Finally, the risk of encountering years whose data differs exceptionally from those used to train the model could render it totally ineffective.

## Conclusion

Burkina Faso has been regularly facing DF outbreaks since 2016, when this pathology was included in the list of diseases subject to weekly monitoring. In addition to the loss of human life it causes, this disease deals a serious blow to the country's already battered healthcare system and economy. This is why it is imperative that tools or solutions be put in place to enable better monitoring and management of DF. With this in mind, our study set out to improve knowledge of the relationships between DF and climate in three regions of Burkina Faso: Centre region, Hauts-Bassins region, and Sahel region. To achieve this, the aim of each study area was to highlight the seasonal evolution of the DF, then to identify the climatic parameters best correlated with DF, and finally, to establish a model based on climatic parameters for the prediction of DF cases.

The descriptive analysis revealed that the DF is more or less present in the 3 regions throughout the year. However, the rainy season is the preferred period for this disease, with a peak occurring 1–3 months after the end of the rains, i.e., in November for the Centre and Hauts-Bassins regions and in December for the Sahel region.

The results of the correlation analysis in each region showed that the majority of the climatic parameters presented are correlated with DF, at different time lags. However, 20 of them in the Centre region, 22 in the Hauts-Bassins region, and 6 in the Sahel region were retained because they were the most significantly correlated with the disease and the least correlated with the other climatic variables. For the Sahel region, these parameters were wind (at lags 1, 4 and 6), minimum humidity at lag 3, average temperature at lag 6, and rain at lag 4. In the Hauts-Bassins region, the minimum humidity at lag 1, pressure at lag 6, vapor pressure at lag 5, rain (lag 2, 3, 4), minimum humidity (lag 2, 3, 4), maximum temperature (lag 2, 3, 4), sunshine duration (lag 2, 3, 4, 5), wind (lag 0, 1, 2, 6), and minimum temperature (lag 0, 6) were retained. For the Centre region, the evaporation pan at lag 6, minimum humidity at lag 2, wind (lag 0, 1, 4, 5, 6), vapor pressure (lag 1, 5, 4), sunshine duration (lag 2, 3, 4), rain (lag 3, 4), minimum temperature (lag 4, 5, 6), and atmospheric pressure (lag 5, 6) were retained.

In addition, at the end of the modelling process, Im_lag2, Pm_lag3, Em_lag4, and Pam_lag6 constituted the explanatory variables of the Centre region model. The Hauts-Bassins region included predictors Txm_lag4, Tnm_lag0, Vm_lag0, and Pam_lag6. The Sahel region had Vm_lag4 and Pm_lag4 as explanatory variables. Apart from the model for the Hauts-Bassins region, these models showed good representation of the evolution of IDF for the period used for validation.

Ultimately, despite the limitations noted, the results obtained could improve the resilience of the country in general and of health systems in particular through the establishment of effective mechanisms for the management and surveillance of DF. Our study could be improved by considering other non-climatic parameters involved in the epidemiology of DF, the determination of epidemiological thresholds, and the improvement of data collection systems.

## References

Akaike H (1974) A new look at the statistical model identification. IEEE Trans Autom Control 19:716–723

Amarakoon D, Chen A, Rawlins S, Chadee DD, Taylor M, Stennett R (2008) Dengue epidemics in the Caribbean-temperature indices to gauge the potential for onset of dengue. Mitig Adapt Strat Glob Change 13:341–357

Amarasinghe A, Kuritsky JN, Letson GW, Margolis HS (2011) Dengue virus infection in Africa. Emerg Infect Dis 17:1349

Arsin AA, Istiqamah SNA, Elisafitri R, Nurdin MA, Sirajuddin S, Pulubuhu DAT, Usman AN, Yani A (2020) Correlational study of climate factor, mobility and the incidence of Dengue Hemorrhagic Fever in Kendari, Indonesia. Enferm Clin 30:280–284

Beserra EB, Castro Jr FPd, Santos JWd, Santos TdS, Fernandes CR (2006) Biologia e exigências térmicas de Aedes aegypti (L.) (Diptera: Culicidae) provenientes de quatro regiões bioclimáticas da Paraíba. Neotrop Entomol 35:853–860

Breusch TS, Pagan AR (1979) A simple test for heteroscedasticity and random coefficient variation. Econometrica: J Econ Soc 1287–1294

Cazelles B, Chavez M, McMichael AJ, Hales S (2005) Nonstationary influence of EI Nino on the synchronous dengue epidemics in Thailand. PLoS Med 2:313

Christophers SR (1960) Aedes aegypti: the yellow fever mosquito. CUP Archive

Collenberg E, Ouedraogo T, Ganamé J, Fickenscher H, Kynast-Wolf G, Becher H, Kouyaté B, Kräusslich HG, Sangaré L, Tebit DM (2006) Seroprevalence of six different viruses among pregnant women and blood donors in rural and urban Burkina Faso: a comparative analysis. J Med Virol 78:683–692

Dubé C, Stevenson M, Garner M, Sanson R, Harvey N, Estrada C, Corso B, Wilesmith J, Griffin J (2007) Foot-and-mouth disease model verification and validation through a formal model comparison. In: Proceedings of a meeting held in the theatre of the institution, on Friday Evening, the 12th November, 1858

Durbin J, Watson GS (1971) Testing for serial correlation in least squares regression. III. Biometrika 58:1–19

Failloux A, Hervé J-P (2003) Bio-écologie et competence vectorielle d'Aedes aegypti. La dengue 3

Fox J, Weisberg S (2018) An R companion to applied regression. Sage Publications

Gagnon AS, Bush AB, Smoyer-Tomic KE (2001) Dengue epidemics and the El Niño southern oscillation. Clim Res 19:35–43

Gama ZP, Nakagoshi N (2013) Climatic variability and dengue haemaorrhagic fever incidence in Nganjuk district, East Java, Indonesia. Acta Biologica Malaysiana 2:31–39

Gharbi M, Quenel P, Gustave J, Cassadou S, Ruche GL, Girdary L, Marrama L (2011) Time series analysis of dengue incidence in Guadeloupe, French West Indies: forecasting models using climate variables as predictors. BMC Infect Dis 11:1–13

Gubler DJ (1998) Dengue and dengue hemorrhagic fever. Clin Microbiol Rev 11:16

Islam MA, Hasan MN, Tiwari A, Raju MAW, Jannat F, Sangkham S, Kumar M et al (2023) Correlation of dengue and meteorological factors in Bangladesh: a public health concern. Int J Environ Res Public Health 20(6):5152

Jolion J (2003) Probabilité et Statistique. Cours de l'INSA

Li TG, Yang ZC, Lei L, Biao D, Ming W (2013) Dengue fever epidemiological status and relationship with meteorological variables in Guangzhou, Southern China, 2007–2012. Biomed Environ Sci 26:994–997

Lozano R, Naghavi M, Foreman K, Lim S, Shibuya K, Aboyans V, Abraham J, Adair T, Aggarwal R, Ahn SY (2012) Global and regional mortality from 235 causes of death for 20 age groups in 1990 and 2010: a systematic analysis for the Global Burden of Disease Study 2010. The Lancet 380:2095–2128

Ministry of health of Burkina Faso (2017) Guide de diagnostic et de traitement de la dengue au Burkina Faso

Morin CW, Comrie AC, Ernst K (2013) Climate and dengue transmission: evidence and implications. Environ Health Perspect 121:1264–1272

National Institute of Statistics and Demography (NISD) (2022) Annuaire_Statistique_National_2021 (RGPH-2019). http://www.insd.bf/contenu/pub_periodiques/annuaires_stat/Annuaires_stat_nationaux_BF/Annuaire_Statistique_National_2021.pdf

Nakhapakorn K, Tripathi NK (2005) An information value based analysis of physical and climatic factors affecting dengue fever and dengue haemorrhagic fever incidence. Int J Health Geogr 4:1–13

Natal D (2002) Bioecologia do Aedes Aegypti. Biológico 64:205–207

Ouattara CA, Traore TI, Traore S, Sangare I, Meda CZ, Savadogo LGB (2022) Climate factors and dengue fever in Burkina Faso from 2017 to 2019. J Public Health Afr 13(1):2145. https://doi.org/10.4081/jphia.2022.2145

Pokona EP, Pascal Y, Ndione JA (2021) Modélisation épidémiologique des pathologies climato-dépendantes au nord-Togo: Cas de la Fièvre Aphteuse dans la région des Savanes. Revue Marocaine des Sciences Agronomiques et Vétérinaires 9:675–682

Poveda G, Graham NE, Epstein PR, Rojas W, Quiñones ML, Velez ID, Martens WJ (2000) Climate and ENSO variability associated with vector-borne diseases in Colombia. El Niño and the Southern Oscillation, Multiscale Variability and Global and Regional Impacts 1:183–204

Seogo PH, Bicaba BW, Yameogo I, Moussa G, Charlemangne KJ, Ouadraogo S, Sawadogo B, Yelbeogo D, Savadogo Y, Sow H-c (2021) Ampleur de la dengue dans la ville de Ouagadougou, Burkina-Faso, 2016. J Interval Epidemiol Public Health

Singh S, Herng LC, Sulaiman LH, Wong SF, Jelip J, Mokhtar N, Gill BS et al (2022) The effects of meteorological factors on dengue cases in Malaysia. Int J Environ Res Public Health 19(11). https://doi.org/10.3390/ijerph19116449

Tarnagda Z, Congo M, Sagna T, Ouédraogo C, Nikiéma V, Cissé A, Sanou MA, Ouédraogo A, Bosco J (2014) Outbreak of dengue fever in Ouagadougou, Burkina Faso, 2013. Int J Microbiol Immunol Res 2:101–108

Taylor KE (2001) Summarizing multiple aspects of model performance in a single diagram. J Geophys Res Atmosph 106:7183–7192

Tomassone R, Lesquoy E, Millier C (1983) La régression. Nouveaux regards sur une ancienne méthode statistique

Wang L (2018) Résistance aux insecticides: importance dans la transmission du virus chikungunya par les moustiques Aedes aegypti. Université de Guyane

Wu J-Y, Lun Z-R, James AA, Chen X-G (2010) Dengue fever in mainland China. Am J Trop Med Hyg 83:664

Zambrano L, Sevilla C, Reyes-Garcia S, Sierra M, Kafati R, Rodriguez-Morales A, Mattar S (2012) Potential impacts of climate variability on dengue hemorrhagic fever in Honduras, 2010. Trop Biomed 29:499–507

# Chapter 6
# Climate Change and the Health System in Vietnam: Impacts, Progress, and Future Priorities

**Le Thai Ha, Do Vu Khanh Huyen, Do Manh Cuong, Le Ha Thanh, and Do Phuong Hien**

**Abstract** Vietnam is among the most vulnerable countries in the world regarding climate change. Climate-related hazards already affect human health and the health system in Vietnam. As the health system plays an important role in individual and community-level resilience, their components—such as healthcare facilities, health services, health workforces, and health supply chains—must continue to adapt to current and future challenges including climate change. This paper investigates the impacts of climate change on the Vietnamese health system and suggests strategies for health system practitioners to apply a climate lens and build a comprehensive approach to promoting resilient primary healthcare and health security. Analysis involved systematic literature and policy reviews, supplemented by two focus group discussions. The results showed that climatic factors such as temperature and rainfall show complex patterns of change across Vietnam. Natural disasters and extreme weather events occur every year across the country, causing human and physical damage in general and to the health sector in particular. Given the context, efforts have been made to develop and promote measures that can help the health sector in Vietnam overcome climate change challenges, evidenced in several national regulations and policies of Vietnam. Despite increased efforts to focus on climate and health, development challenges persist. Policy-related recommendations are proposed to further enhance the health sector to more effectively respond to climate change. They include establishing an early warning system, developing a climate-health database, vulnerability and adaptation mapping, and the scaling up of successful models on water,

L. T. Ha (✉) · D. M. Cuong
Vietnam Administration of Disease Prevention, Ministry of Health, Hanoi, Vietnam
e-mail: halt.pb@moh.gov.vn; lethaiha.nioeh@gmail.com

D. V. K. Huyen · D. P. Hien
National Institute of Occupational and Environmental Health, Ministry of Health, Hanoi, Vietnam

L. H. Thanh
National Economics University, Hanoi, Vietnam
e-mail: thanhlh@neu.edu.vn



W. Leal Filho and F. Wolf (eds.), *Climate Change and Health: Perspectives from Developing Countries*, Climate Change Management,
https://doi.org/10.1007/978-3-031-93177-2_6

sanitation, and hygiene. Vietnam's case is likely, in many aspects, to be also relevant to other developing countries.

**Keywords** Vietnam · Health system · Climate change resilience · Climate change adaptation · Low-carbon health care facilities

## Introduction

Vietnam is situated in the eastern part of the Indochinese Peninsula bordering China, Laos, Cambodia, the East Sea, and the Pacific Ocean. The total area is 331,345.7 km$^2$. As of 2022, it is home to 99,474,400 persons, of which 50.15% are female, and 37.55% are urban residents (General Statistics Office 2023). Vietnam has an S-shaped landform and is located in the tropical zone of Southeast Asia. However, with the mountainous and hilly topography and a long coastal line of 3260 km, Vietnam is affected by not only tropical monsoons but also cold northeast winds. Vietnam is one of the nation's most seriously impacted by climate change (World Bank and Asian Development Bank 2020). Over the past few decades, Vietnam has encountered elevated temperatures and a significant increase in the frequency of extreme weather, including droughts, floods, and tropical cyclones (Son et al. 2018). These present multiple risks to both the economy and the health system of Vietnam. Evidence suggests that health risks caused by climate change are growing in many parts of Vietnam. Not only do people face the risk of death, injury, or financial loss, but hospitals and clinics may be physically damaged by storms and floods, doctors and staff may be immobilised, and access to and from health facilities may be cut off by damage in communication and roads.

To respond, several national strategies, action plans, and programmes have been issued by the Government of Vietnam to strengthen the adaptation capacity and resilience of the country to climate change. At COP26, Vietnam committed to net-zero carbon emissions by 2050. It is estimated that the health sector produces 5% of Vietnam's greenhouse gas emissions (Pichler et al. 2019). Given the importance of the health sector as both the victim and the cause of climate change, Vietnam has made efforts to enhance climate change resilience and build a low-carbon health system.

This paper reviews and summarises the impacts of climate change on Vietnam's health system, as reported in the literature, emphasising the sector's adaptation and mitigation actions. We offer policy recommendations for health system policymakers and practitioners to embrace an integrated strategy and climate perspective so that the capacity of the sector to respond to climate challenges is strengthened.

## Methodology

This study uses a systematic literature and policy review and focus group discussion to answer the research questions. These methods are mutually complementary for the following reasons: (i) they provide an overview of the recent and current literature and their focus (literature review and analysis); (ii) they enable the identification of policy directions and efforts (policy review), and (iii) they cater to the provisions of expert's assessments illustrating adaptation and mitigation actions in the health sector of Vietnam, as well as attempts to incorporate them into the policy-making process (focus group discussion).

### Literature Review

A literature review was performed between January 2024 and July 2024 on a set of peer reviewed papers published in scientific journals and referenced in Web of Science, Scopus, and Science Direct. References published between 2010 and 2024 were identified using both Google Scholar and Google search. The key words and phrases of climate change, health impacts, public health, and Vietnam were used in various combinations. The most crucial criterion for selection of references for this paper was the level of relevance to Vietnam and other countries with similar climatic conditions. Other criteria, such as publication date, outlet type, and authors, were then considered for the selection. To supplement the findings of the systematic review and address the relevant research questions, other scientific data on climate change were also collected, such as health impacts reported in official publications, guidelines, and technical and/or monitoring reports issued by responsible institutions and organisations, e.g., WHO, UNDP, ADB.

### Policy Review

For the policy review, we collected information from the official legal databases of Vietnam and available sources from the website of the Vietnam Ministry of Health, Ministry of Natural Resources and Environment, and General Statistics Office to compile a comprehensive list of country climate change adaptation and mitigation policies.

## *Focus Group Discussion*

To gain further insights, we conducted two focus group discussions with participants from government health agencies, academia, civil society, and selected national and provincial health care facilities. Participants were invited to express their views on the impacts of climate change on the health sector, health policy's response to climate change, and ways to enhance the sector's adaptation and resilience capacity to climate change.

Selective data, information, and findings gathered from the different sources were reviewed and analysed to answer the following questions:

- How have climatic factors evolved in the country over time?
- What are the main impacts of climate change on Vietnam's health system?
- How has the health sector responded to climate change?
- What are the policy priorities of Vietnam's health sector to effectively respond to climate challenges?

# Results

## *Climate Change in Vietnam*

Vietnamese climate is characterised by tropical monsoons with an average temperature of 15–27.5 °C in northern regions and 26–29 °C in southern regions, depending on topographical features. Various sources indicate the rising trend of average temperature in Vietnam over the decades, with the largest increase recorded in 2011–2018. According to MONRE (2020), from 1958 to 1985, the mean annual temperature increased by 0.059 °C per decade. In the next 33 years (1986–2018), it increased by 0.74 °C or by 0.22 °C per decade. International organisations reported a rising trend of mean annual temperature in the whole country, including tropical islands, up to 2020, with a rate of 0.22 °C per decade (Phan and Nguyen 2023; World Bank and the Asian Development Bank 2020). The World Bank (2022) further noted that, from 2011 to 2022, the mean annual surface temperature rose even faster at the rate of 0.09 °C per year (Fig. 6.1).

Under the scenario of increasing global surface mean temperature by 2 °C, the mean temperature in Vietnam may increase up to 3 °C by the end of the twenty-first century (IPCC 2018; GIZ et al. 2023). There is even the possibility of an increase of 4.4 °C under the scenario of a very high greenhouse gas emission rate (IPCC 2023).

When looking at maximum and minimum temperatures, the situation is quite complex in Vietnam. Generally, both maximum and minimum temperatures tend to increase over different regions of any nation at the rate of 0.2–0.3 °C per decade and 0.1 °C per decade, respectively (VNU University of Science 2015). In Vietnam, the maximum temperature increased significantly (up to 2.1 °C) in the Northern plains and North-eastern and Southeast regions, while it decreased in some parts

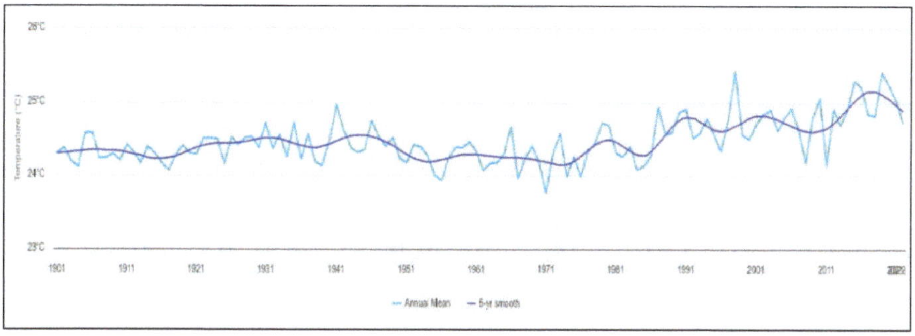


**Fig. 6.1** Annual average mean surface air temperature in Vietnam: 1901–2022 (*Source* World Bank 2022). CC BY 4.0

of the Northwest and Central Highlands regions by 0.2–0.6 °C over 58 years. By contrast, the minimum temperature increased greatly (up to 1.8 °C) in the Central Highlands, Northwest, and North Central regions, while the increase was small in the Northern plains and South-Central region (MONRE 2020). Despite such regional disparities, the overall increase in maximum and minimum temperatures contributes to an increase in the mean temperature of the nation and the increase in hot or even extremely hot days.

Furthermore, variation in precipitation is large in Vietnam across different months, years, and locations. Average precipitation is 700–5000 mm per year, and the common range is 1400–2400 mm per year. Up to the year 2010, annual precipitation had decreased in the Northern region and increased in the Southern region with a rate of about 1.5 mm per year each (VNU University of Science 2015). From 1951 to the year 2100, mean precipitation in Vietnam is predicted to increase by 5% in the South and decrease by 5% in the North. However, heavy precipitation, especially maximum precipitation per day, might increase by 10% (IPCC 2023). Nevertheless, past trends in precipitation exhibited high variation between provinces and even within the same province (Fig. 6.2).

## *Natural Disasters and Extreme Weather Events*

The change in temperature and rainfall can lead to changes in weather patterns, such as extremely cold or high temperatures, tropical storms, and tropical depressions. They are likely to have serious consequences on public health and health infrastructure. Different types of natural disasters hit Vietnam every year. Flooding and inundation are the most prevalent and damaging ones (Table 6.1), while the frequency of others depends on the topographical characteristics of each region (Fig. 6.3).

From 1999 to 2018, Vietnam was one of the ten nations most affected by climate change and natural disasters, which included severe floods, storms or tropical cyclones, and drought, especially in 2016 and 2017. In the past ten years, 34

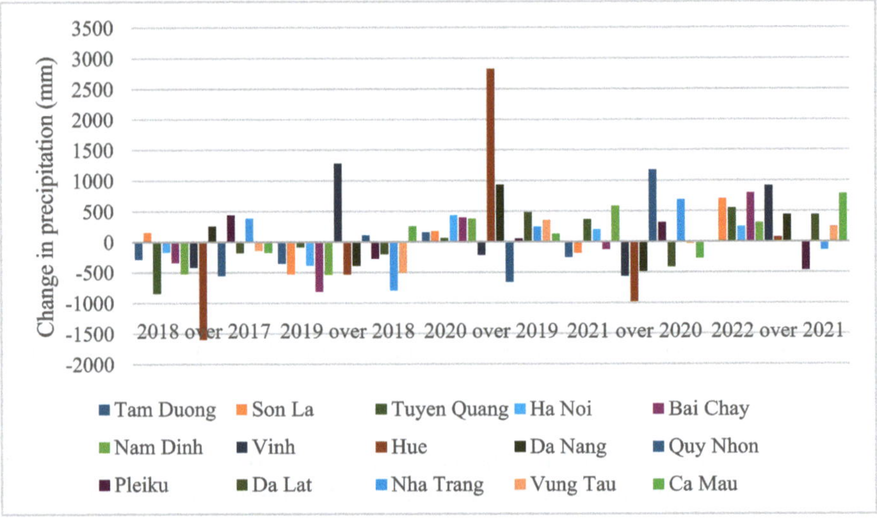


**Fig. 6.2** Change in precipitation in Vietnam at weather stations from 2018 to 2022 2023 (*Source* Author's illustration, adapted from General Statistics Office)

**Table 6.1** Natural disasters in Vietnam in the period from 2013 to 2022 (classified based on the level of damage)

| The level of damage | | |
|---|---|---|
| High | Medium | Low |
| Flood, storm, inundation | Hail rain | Frost |
| Typhoon | Landslide | Damaging cold |
| Flash flood | Saltwater intrusion | Extreme heat |
| Tornado | Drought | |

*Source* Own illustration, adapted from EM-DAT—the international disaster database. Available at: https://www.emdat.be/; National Steering Committee for Natural Disaster Prevention and Control (2024)

floods, 35 tropical cyclones, and other natural disasters occurred throughout the country (EM-DAT—The international disaster database. Available at: https://www.emdat.be/).

These extreme events wreaked great damage not only on the economy but also on public health. Based on the Global Climate Risk Index (CRI), Vietnam was ranked as the fifth and sixth country most affected by extreme weather events in 2016 and 2017, respectively (Eckstein et al. 2017, 2018).

In 2015 and 2016, Vietnam was impacted by the strongest drought in 60 years and the most massive saltwater intrusion in 90 years, induced by the peaking of El Niño, affecting 52 out of the 63 provinces. Approximately two million people lacked access to clean water from March to July 2016 (UNDP 2016).

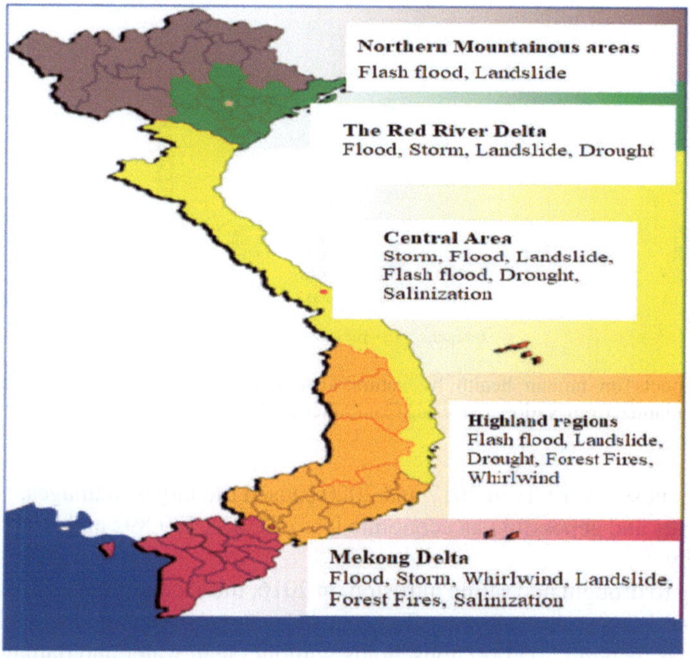


**Fig. 6.3** Map of hazard zones in Vietnam (*Source* Ministry of Education and Training 2012, as cited in Vu et al. 2023). CC BY

Severe floods also occur in the central part of Vietnam. In 2017, two typhoons, Damrey and Doksuri, hit the northern and central provinces, bringing strong winds and heavy rainfall, with the largest precipitation in a very short period over the decade. The typhoons caused floods and landslides and produced enormous human and economic losses.

Natural disasters in 2020 were also historic with 14 storms in the East Sea and 120 floods, flash floods, and landslides, which affected 49 out of the 63 provinces of Vietnam (National Steering Committee for Natural Disaster Prevention and Control 2024).

## *Impacts of Climate Change on Public Health*

According to the National Steering Committee for Natural Disaster Prevention and Control, over the past ten years, the most frequent natural disasters were heavy rains (including heavy rains after storms), storms, floods (including flash floods), landslides, and inundation. As a result, 2059 people died, 3627 people were injured and 513 people remained missing (Fig. 6.4).

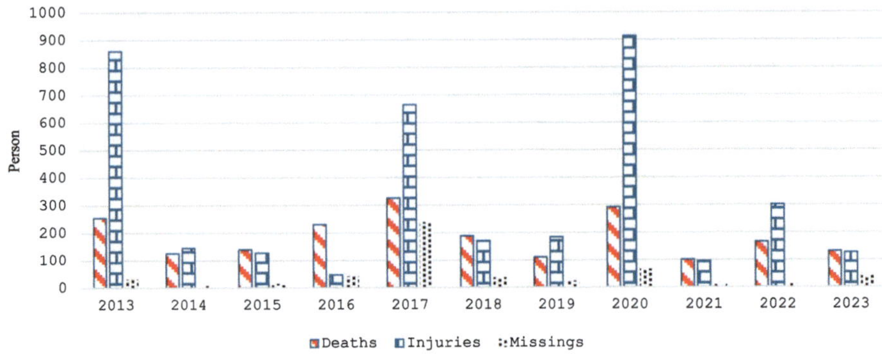


**Fig. 6.4** Impacts on human health by natural disasters in Vietnam 2024 (*Source* Author's illustration, adapted from National Steering Committee for Natural Disaster Prevention and Control)

Among these, disasters in 2017 and 2020 caused the largest damages, taking 325 and 291 lives and generating an economic loss of US$2,539,892 and US$1,691,530, respectively.

Turning to drought and saline intrusion, in 2016, these conditions proved to be the most serious in more than 60 years. They caused water shortage in 114,000 ha of residential areas, leaving 505,000 households without clean water and damaging many crops (National Steering Committee for Natural Disaster Prevention and Control 2024).

Moreover, 520,000 children and 1,000,000 women suffered undernutrition, and 400,000 individuals were at risk of water-related epidemics (UNICEF Viet Nam 2019).

Other climate-related diseases generate suffering and burden on many households and society. The occurrence of mosquito-borne diseases, such as dengue fever and malaria, is sensitive to variability in temperature and humidity. These diseases are often endemic to specific countries or regions, with an estimated 100 million people infected each year worldwide (Bhatt et al. 2013). In Vietnam, though 40 out of the 63 provinces were proclaimed malaria-free, it is still an issue in the Central Highlands, Mekong Delta, and South Central Coast (Tran et al. 2023). As a consequence, mosquito-borne diseases continue to put a great burden on the economy, society, and families.

With rising mean annual temperature and the increasing frequency of heavy rains in a short period, the behaviour and reproduction methods of pathogens may change (Levy et al. 2016). This, in turn, may affect human health (Bartram and Cairncross 2010; GBD Diarrhoeal Diseases Collaborators 2017). Disease outbreaks are likely to increase as water is contaminated during storms, floods, or droughts (Phung et al. 2015).

Young children are susceptive to hand, foot, and mouth disease (HFMD), which is caused by enteroviruses. It can spread rapidly and become an epidemic unless early detection and treatment are implemented. This infectious disease is influenced by meteorological variables, temperature in particular, and may reach a peak in summer

(Coates et al. 2019). In Vietnam, there are few scientific studies addressing the link between climate change and HFMD transmission. However, the observed trend is a sharp increase in the disease and a seemingly positive correlation between the occurrence of HFMD and climate factors (Nguyen et al. 2017, 2014).

An unusual increase in dengue fever was recorded in recent years. For example, in 2020 (mostly in September), 65,046 dengue cases and seven deaths were reported, though these figures were lower than those reported in 2019 (General Department of Preventive Medicine 2020). Previous studies conducted in different parts of Vietnam have shown that an increase of 1 °C in temperature may cause an increase of 5% in dengue cases (Vu et al. 2014; Do et al. 2014) and increases of 0.4–2.5% in other infectious, intestinal, or respiratory diseases, including influenza, mumps, chickenpox, and tuberculosis (Tran et al. 2022; Phung et al. 2018).

The study by Barrak et al. (2023) collected information from 567 cities in 27 countries around the world and concluded that extreme temperature change could increase the risk and burden of heart failure, ischaemic heart disease, stroke, and all-cause cardiovascular disease (CVD). The impact of cold weather on the incidence of CVD among elderlies in Northern Vietnam was found statistically more significant than the impact of warm weather (Giang et al. 2014). This result was similar to those of Chen and Zhang (2023) and Yoneyama et al. (2021), which confirmed the negative association between low temperature and the prevalence of CVD among elderlies in China and Japan.

## *Impacts of Climate Change on the Health System*

The health system of Vietnam is also strongly affected by climate change. According to the Law on Medical Examination and Treatment (Law No. 15/2023/QH15 issued on 09/1/2023), the health system consists of four levels: national, provincial, district, and commune levels (Article 48). Within each level, there are general hospitals, specialised hospitals, clinics, health centres, and health stations at the commune level. They collectively provide all types of medical services from first aid to special treatments (Article 104). Vietnam has more than 13,000 public medical facilities and more than 12,000 private ones (National Institute of Occupational & Environmental Health and Hanoi University of Public Health 2019), with approximately 300,000 patient beds and about 100,000 doctors in 2022. This is equivalent to 10 doctors and 31.7 beds per 10,000 persons (General Statistics Office 2023). However, 32% of the population has no health coverage (WHO 2023).

As heavy rains (especially in a short period) and storms become more frequent and powerful under climate change, once a natural disaster or an extreme weather event occurs, there is a high likelihood that health facilities in affected areas are damaged or become isolated. A study by Tran et al. (2020), using WHO Guidelines for the implementation of the Vulnerability and Adaptation Assessment (V&A), showed that local healthcare facilities in Vietnam are highly sensitive to climate change and natural disasters. Under the WHO Guidelines, each indicator ranges from 1 to 5, from very low risk to very high risk or from very high capacity to very low capacity, with a higher indicator showing greater vulnerability.

Many commune health stations face the risk of flooding, collapse, power outage, water shortage, and isolation during storms and floods. According to the "health sensitivities" index that ranges from 1 (very low sensitivities) to 5 (very high sensitivities), the level of sensitivities over the past five years decreased slightly from 3.8 in 2013 to 3.4 in 2017. Nevertheless, the overall health risks remain high.

As for the "adaptive capacity" index that ranges from 1 (very high capacity) to 5 (very low capacity), the overall score was 4.5 in 2013 and 2014 and remained at 4.4 in the following three years. Against climate change-related risks and high health sensitivities, the current capacity of Vietnam's health sector is assessed to be extremely low.

The overall score for the Vietnamese health sector, ranging from 1 (very low risk) to 5 (very high risk), was 4.2 in 2013, 4.0 in 2014, 3.8 in 2015, 3.9 in 2016 and 3.8 in 2017. Though a decline (improvement) was observed from 2013 to 2015, the score remained high thereafter.

While assessments beyond 2017 were not produced by this study, high risk is likely to have remained or even intensified in more recent years. For instance, storms and floods in 2020 unroofed, inundated, isolated, or damaged 459 hospitals, medical centres, and commune health stations, with medical equipment ruined or impaired. Moreover, many medical facilities in the affected areas that remained physically intact were isolated and unable to receive and treat patients (Ministry of Health 2020).

## *Progress in Adaptation and Mitigation*

In 2018, the Ministry of Health issued the Health National Action Plan to respond to climate change for the period of 2019–2030 with a vision to 2050. The plan aimed to (i) develop and finalise a mechanism for climate change adaptation within the health sector; (ii) increase the awareness and capacity of medical staff in protecting human health and responding to climate change; and (iii) strengthen the capacity of the health sector, especially commune health stations, to adapt to and mitigate the damage of climate change and extreme weather events. In early 2024, the national strategy for protection, care, and improvement of the people's health by 2030, developed by the Ministry of Health, was approved by the prime minister. The objective of the strategy was to provide high-quality health services and improve the quality of life of the citizens by strengthening the capacity of the health system concerning disease prevention, health supply chains, skillful medical staff, scientific research, and AI application.

A tool to assess vulnerabilities and adaptation (V&A assessment) in the health sector was developed in the research on climate change and health vulnerability in 2018 that conformed to the WHO guidelines (Tran et al. 2020). The tool has subsequently been reviewed, updated, and adjusted annually to adapt to changes in climatic conditions and the policy framework. In 2023, the tool consisted of 31 indicators (the WHO Guidelines and the first version of the tool had 45 indicators), including 8 indicators on climate change exposure, 12 on health sensitivities, and 11 on adaptation capacity using the automatic scoring system. To implement the tool, about 800 health staff have been trained on V&A assessment and national and provincial health action planning and implementation.

The Ministry of Health also issued technical guidelines on health protection during heatwaves and cold seasons, emergency water treatment, water and environmental sanitation treatment during floods and storms, and health protection from outdoor air pollution.

For capacity building on climate change response and adaptation, Training of Trainers (ToT) courses were provided to targeted medical staff. By the end of 2023, 21 ToT courses were organised for more than 800 provincial and district health staff. However, the training needs are still very high, as Vietnam has more than 13,000 healthcare facilities. ToT courses will be organised for these facilities in the coming years.

Additionally, several climate change-related models have been piloted and applied to health care facilities (HCFs), which produced good results up to now. For example, the model of Green-Clean-Beautiful Hospital has been deployed nationwide since 2016, which focuses on environmental sanitation, clean water, and medical waste management to create a friendly, green, and beautiful environment in HCFs. This is expected to improve the quality of health services and the satisfaction of patients, visitors, and health staff. The programme of reducing medical plastic waste was applied to all HCFs in 2019 and continues to be implemented to this date.

To build climate-resilient healthcare facilities, the Water and Sanitation for Health Facility Improvement Tool (WASH-FIT) was introduced and tested in Vietnam in 2018, contributing to improvements in clean water and environmental sanitation in targeted HCFs. In 2021, the models of climate change resilient and environmentally sustainable HCFs were developed and piloted in three provinces. The models were based on the WHO framework with the following four main pillars: (i) skilled health workforce; (ii) sustainable and safe management of water, sanitation, hygiene, and medical waste; (iii) green and sufficient energy usage; and (iv) appropriate infrastructure, technologies and products.

Up to now, the following related actions have been made in three piloting HCFs. In 2021, water supply systems were installed to provide clean water for medical health and hygiene activities, and medical staff were trained in general WASH-FIT and health impacts from climate change. In 2023, training in V&A assessment and green energy was conducted, hand-washing facilities were provided, and guidelines for developing and implementing models for climate change-resilient HCFs were created.

Based on these initial achievements, these models are to be scaled up to other HCFs in 2024 and the coming years. Recently, a community-based pilot model with the support of WHO was implemented in three provinces of Vietnam and is expected to be replicated in other provinces in the future.

The Ministry of Health also plans to increase awareness at the community level on climate change, clean water, and environmental sanitation during flood and storm seasons through various means, including leaflets, local mass media, and social networking services.

At the national level, the Ministry of Health is now a member of the Alliance for Transformative Action on Climate and Health (ATACH) and the ATACH working group. In collaboration with UNDP and WHO, the Ministry developed a proposal for the "Climate and Health Co-Investment Facility Coordination Programme" to receive funding from the Green Climate Fund (GCF). The programme is expected to be approved at the end of 2024.

## Discussion

### *Climate Change in Vietnam*

The mean annual temperature in Vietnam is gradually increasing. Findings showed an increase of 0.22 °C per decade (Phan and Nguyen 2023). In 2023, the mean annual temperature in various locations in Vietnam ranged from 20 to 28.5 °C, which was an increase of 0.5–1.5 °C above the average temperature over many years, and the year was the second hottest between 2014 and 2023 with 20 heatwaves (Pham et al. 2024). With this trend continuing, the scenario of increasing heatwaves in Vietnam, predicted by the Ministry of Natural Resources and Environment, may be inevitable (MONRE 2020).

Similarly, according to Pham et al. (2024), the total annual precipitation in Vietnam in 2023 was the fourth lowest in the most recent decade (2014–2023). The study also presented the forecast of mean precipitation in Vietnam up to 2100 with an increase of 5% in the South and a decrease of 5% in the North (IPCC 2023). These trends will lead to a change in weather patterns, with the number of extremely cold days decreasing in the North, the number of hot and extremely hot days increasing across the country, and storms becoming more powerful and complex (MONRE 2020). As a consequence, the number of natural disasters such as storms, typhoons, heavy rains, etc. will increase. In 2023, the country experienced more than 1000 events of natural disasters of 21 types when compared to 600 such events in 2019 (Saigon Liberation Newspaper 2020; National Steering Committee for Natural Disaster Prevention and Control 2023; Viet Nam Meteorological and Hydrological Administration 2023).

### *Climate Change and Health*

The studies and data presented above verify that climate change has put a heavy burden on the health system of Vietnam, including, but not limited to, the increasing number of hospitalisations due to heatwaves or extreme cold in the North, the rising cases of infectious diseases such as dengue, influenza, mumps, tuberculosis, etc. (Tran et al. 2022; Phung et al. 2018), and injuries and deaths due to natural disasters (MONRE 2020). However, the capacity of the health system to provide health services to the population is limited and sometimes disrupted, as the means of communication and transportation are lost during a flood, infrastructure is damaged by long-inundation, salt intrusion, or tornadoes (MONRE 2020), and the supply of water for medical treatment is reduced in both quantity and quality.

## *Adaptation and Mitigation Actions of the Health Sector*

Since the issuance of the Health National Action Plan to respond to climate change in the period of 2019–2030 with a vision to 2050, the Ministry of Health has implemented several projects and programmes for creating a climate-resilient health system with technical and financial support from the government, international organisations such as ADB, WHO, UNDP, and UNICEF, and national experts. They include capacity building, V&A assessment tools, Green-Clean-Beautiful Hospital, and WASH application, as discussed in this paper. However, gaps still remain between current achievements and prevailing challenges, which call for broader and stronger actions, strategies, and support from all stakeholders.

Some of the remaining challenges are discussed below.

## *The Expansion of the V&A Assessment*

As explained above, the V&A Assessment tool was developed in 2018 and subsequently revised and updated. The 2024 version includes 31 indicators and was applied to all 63 provinces of Vietnam. However, the tool is for provincial and district levels only and does not obtain information on vulnerability and adaptive capacity at the lowest levels of administration (commune and grass-root levels). As a result, policy intervention to improve and strengthen the health system has not reached the "last-mile" HCFs. It is necessary to continue to develop the tool to encompass the V&A assessment of HCFs at these levels as well.

## *Assessment of Greenhouse Gas (GHG) Emission from HCFs*

At COP26, the prime minister of Vietnam made a strong commitment to net-zero emissions of GHG by 2050. Under this commitment, the Ministry of Health plans to build low-carbon health systems. However, Vietnam lacks an appropriate instrument and system to measure GHG emissions from its health system, including from HCFs and the health supply chain.

## An Early Warning System on Climate and Health

With available data on climatic factors and climate change in the last several decades on the one hand, and accumulated data on public health and climate-related diseases on the other, Vietnam should be able to develop health impact thresholds for climatic factors. This will be the starting point for creating an early warning system on climate change impacts and health, which is crucial in decreasing deaths and damages by natural disasters and extreme weather events. However, this remains a challenging task for Vietnam, as it requires high expenditure as well as scientific and systematic research to construct and analyse a relevant database.

## Future Priorities for Policy and Programming

Ongoing health system efforts can be strengthened by developing new initiatives at the country level. The priorities outlined below should be considered as the entry points for better policy intervention.

Establishment of short-term and/or medium-term early warning systems on climate risks to human health and health care facilities: This is important for avoiding and reducing the impacts of climate change and increasing the preparedness of the community and health systems for effective response.

Developing a climate and health database: This will help to develop and maintain early warning systems on climate-sensitive health risks as well as to plan effective interventions. Such a database can expedite the decision-making process and reduce the cost of policy implementation. It also helps the health sector to be better prepared for the emergence and expansion of diseases and other health-risk incidences.

## Mapping of Vulnerability and Climate Risks to Health

This mapping, based on V&A assessments at national, provincial, and district levels, should give us a detailed picture of the geographical scope of climate change impacts and facilitate the mobilisation of resources needed to deal with possible consequences.

### *Scaling up of Successful Healthcare Facility-Based and Community-Based Models*

Success stories that demonstrate the benefit of healthcare facility-based and community-based models on climate resilience should be collected and publicised widely. Based on these showcases, a large-scale introduction of successful models should be designed and executed nationwide. This will enable both the central and local governments to mobilise limited resources in an efficient way to improve the risk management capacity of the health sector.

#### Collection of Location-Specific Information

Finally, as adaptation and resilience to climate change is largely a location-specific issue, knowing the characteristics and constraints of the health sector of each location is critical. This will enable each district and commune to formulate and execute policies effectively to cope with specific problems that may arise in the community.

## Conclusion

The objective of this paper was to summarise and evaluate the impacts of climate change on the health sector of Vietnam, and to identify possible future interventions.

Data indicates that climatic factors such as temperature and rainfall show complex patterns of change in both northern and southern Vietnam. These lead to an increase in the number of hot or extremely hot days. Heavy precipitation is projected to increase by 10% by 2100.Natural disasters and extreme weather events occur every year across the country. In recent decades, historic events were frequently recorded, especially in 2016, 2017, and 2020, causing human and physical damage in general and to the health sector in particular. Hundreds of HCFs were inundated, destroyed, and/or failed to provide health services to local people due to the lack of access when they were most needed.

Vietnam has taken several policy actions to strengthen the capacity of HCFs in climate change adaptation and resilience, including the development of the V&A tool, ToT courses for medical staff, the Green-Clean-Beautiful hospital programme, plastic waste reduction movement in HCFs, and the community-based model. Their initial results are encouraging.

To further enhance the health sector to more effectively respond to climate change, Vietnam needs a concrete and transparent roadmap with sufficient data. For this, consideration and priorities should be given to the establishment of an early warning system, the development of a climate-health database, V&A mapping at the district level, and the scaling up of successful models in water, sanitation, and hygiene.

This paper is not without limitation. The data for the study were collected mainly from secondary data sources. Future research would benefit from the examination of another research design such as longitudinal experiments, case studies, and other impact analysis methods. Also, the study only looked at the impacts of climate change on the Vietnamese health system. Further research should be carried out in multiple climate regions of Vietnam with diversified environmental, socioeconomic, and population characteristics.

## References

Barrak A, Haitham K, Dominic R (2023) Associations between extreme temperatures and cardiovascular cause-specific mortality: results from 27 countries. Circulation 147(1):35–46. https://doi.org/10.1161/CIRCULATIONAHA.122.061832

Bartram J, Cairncross S (2010) Hygiene, sanitation, and water: forgotten foundations of health. PLoS Med 7(11):e1000367. https://doi.org/10.1371/journal.pmed.1000367

Bhatt S, Gething PW, Brady OJ, Messina JP, Farlow AW, Moyes CL, Drake JM, Brownstein JS, Hoen AG, Sankoh O, Myers MF, George DB, Jaenisch T, William Wint GR, Simmons CP, Scott TW, Farrar JJ, Hay SI (2013) The global distribution and burden of dengue. Nature 496(7446):504–507. https://doi.org/10.1038/nature12060

Chen H, Zhang X (2023) Influences of temperature and humidity on cardiovascular disease among adults 65 years and older in China. Front Public Health 10:1079722. https://doi.org/10.3389/fpubh.2022.1079722

Coates SJ, Davis MDP, Andersen LK (2019) Temperature and humidity affect the incidence of hand, foot, and mouth disease: a systematic review of the literature—a report from the International Society of Dermatology Climate Change Committee. Int J Dermatol 58(4):388–399. https://doi.org/10.1111/ijd.14188

EM-DAT (2024) The international disaster database. Available at: https://www.emdat.be/. Accessed: 15 Mar 2024

Eckstein D, Künzel V, Schäfer L (2017) Global climate risk index 2018. In: Who suffers most from extreme weather events? Weather-related loss events in 2016 and 1997 to 2016. Available at https://www.germanwatch.org/en/node/14987. Accessed: 10 Jan 2024

Eckstein D, Hutfils ML, Winges M (2018) Global climate risk index 2019. In: Who suffers most from extreme weather events? Weather-related loss events in 2017 and 1998 to 2017. Available at https://www.germanwatch.org/en/16046. Accessed: 12 Jan 2024

Do TT, Martens P, Luu NH, Wright P, Choisy M (2014) Climatic-driven seasonality of emerging Dengue fever in Hanoi, Viet Nam. BMC Public Health 14:1078

GBD Diarrhoeal Diseases Collaborators (2017) Estimates of global, regional, and national morbidity, mortality, and aetiologies of diarrhoeal diseases: a systematic analysis for the Global Burden of Disease Study 2015. Lancet Infect Dis 17(9):909–948. https://doi.org/10.1016/S1473-3099(17)30276-1

General Department of Preventive Medicine (2020) Tình hình dịch bệnh sốt xuất huyết và các biện pháp phòng chống trọng tâm Hà Nội. Available at: http://vncdc.gov.vn/files/article_attachment/2020/9/2-cuc-ytdp-bai-trinh-bay-hoi-nghi-sxh-1992020-final-1(1).pdf). Accessed: 16 July 2024

General Statistics Office (2023) Statistical yearbook of 2022. Statistical Publishing House. Available at: https://www.gso.gov.vn/en/data-and-statistics/2023/06/statistical-yearbook-of-2022/. Accessed: 15 Mar 2024

Giang PN, Dung DV, Giang KB, Vinh HV, Rocklov J (2014) The effect of temperature on cardiovascular disease hospital admissions among elderly people in Thai Nguyen Province, Vietnam. Glob Health Action 7:23649. https://doi.org/10.3402/gha.v7.23649

GIZ, UNEP FI, and CDSB (2023) Vietnam's climate risks and best practices for TCFD disclosure. Available at: https://www.unepfi.org/themes/climate-change/vietnams-climate-risks-and-best-practices-for-tcfd-disclosure/. Accessed: 15 Mar 2024

IPCC (2018) Global Warming of 1.5°C. An IPCC Special Report on the impacts of global warming of 1.5°C above pre-industrial levels and related global greenhouse gas emission pathways, in the context of strengthening the global response to the threat of climate change, sustainable development, and efforts to eradicate poverty [Masson-Delmotte V, Zhai P, Pörtner H-O, Roberts D, Skea J, Shukla PR, Pirani A, Moufouma-Okia W, Péan C, Pidcock R, Connors S, Matthews JBR, Chen Y, Zhou X, Gomis MI, Lonnoy E, Maycock T, Tignor M, Waterfield T (eds)] Cambridge University Press, Cambridge, UK and New York, USA, 616 pp. Available at: https://doi.org/10.1017/9781009157940. Accessed: 15 Feb 2024

IPCC (2023) Summary for policymakers. In: Climate change 2023: synthesis report. Contribution of Working Groups I, II and III to the Sixth Assessment Report of the Intergovernmental Panel on Climate Change IPCC, Geneva, Switzerland, pp 1–34. Available at: https://doi.org/10.59327/IPCC/AR6-9789291691647.001. Accessed: 20 Mar 2024

Levy K, Woster AP, Goldstein RS, Carlton EJ (2016) Untangling the impacts of climate change on waterborne diseases: a systematic review of relationships between diarrheal diseases and temperature, rainfall, flooding, and drought. Environ Sci Technol 50(10):4905–4922. https://doi.org/10.1021/acs.est.5b06186

Ministry of Health (2020) Report on Results in natural disasters prevention and searching for victims in storm seasons (October, November 2020) in the Central Region of Vietnam

Ministry of Natural Resources and Environment (MONRE) (2020) Kịch bản biến đổi khí hậu. Viet Nam Publishing House of Natural Resources, Environment and Cartography (NARENCA)

National Institute of Occupational & Environmental Health and Hanoi University of Public Health (2019) Báo cáo tổng kết nhiệm vụ Tổng điều tra toàn quốc về nguồn thải từ hoạt động y tế và đánh giá mức độ ô nhiễm từ các nguồn thải tại các cơ sở y tế

National Steering Committee for Natural Disaster Prevention and Control (2023) Thiên tai gây thiệt hại hơn 8.000 tỷ đồng. Available at: https://phongchongthientai.mard.gov.vn/Pages/2023-thien-tai-gay-thiet-hai-hon-8-000-ty-dong.aspx. Accessed: 30 July 2024

National Steering Committee for Natural Disaster Prevention and Control (2024) Published data on disasters' damages in Vietnam. Available at: https://phongchongthientai.mard.gov.vn/Pages/Thong-ke-thiet-hai.aspx. Accessed: 15 Mar 2024

Nguyen DQ, Renwick J, McGregor J (2013) Variations of surface temperature and rainfall in Vietnam from 1971 to 2010. Int J Climatol 34(1):249–264. https://doi.org/10.1002/joc.3684

Nguyen HX, Chu C, Nguyen HLT, Nguyen HT, Do CM, Rutherford S, Phung D (2017) Temporal and spatial analysis of hand, foot, and mouth disease in relation to climate factors: a study in the Mekong Delta region, Vietnam. Sci Total Environ 581–582:766–772. https://doi.org/10.1016/j.scitotenv.2017.01.006

Nguyen NT, Pham HV, Hoang CQ, Nguyen TM, Nguyen LT, Phan HC, Tran Minh NN (2014) Epidemiological and clinical characteristics of children who died from hand, foot and mouth disease in Vietnam, 2011. BMC Infect Dis 14:341. https://doi.org/10.1186/1471-2334-14-341

Pichler PP, Jaccard IS, Weisz U, Helga Weisz H (2019) International comparison of health care carbon footprints. Environ Res Lett 14:064004. https://doi.org/10.1088/1748-9326/ab19e1

Phan TTH, Nguyen HA (2023) Spatial and temporal distributions of temperature and rainfall on tropical islands of Vietnam. J Water Clim Change 14(5):1395–1412. https://doi.org/10.2166/wcc.2023.342

Pham THY, Vu VT, Nguyen TT, Tran TT, Truong TTT, Phung TML, Le MD (2024) Climate characteristics of Viet Nam in 2023. J Clim Change Sci 29. https://doi.org/10.55659/2525-2496/29.94651

Phung D, Huang C, Rutherford S, Chu C, Wang X, Nguyen M (2015) Climate change, water quality, and water-related diseases in the Mekong Delta Basin: a systematic review. Asia Pac J Public Health 27(3):265–276. https://doi.org/10.1177/1010539514565448

Phung D, Nguyen HX, Nguyen HLT et al (2018) The effects of socioecological factors on variation of communicable diseases: a multiple-disease study at the national scale of Viet Nam. PLoS ONE. https://doi.org/10.1371/journal.pone.0193246

Saigon Liberation Newspaper (2020) Thiên tai gây thiệt hại về kinh tế trên 7.000 tỷ đồng trong năm 2019. Available at: https://www.sggp.org.vn/thien-tai-gay-thiet-hai-ve-kinh-te-tren-7000-ty-dong-trong-nam-2019-post555870.html. Accessed: 30 May 2024

Son NH, Yen BT, Sebastian L (2018) Development of climate-related risk maps and adaptation plans (Climate Smart MAP) for rice production in Vietnam's Mekong River Delta. CCAFS Working Paper. Available at: https://cgspace.cgiar.org/rest/bitstreams/148994/retrieve. Accessed: 30 Apr 2024

Tran NQL, Huynh TCH, Pham CT, Nguyen XH, Tran ND, Tran TTH, Nghiem S, Luong TML, Bui V, Nguyen HT, Don VQ, Dang KA, Do THT, Ngo TKH, Nguyen VT, Nguyen NH, Do MC, Ton TN, Dang TAT, Nguyen K, Tran XB, Thai P, Phung D (2023) Climate change and human health in Vietnam: a systematic review and additional analyses on current impacts, future risk, and adaptation—The Lancet Regional Health—Western Pacific. https://doi.org/10.1016/j.lanwpc.2023.100943

Tran QA, Le VTH, Le TH, Phung DT, Berman JD, Nguyen HLT (2022) The association between ambient temperatures and hospital admissions due to respiratory diseases in the capital city of Viet Nam. Front Public Health 10:903623. https://doi.org/10.3389/fpubh.2022.903623

Tran TTH, Le TTH, Nguyen TLH, Tran NQL, Nguyen HQ, Nguyen TTN, Ebi K, Nguyen DC, Ha VN, Tran MK, Hales S, Do MC, Nguyen TTT, Luu QT, Nguyen NB, Hoang VM (2020) Vietnam climate change and health vulnerability and adaptation assessment, 2018. Environ Health Insights 14:1–11. https://doi.org/10.1177/11786302209246

UNDP (2016) Viet Nam drought and saltwater intrusion: transitioning from emergency to recovery—analysis report and policy implications. Available at: https://www.undp.org/sites/g/files/zskgke326/files/migration/vn/Recovery-draft-Sep-2016_final.pdf. Accessed: 1 May 2024

UNICEF Viet Nam (2019) Viet Nam child centred risk assessment. Available at: https://www.unicef.org/vietnam/reports/viet-nam-child-centred-risk-assessment. Accessed: 18 Mar 2024

Viet Nam Meteorological and Hydrological Administration (2023) Thiên tai năm 2022 diễn biến phức tạp, khốc liệt. Available at: http://vnmha.gov.vn/kttv-voi-san-xuat-va-doi-song-106/thien-tai-nam-2022-dien-bien-phuc-tap-khoc-liet-13921.html. Accessed: 30 May 2024

VNU University of Science (2015) Climate change-induced water disaster and participatory information system for vulnerability reduction in North Central Vietnam (CPIS). Available at: http://danida.vnu.edu.vn/cpis/vn/content/chuong-3-bdkh-o-vn.html. Accessed: 15 Mar 2024

Vu HH, Okumura J, Hashizume M, Tran DN, Yamamoto T (2014) Regional differences in the growing incidence of Dengue fever in Viet Nam explained by weather variability. Trop Med Health 42:25–33. https://doi.org/10.2149/tmh.2013-24

Vu BD, Nguyen HT, Dinh HVT, Nguyen QAN, Ha XV (2023) Natural disaster prevention literacy education among Vietnamese high school students. Educ Sci 13:262. https://doi.org/10.3390/educsci13030262

WHO (2023) Index of service coverage—data by country. Available at: https://apps.who.int/gho/data/view.main.INDEXOFESSENTIALSERVICECOVERAGEv?lang=en. Accessed: 18 Mar 2024

World Bank (2022) Climate change knowledge portal. Available at: https://climateknowledgeportal.worldbank.org/. Accessed: 15 Mar 2024

World Bank and the Asian Development Bank (2020) Climate risk country profile: Vietnam, Hanoi, Vietnam. Available at: https://climateknowledgeportal.worldbank.org/sites/default/files/2021-04/15077-Vietnam%20Country%20Profile-WEB.pdf. Accessed: 29 May 2024

Yoneyama K, Nakai M, Higuma T et al (2021) Weather temperature and the incidence of hospitalization for cardiovascular diseases in an aging society. Sci Rep 11:10863. https://doi.org/10.1038/s41598-021-90352-x

# Chapter 7
# The Health-Related and Socio-economic Impacts of Climate Change on Women in India

**K. S. Navya and P. B. Shankar Narayan**

**Abstract** *Background.* Climate change refers to the average of weather patterns in a particular area over a long period, representing the climate system's general state (UNDP 2011). The consequences of the climatic changes are experienced unevenly by people. Concerning it, there are gender disparities as well, and women are disproportionately affected by climatic variations. *Objective.* The current research paper explores climate change's health-related and socio-economic impact on women and its probable causes through the existing literature. *Methodology.* A systematic literature review was used, and 45 articles were identified for thematic analyses with respect to the objectives and inclusion criteria of the study. *Results.* The study points out that health and well-being, gender-based violence, social roles and responsibilities, livelihood, and survival are some of the major areas that require attention in considering the impact of climate change on women in India. These impacts were found to be deeply rooted in their socio-cultural contexts that perpetuate different kinds of vulnerabilities. *Conclusion.* As gender can be an important dimension of adaptability as well, the research study points out recommendations for policymakers to integrate the needs of women and enhance their coping strategies and overall health.

**Keywords** Climate change · Women in India · Health impacts · Socio-economic impacts · Vulnerabilities

## Introduction

Climate change is the change in weather patterns that have evolved through generations and decades and has the potential to influence the biotic process through radiation, tectonic plates, volcanic eruptions, and natural disasters (Dimitrov 2019). Climatic variations on planet Earth have been happening throughout history. Apart

K. S. Navya (✉) · P. B. Shankar Narayan
Department of Social Work, Pondicherry University, Kalapet, Puducherry, India
e-mail: navyaksujith@gmail.com



W. Leal Filho and F. Wolf (eds.), *Climate Change and Health: Perspectives from Developing Countries*, Climate Change Management,
https://doi.org/10.1007/978-3-031-93177-2_7

from the periodic and episodic natural changes, human activities have significantly altered the Earth's climate since the Industrial Revolution. However, present-day climate change is happening rapidly compared to natural variations alone (Krishnan et al. 2020). As a result of climate change, temperatures are rising and there are more unpredictable weather patterns, such as unusually heavy rainfall (George et al. 2022; Pailler and Tsaneva 2018). The 2024 Wayanad and Himachal Pradesh landslides in India are a testimony to it. According to the Intergovernmental Panel on Climate Change (IPCC) Synthesis Report of 2023, human-caused climate change is influencing weather patterns and climate extremes worldwide (Owen-Burge 2023). Approximately 3.6 million people currently reside in regions that are extremely vulnerable to climate change, and between 2030 and 2050, it is predicted that undernutrition, malaria, diarrhoea, and heat stress will contribute to an additional 25,000 fatalities per year (WHO 2023).

Extreme climate variability is causing changes in disaster patterns and an increase in the extent, magnitude, intensity, and frequency of disasters, which affect millions of people worldwide (Chisty et al. 2022). According to estimates based on a climatic model, between 920 and 3,400 million people globally are expected to be exposed to and impacted by the growing number of natural disasters by 2050 as a result of greenhouse gas emission scenarios (George et al. 2022).

Even though climate change is global, its impact might not be uniform across the planet. Concerning the Indian subcontinent, there are complex interactions of the atmosphere and the evidence of various anthropological activities that have influenced the regional climatic conditions in recent decades. Some of the human-induced climate changes such as the rising sea levels, changes in the monsoon patterns, and increasing heat extremes are considered a threat to lives and livelihoods on the subcontinent (Krishnan et al. 2020). The significant events caused by climate change are so much that recovery from it can take decades, with long-lasting effects on the communities and their healthcare services. In comparison to developed countries, the developing countries are less resilient to these extreme climatic events as they take longer to recover from them (Ebi and Bowen 2016). The IPCC's sixth assessment report has also pointed out that the vulnerable communities are those that have least contributed to the current climate change and yet are disproportionately affected by it (IPCC 2023).

The nations and areas with the lowest levels of greenhouse gas emissions are those that are most impacted by climate change and those that find it difficult to adapt or deal with it (Pinchoff et al. 2024). Due to several social, physical, political, and environmental factors, India's low-middle-income population and especially the women in this group are more susceptible to the effects of climate change (Sorensen et al. 2018). The effects of climate change are gendered; women are disproportionately affected, and the severity of the effects varies with the degree of susceptibility (Erman et al. 2021). Men's and women's experiences and adaptation skills are different because of cultural norms, resource availability, and social hierarchy (Singh et al. 2021).

Climate change has a structural impact on the vulnerable population, as they have limited capacity to cope with the consequences of climate change; women are vulnerable considering their marginalised position in society (Parikh et al. 2012).

Disasters expose the already existing vulnerabilities of women, which are a result of various socio-political factors (Bhadra 2017). As climate change can amplify the existing vulnerabilities of an individual, there is a need to consider the differential vulnerability and to tailor the climate adaptation policies concerning the diverse needs of the population (Pinchoff et al. 2024).

## *Significance of the Study*

Different effects of climate change are felt by all life forms on Earth. In its second section of the Sixth Assessment Report, "Climate Change 2022: Impacts, Adaptation and Vulnerability", the IPCC noted how various climate change consequences on ecosystem and human susceptibility depend on the region. India has been identified as one of the world's hotspots by the report and other research because of its socioeconomic and geographic vulnerabilities. The disproportionate effects of climate change on women are also mentioned in the field of climate research, and other studies support this claim (World Meteorological Organisation 2024).

The main research gap is that the gendered impact of climate change is under-researched. There is also a dearth of studies that address the socio-economic and health impacts of climate change on women in India, as well as its underlying causes. A comprehensive understanding of these impacts and causes is an urgent need for effective climate action based on equity, and a systematic literature review can bridge this gap. Hence, this study focuses on exploring the impact of climate change on women in India and analysing the rationale for the gendered impact of climate change through the existing literature.

## Methodology

A systematic literature review was conducted for this study. Considering the IPCC's Third Assessment Report (TAR) from 2001 focusing on the impact of climate change and the need for adaptation, literature from the period between 2001 and 2024 from the electronic database Google Scholar and Scopus was used for the study. As a first step, the research questions were formulated to understand how climate change impacts the health and socio-economic aspects of women in India and explore the rationale behind it. After this, the inclusion and exclusion criteria were defined. The inclusion criteria included studies related to climate change and natural disasters concerning women based in India, and exclusion criteria included studies concerning climate change and natural disasters based outside India and not related to women, as well as those that focused on man-made disasters. Based on this, the keywords 'climate change', 'women' and 'India' were used for the search. Articles and grey literature in English were identified, and a database was created under relevant heads.

A total number of 67 studies were included after screening the title, and after examining the abstract concerning the set criteria of the study, 45 articles were accessed in full text. As a final step, a thematic analysis was later undertaken to identify the major themes by recording the common experiences from the findings concerning the objectives of the study.

## Results

It has been noted that the effects of climate change, including water scarcity, natural disaster impacts, and population displacement, have resulted in the double victimisation of young girls and women worldwide (Anand et al. 2022). India is a low-middle-income nation with a variety of social, geographic, political, and environmental factors that increase the susceptibility of its people—especially women—to the effects of climate change (Sorensen et al. 2018). Women are the most vulnerable group, along with children and the elderly, according to the Indian government's National Action Plan on Climate Change (NAPCC), which highlights the importance of addressing all facets of gender (Arora-Jonsson 2011).

There is a need to look beyond women as a homogeneous group and consider the socio-demographic factors that define their identity. As women are seen as the primary caregivers in society with the responsibility for the provision of basic needs for the family, climate change adversely affects them as it has a direct bearing on their responsibilities (Panda et al. 2014). The primary domains of the socio-economic and health consequences of climate change on women in India are illustrated in Fig. 7.1.

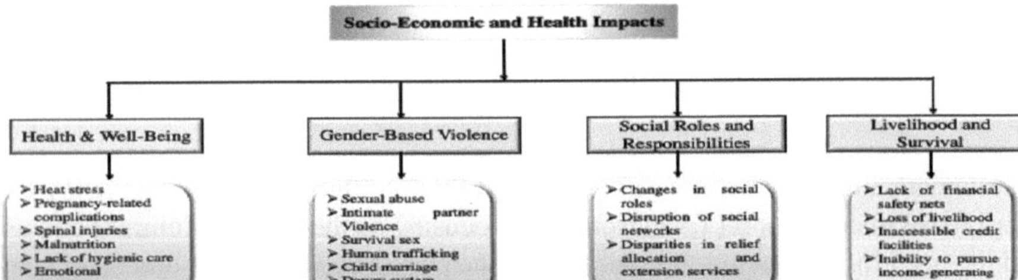


**Fig. 7.1** The socio-economic and health impacts of climate change on women in India (*Source* Compiled by the authors)

### *Health and Wellbeing*

Climate change has a significant impact on women's health, as established through various studies undertaken in India. The key issues identified include increased exposure to the ill effects of high temperatures, pregnancy-related complications, spinal injuries, malnutrition, health hazards due to lack of basic hygienic care, and increased emotional disturbances, including mental illness such as depression, anxiety, somatic complaints, and traumatic stress.

### *Mental Health Issues*

Climatic fluctuations and individual psychological wellbeing are linked; however, the psychological effects of climate events may differ between men and women (Goh 2012; Pailler and Tsaneva 2018). Women are more prone to depression and other emotional disturbances during disasters because they are obliged to care for their family members while unable to process their grief or loss, making them exposed to many types of stress (Gokhale 2008). In a case study conducted in the Himalayan state of Uttarakhand, livelihood, financial, and personal losses were discovered to be closely related to depression, anxiety, somatic symptoms, and general and traumatic stress (Parida 2015). Women's caretaking roles remain even after a disaster, as they must fend for food and shoulder such tasks, which adds to their agony. Because of their limited access to trauma recovery options, less privileged survivors may face long-term consequences in addition to developmental issues (Bhadra 2017).

Indian women score the lowest on some development indices, which can have major consequences for their wellbeing (Parikh et al. 2012). Isolation from social networks, loss of livelihood, forced dependency on high-risk activities to sustain oneself, and intimate partner violence can all contribute to women's emotional issues. The sense of loss that women feel after a disaster renders them subject to numerous emotional disturbances, which in turn makes them more vulnerable when the strain of the stressful environment increases (Bhadra 2017).

### *Physical Health Issues*

Heat exposure is related to a significant burden of disease and so requires consideration in the context of climatic changes (Deshpande et al. 2024). Men and women have different physiological responses to high temperatures. Culturally imposed heavy clothing garments, as well as limited access to healthcare and cooling facilities, all contribute to women's heat-related health issues. The situation is much more difficult for pregnant women, since extended heat exposure can lead to a variety of issues, including stillbirth, congenital birth abnormalities, and preterm delivery (Sorensen et al. 2018).

In Indian states such as Gujarat and Rajasthan, illiteracy among women is high and they are largely given the responsibility for household work, harvesting, and carrying fodder, in addition to farming and related activities. The increased occurrence of adverse weather conditions has a direct impact on these women. Water fetching, fuel wood collection, and long-distance carrying of huge loads on their heads can result in severe backaches and spinal damage. In the case of pregnant women, it may cause pregnancy-related problems and maternal death.

Even though women are responsible for gathering household necessities in most Indian village homes, their access to food is limited due to scarcity and priority hierarchy, harming their health (Yadav and Lal 2018). According to OXFAM India, women and young girls, accounting for nearly half of the agricultural workforce, consume less food (Anand et al. 2022). Pregnant and nursing women face an additional layer of challenges because of food scarcity and poor food intake (Nwoke and Ibe 2014 as cited in Yadav and Lal 2018). The water crisis and food shortages caused by climate change influence malnutrition and a lack of sanitation, exposing women to serious health risks. In India, impoverished households already have higher nutritional inadequacies, and climate-driven agricultural production reductions exacerbate their situation (Bhuyan et al. 2020; Singh et al. 2021). Inadequate access to clean water and sanitation can also pose several health risks for women, particularly during menstruation (Birch et al. 2012 as cited in Anand et al. 2022).

Women encounter hygiene and sanitation concerns in disaster relief camps as well. The unsanitary circumstances in the camps, as well as cultural traditions that permit certain forms of female hygiene care in private, make it impossible for women to receive sufficient hygiene treatment (Neumayer and Plümper 2007). In a case study of Majuli Island in Assam, many young women reported chronic rashes and urinary tract infections while in camps (Nongmaithem 2024). Extreme heat events, poor air quality, shifting rainfall and temperature patterns, and water insecurity, among other things, have a direct impact on women's health because they can cause adverse reproductive outcomes, macro and micronutrient deficiencies, bacterial, viral, and protozoan infections through contaminated water, and infection during pregnancy, among other things (Sorensen et al. 2018).

## *Gender-Based Violence*

Extreme weather patterns are linked to increasing gender-based violence among women. Some of the most frequently reported cases were various forms of sexual abuse and intimate partner violence. People are forced to leave their homes during a disaster to seek relief in shelters and relief camps. Sexual harassment and violence have been reported even in places of relief (Anand et al. 2022). Unaccompanied women and girls are more vulnerable to sexual abuse and rape in overcrowded camps, as well as unwanted physical touch and harassment (Alam and Rahman

2014; Neumayer and Plümper 2007). The shelters and temporary camps are mostly located along highways or railway tracks, where women lack privacy because they are in public view, and the tents, which are set up very close to each other, expose them to harassment and abuse (Ministry of Women and Child Development New Delhi 2015). Aside from that, women are forced to go to isolated regions at odd hours to gather fuelwood and fodder, exposing them to unsafe situations (Yadav and Lal 2018). Women and young girls are also exposed to high-risk behaviours and survival sex because they are often left with no other way to support themselves. They are frequently prone to criminal human trafficking, with 70% of them being coerced into the sex trade and other forms of sexual exploitation (Nellemann et al. 2011). Children exposed to climate hazards are also at risk of early pregnancies (Porecha 2024).

Climate change, with its impact on resource availability, influences the practice of several social evils in society, such as child marriage and the dowry system. The male preference, considering their conditioned responsibility to provide financial support for their parents, leaves adolescent girls feeling less cared for; marrying the girls off is considered a way to minimise the number of people to feed and prevent high dowries. The floods in Bihar resulted in a rise in dowry and child marriage. It has also been noticed that failing to meet expectations in terms of household necessities, such as water security, is connected with physical violence by the men in the family (Anand et al. 2022). Women are more likely to experience domestic violence and sexual harassment during climate-related disasters (Nellemann et al. 2011). In terms of intimate partner violence (IPV), some research has identified a link between the two. From 2005 to 2015, IPV rose in all Indian states except Kerala. A long-term analysis also shows that the rate of violence was high in the months following the tsunami and remained high even a decade later (Rao 2020).

## *Social Roles and Responsibilities*

Climatic variations have a greater impact on women's socially defined roles, exacerbating their socioeconomic situations. Changes in socioeconomic conditions are linked to several kinds of family and societal issues that women encounter (Dimitrov 2019). Changes in social roles, disruption of social networks, disparities in relief allocation, and further social inequities are among the many social inequalities.

Women face increased inequality and inequities because of extreme climatic events. Women bear most of the responsibility of household management, yet females have restricted access to health care, employment, economic possibilities, and political and decision-making processes (Parikh et al. 2012). Following a disaster, women may be forced to take on new societal roles such as single parent, widow, or head of household, in addition to other socioeconomic consequences such as disruption of life routines, feelings of isolation, stigma, and withdrawal from external life, among others (Gokhale 2008). Considering post-disaster relief operations, even though they are intended for the entire population of the affected region, they primarily rely on the existing patriarchal resource allocation arrangements, leaving women with limited access to resources. Rather than receiving preferential treatment because of their vulnerability, they are frequently marginalised in terms of resource access (Neumayer and Plümper 2007).

A study comparing men's and women's responses to climate variability during the drought in villages in Andhra Pradesh found a clear difference in men's and women's experiences based on their responsibilities in different situations. Some of the major differences identified were in government assistance, extension services, remuneration for their participation in National Rural Employment Guarantee initiatives, and available services (Arora-Jonsson 2011).

## *Livelihood and Survival*

Climate-related extreme weather events also have an impact on the social structure, as women and children are often left unattended, separated, or orphaned (Nellemann et al. 2011), putting them under financial strain. Climate change and its consequences affect women's livelihoods, often deteriorating their financial situation due to a lack of financial safety nets. Some of the primary challenges include a lack of economic capital to deal with the climatic occurrences, an inability to pursue income-generating activities, inaccessible credit facilities, a negative impact on agriculture activities, and land rights as well as ownership concerns.

As a result of the gendered division of labour, women are represented more in agriculture and informal sectors such as fishing, paddy cultivation, and the collection of non-timber forest produce, as well as the provision of water, food, and firewood to households (Anand et al. 2022; Parikh et al. 2012). Climate change has a significant impact on these activities. Furthermore, loss of livelihood due to floods impacts the livelihood of women who are dependent on livestock and paid labour (George et al. 2022). Women-headed households are particularly vulnerable because they have limited access to natural, social, and economic capital to deal with climate-related risks (Giri 2021). Furthermore, the variety of roles that women must perform limits their ability to pursue income-generating activities. The additional burden of social and household responsibilities, increased responsibilities as a result of male members' migration, and the responsibility to care for children and the elderly, given their increased risk of becoming ill due to heat stress, among many other factors, create a barrier for women to engage in other income-generating activities (Parikh et al. 2012).

In the case of Mahanadi Delta situated in the State of Odisha, agriculture and fishing are the major sources of livelihood, but the rise in sea levels and salinisation of the agricultural fields left them with no livelihood. In the planned relocation of these coastal communities, even though it had its positive impacts, women had to suffer from several issues. As there was an irregular supply of money in terms of the construction of houses and toilets, only a few people with additional income were able to complete it, while the others, especially the female-headed households, had to live in houses that were just half-built (Hans et al. 2019).

Even though they make up almost half of the agricultural workforce in India, rural women still experience discrimination due to their limited land rights, difficulties obtaining ownership, inability to access finance facilities, and unpaid labour (Ravi 2021). Despite their substantial contribution to the industry, women growers typically do not have access to assets, arable land, or, in certain situations, the legal right to own the land they cultivate (Gupta 2015). As education—which has the potential to increase their opportunities—remains limited or non-existent to a majority of rural women, they are left with very few options, such as with the loss of usufruct rights to community land (Roy and Venema 2002).

## Discussion

The rationale for the gendered impact of climate change on women is rooted in the multiple layers of inequality and discrimination that place women in a disadvantaged position to their male counterparts. The intersectionality approach provides insight into the rationale behind it. Intersectionality focuses on how the multiple forms of inequality or disadvantage can compound themselves and create obstacles for an individual (Crenshaw 1991). Understanding the vulnerabilities of women enhances the understanding of the existing inequalities of women. Figure 7.2 portrays the major vulnerabilities of women in India that perpetuate inequality in various domains.

Socio-cultural vulnerabilities, the factors that are culturally and socially rooted in the community (such as certain beliefs and customs) make women a vulnerable

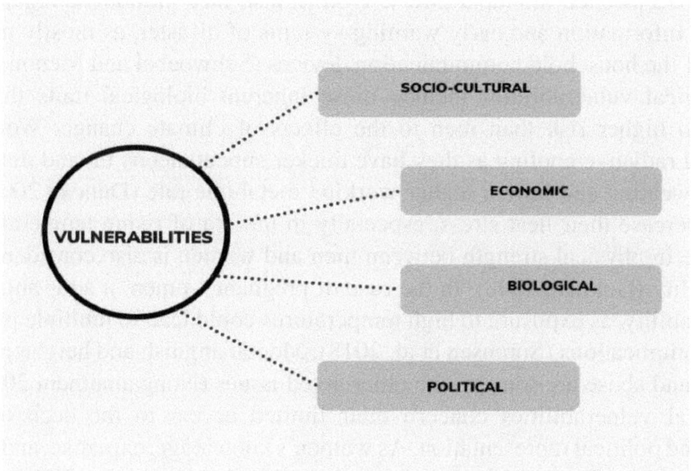


**Fig. 7.2** The major vulnerabilities of women in India (*Source* Compiled by the authors)

group. These conditioned beliefs hold women in a disadvantaged position and affect their ability to cope with extreme climate events. Traditional gender roles and social norms that exist act as major factors that increase women's risk of the adverse effects of climate change. In the case of a disaster, often, the kind of work they engage in has been a factor that has influenced their mortality. For instance, in the earthquake that occurred in the western state of Maharashtra, more women died because they were at home while the men were either harvesting crops, preparing for festivals, or in other districts for work (Schwoebel and Menon 2004). In the case of tsunamis in India, many women are at home and on the seashore waiting for their husbands, and because a tsunami has the most fatal impact on the coast, more women are affected than men (Neumayer and Plümper 2007).

Numerous health problems among women are a result of culturally prescribed attire that may restrict evaporative cooling, poor access to cooling facilities and healthcare, and a lack of personal transportation (Sorensen et al. 2018). Not only does culturally prescribed clothing contribute to heat stress, but it can also hinder mobility during disaster (Neumayer and Plümper 2007). Furthermore, the fundamental cultural convictions of some groups, who view women as impure and saltwater as sacred, have limited their exposure to the sea, preventing them from receiving life-saving instruction like swimming (Jayarathne and Babu 2017).

Economic vulnerabilities—such as the lack of access to resources—are a major risk factor for women in terms of their wellbeing, and this is even more evident in the context of extreme climatic events. Resources that range from land ownership to educational opportunities are deprived for many, especially rural women. As boys are expected to provide for their parents in their old age, the girls in the family get lesser priority for educational opportunities; reducing the financial burden by marrying them off is the priority (Anand et al. 2022). Women farmers face disparities in land rights and credit facilities even when they significantly contribute to food production (Yadav 2013 as cited in Gupta 2015). Women also face limitations regarding their access to information and early warning systems of disaster, as mostly men are in control of the household communication devices (Schwoebel and Menon 2004).

Biological vulnerabilities include those inherent biological traits that expose women to higher risk than men to the effects of climate change. Women have decreased radiative cooling as they have thicker subcutaneous fat and disperse less heat by sweating and have a higher working metabolic rate (Duncan 2006). These factors increase their heat stress, especially in the era of rising temperatures. The difference in physical strength between men and women is also considered to be a vulnerability (Gokhale 2008). In the case of pregnant women, it adds another layer of vulnerability, as exposure to high temperatures could lead to multiple pregnancy-related complications (Sorensen et al. 2018). Mental anguish and heightened risk of violence and abuse are some of the other noted issues (Nongamaithem 2024).

Political vulnerabilities concern their limited access to the decision-making process and political representation. As women's knowledge, expertise, and needs are not adequately reflected in the policies and programmes, it is not sufficient to reduce the risk factors and improve the protective factors of women. Women, especially those who belong to poorer communities, do not have much say in decisions that are made even at the local or national levels (Gupta 2015). Even in disaster management, women's participation is limited because of gender role perceptions (Schwoebel and Menon 2004). As the needs of women in both health and socio-economic aspects are unique, their lack of representation influences the interventions as well.

## Conclusion

Today, climate change is a pressing issue, as it has innumerable effects on planet Earth and all the forms of life in it. It has severe implications for human security, closely linked to the vulnerabilities associated with climate change (Parikh et al. 2012). In terms of gender, climate change and its consequences are likely to increase the existing inequalities in society due to socially constructed gender roles (Nellemann et al. 2011). India is very vulnerable to extreme climate events, and women are severely affected. Cultural norms, varied resource access, and the position in social hierarchies mediate the capacities of men and women, and hence they experience climate change differently (Singh et al. 2021). Gender along with socio-demographic factors significantly shape how climate change is experienced by an individual.

In India, climate change has a multitude of impacts on women. The intensity and nature of this impact varies across the women, as they are not a homogeneous group. Healthwise, women's physical and mental health are directly and indirectly affected by climatic variations as physical injuries, mental illness, malnutrition, and other health hazards are reported. The economic impact is linked with the direct impact of extreme climate events on their livelihood options and how the lack of financial safety nets exacerbates their situation for meeting basic needs. Concerning the social impact, climate change has affected the existing social inequalities. In addition, women's societal roles and responsibilities also change, and their social networks often get disrupted. Increased cases of violence against women have also been reported in different contexts that range from intimate partner violence to different forms of sexual abuse. These impacts are deeply rooted in the vulnerabilities and existing inequalities that women face, which include socio-cultural, economic, biological, and political vulnerabilities. Hence, this study highlights the existing disparities that women experience because of climate change and the urgent need for gender sensitivity in climate action.

There are also certain limitations to the current study, stemming from India's diverse geographical locations and cultural customs. Although the effects of climate change on women have similarities throughout India, the severity of these effects varies depending on sociocultural variables, state government policies and initiatives, and geographic vulnerabilities. As women are not a homogeneous group, the impact varies depending on the group of women and sociodemographic variables. Even though the study gives an overall understanding of the health-related and socio-economic impact of climate change on women in India, the nuances and intricacies of differences in geographical locations and experiences of different groups of women are not captured in this.

Further research areas have also been identified. Since women's experiences with climate change are distinct, more research is required to quantify the impact of the phenomenon and concentrate on these experiences through qualitative studies. Studies that concentrate only on women are necessary, and as women are not a homogenous group, investigating the effects on various subgroups of women can help policymakers better understand the solutions that are required for them. Location-specific research is also required, considering the unique characteristics and socio-cultural customs of different regions. Aside from that, as women actively participate in addressing climate change in addition to being its victims, further research is required to fully understand their contributions and the sustainable behaviours they uphold.

## Recommendations and Social Work Implications

Based on the nature of the health-related and socio-economic impact of climate change and the rationale behind it, there is room for further interventions from different stakeholders. Some of them are mentioned as follows.

## Recommendation for Policy Makers

Gender mainstreaming in climate action policies and programmes is an urgent need. India's NAPCC recognises the severe impact of climate change on women and the need to pay special attention to the aspects of gender through policies and programmes, but women are seen as a homogeneous category in it. As there are only a few gender-specific measures, the need to address the concerns of women and to incorporate gender-specific measures has been recognised (Parikh et al. 2012). For an effective risk reduction plan, a comprehensive assessment to identify the vulnerabilities of women along with gender-responsive budgeting is needed (Anand et al. 2022). To avoid gender-based inequalities, especially through institutional means, the inclusion of gender-specific considerations at all stages of policy formulation and implementation is required.

An analysis of the State Action Plan for Climate Action (SAPCC) of the Indian states pointed out that 12 states do not recognise gender in their policies at all, while in the remaining 16 states, most of them recognise women as victims without giving due recognition to their contributions and adaptive capacities (Singh et al. 2021). This highlights the need to focus on women's adaptive capacities and skills, as the strengths and expertise they possess are often overlooked. Hence, focusing on interventions that enhance these capabilities and facilitate their empowerment is crucial.

The inclusion of women in decision-making bodies concerning climate action is essential. As the needs of men and women differ, interventions and other programmes that so not consider these differential impacts will not be effective. A "bottom-up" approach that focuses on the inclusion of women from the grassroots level, the ones who are the most affected, is required to craft policies that appropriately represent the needs of the most affected. Informal institutional reform—empowering the individuals at the community level to be the change makers—is also considered a good option for increasing the ability of the poor to adapt to climate change (Roy and Venema 2002).

Better availability and accessibility to credit facilities is another need that was identified. Considering women's economic vulnerability and how extreme climate events adversely affect their wellbeing, credit facilities can benefit them to a great extent. Even though multiple credit systems are available right from the grassroots level, such as through the SHGs, some women lack or have limited access to credit facilities. Apart from credit facilities, information tools that can enhance their understanding of credit systems, income-generating activities, and skill development could be beneficial. Access to information tools and financial assets is considered helpful for women in the context of a disaster, in addition to efforts to improve their income (George et al. 2022). Currently, microfinance institutions are keen on increasing the saving habits of women, and there have been efforts from the government to support the SHG system for better financial inclusion (Kandpal 2022).

## *Recommendations for Government and Non-Government Organisations*

Given the multitude of vulnerabilities ingrained in societal factors that impact women's wellbeing, social work interventions at the micro, meso, and macro levels may prove advantageous. Crisis intervention and psychosocial assistance may be essential, since prompt interventions could prevent mental health issues from becoming mental illnesses. Aside from being necessary after a disaster, mental health interventions are also a fundamental need in light of the various ways that climate change affects women. Given the rise in health-related and gender-based violence incidents, family and individual-level treatments are desperately needed. It is crucial to determine the likely reasons why they are more susceptible to mental health problems and to build their capacities. Need assessment at the community level can provide a better insight to stakeholders, including the government, for the provision of better amenities such as cooling facilities for pregnant women, better resource accessibility, and provision of health services. Improving access to health care services through medical camps can help identify the issues at an earlier stage and thus provide the needed help immediately. Awareness creation for identifying early symptoms, available credit facilities, intimate partner violence, and other forms of abuse have to be carried out at the community level. At the macro level, an analysis of the existing policies and in-depth research studies, especially longitudinal case studies, are needed to identify the gaps concerning the existing policies in place. Hence, as women are disproportionately affected by climatic events, equity not equality in interventions is needed; this can be achieved only through an assessment of the existing vulnerabilities, working towards addressing them and gender mainstreaming in climate actions.

## References


Alam K, Rahman MdH (2014) Women in natural disasters: a case study from southern coastal region of Bangladesh. Int J Disaster Risk Reduct 8:68–82. https://doi.org/10.1016/j.ijdrr.2014.01.003

Anand A, Nagaveni PL (2022) Climate change adaptation and mitigation in India—a gendered perspective—ProQuest. https://www.proquest.com/openview/cad9751f7ef017f02c88bebb8299831a/1?cbl=2035897&pqorigsite=gscholar&parentSessionId=kFDFsiSZxEHLVa%2FJh51kdcVP0l0BcvkaFfk1%2B2Mz7XI%3D

Arora-Jonsson S (2011) Virtue and vulnerability: discourses on women, gender and climate change. Glob Environ Chang 21(2):744–751. https://doi.org/10.1016/j.gloenvcha.2011.01.005

Bhadra S (2017) Women in disasters and conflicts in India: interventions in view of the millennium development goals. Int J Disas Risk Sci 8(2):196–207. https://doi.org/10.1007/s13753-017-0124-y

Bhuyan B, Sahoo BK, Suar D (2020) Nutritional status, poverty, and relative deprivation among socio-economic and gender groups in India: is the growth inclusive? ScienceDirect. https://www.sciencedirect.com/science/article/pii/S2452292918300936?ref=pdf_download&fr=RR-9&rr=866c7e7358f2936d

Chisty MA, Rahman MM, Khan NA, Dola SEA (2022) Assessing Community disaster resilience in flood-prone areas of Bangladesh: from a gender lens. Water 14(1), Article 1. https://doi.org/10.3390/w14010040

Crenshaw K (1991) Mapping the margins: intersectionality, identity politics, and violence against women of color on JSTOR. https://www.jstor.org/stable/1229039?mag=kimberle-crenshaws-intersectional-feminism&seq=3

Deshpande A, Scovronick N, Clasen TF, Waller L, Wang J, Aravindalochanan V, Balakrishnan K, Puttaswamy N, Peel J, Pillarisetti A (2024) Heat exposure among adult women in rural Tamil Nadu, India. Environ Sci Technol 58(1):315–322. https://doi.org/10.1021/acs.est.3c03461

Dimitrov BG (2019) Effects of climate change on women. Res Rev Int J Multidis 4(5):201–215

Duncan K (2006) Global climate change, air pollution, and women's health. WIT Trans Ecol Environ. https://doi.org/10.2495/RAV060611

Ebi KL, Bowen K (2016) Extreme events as sources of health vulnerability: drought as an example. Weather Clim Extrem 11:95–102. https://doi.org/10.1016/j.wace.2015.10.001

Erman A, De Vries Robbe SA, Thies SF, Kabir K, Maruo M (2021) Gender dimensions of disaster risk and resilience: existing evidence. World Bank. https://doi.org/10.1596/35202

George B, Kumar R, Banerjee S (2022) Vulnerability of women in the face of climate change: a study of Wayanad district of Kerala, India. J Chin Econ Foreign Trade Stud 15(3):279–297. https://doi.org/10.1108/JCEFTS-09-2021-0058

Giri SH, Patel A, Das S, Hans A, Ghosh A, Jasmine (2021) Women-headed households, migration and adaptation to climate change in the Mahanadi Delta, India. In: Engendering climate change. Routledge, India

Goh AHX (2012) A literature review of the gender-differentiated impacts of climate change on women's and men's assets and well-being in developing countries. International Food Policy Research Institute. https://doi.org/10.2499/CAPRiWP106

Gokhale V (2008, October) Role of women in disaster management: an analytical study with reference to Indian society. In: The 14th world conference on earthquake engineering, pp 12–17

Gupta H (2015) Women and climate change: linking ground perspectives to the global scenario—Himangana Gupta. https://journals.sagepub.com/https://doi.org/10.1177/0971521515594278?journalCode=ijgb
Hans A, Hazra S, Das S, Patel A (2019) Encountering gendered spaces in climate change policy in India: migration and adaptation. J Migrat Affair 2(1):1. https://doi.org/10.36931/jma.2019.2.1.1-24
Jayarathne H, Babu VS (2017) Socio-cultural factors influencing women to be more vulnerable in natural disasters: a qualitative review on Tamil Nadu state—R discovery. https://discovery.researcher.life/article/socio-cultural-factors-influencing-women-to-be-more-vulnerable-in-natural-disasters-a-qualitative-review-on-tamil-nadu-state/1af4e35785e233d9b51a41a250531a5a
Kandpal V (2022) Socio-economic development through self-help groups in rural India—a qualitative study. https://www.researchgate.net/publication/359240576_Socio-economic_development_through_self-help_groups_in_rural_India_-_a_qualitative_study
Krishnan R, Gnanaseelan C, Sanjay J, Swapna P, Dhara C, Sabin TP, Jadhav J, Sandeep N, Choudhury AD, Singh M, Mujumdar M, Parekh A, Tewari A, Mehajan R, Chopra R, Joshi A, Nagarajan A, Nivsarkar M, Rajeevan M, Niyogi D et al (2020) Introduction to climate change over the Indian Region. In: Krishnan R, Sanjay J, Gnanaseelan C, Mujumdar M, Kulkarni A, Chakraborty S (eds) Assessment of climate change over the Indian Region: a report of the ministry of earth sciences (MoES), Government of India. Springer, pp 1–20. https://doi.org/10.1007/978-981-15-4327-2_1
Ministry of Women and Child Development, Government of India (2015, June) Report of the high level committee on the status of women in India, vol 4. https://wcd.nic.in/sites/default/files/Vol%204.pdf
Nellemann C, Verma R, Hislop L (2011) (PDF) Women at the frontline of climate change: gender risks and hopes. https://www.researchgate.net/publication/235899714_Women_at_the_Frontline_of_Climate_Change_Gender_Risks_and_Hopes
Neumayer E, Plümper T (2007) The gendered nature of natural disasters: the impact of catastrophic events on the gender gap in life expectancy, 1981–2002. Ann Assoc Am Geogr 97(3):551–566. https://doi.org/10.1111/j.1467-8306.2007.00563.x
Nongmaithem J (2024) Gendered vulnerabilities in disaster responses: a case study of Majuli Island, Assam. https://www.researchgate.net/publication/379430131_Gendered_Vulnerabilities_in_Disaster_Responses_A_Case_Study_of_Majuli_Island_Assam
Owen-Burge C (2023, March 20) The IPCC just published its summary of 5 years of reports—here's what you need to know. Climate Champions. https://climatechampions.unfccc.int/the-ipcc-just-published-its-summary-of-5-years-of-reports-heres-what-you-need-to-know/
Pailler S, Tsaneva M (2018) The effects of climate variability on psychological well-being in India. World Dev 106:15–26. https://doi.org/10.1016/j.worlddev.2018.01.002
Panda GR, Shrivastava S, Kapoor A (2014) Climate change and gender: study of adaptation expenditure in select states of India. SpringerLink. https://link.springer.com/referenceworkentry/https://doi.org/10.1007/978-3-642-38670-1_127
Parida PK (2015) Natural disaster and women's mental health. Social Change 45(2):256–275. https://doi.org/10.1177/0049085715574189
Parikh J, Upadhyay DK, Singh T (2012) Gender perspectives on climate change & human security in India: an analysis of national missions on climate change. Cadmus J. http://www.cadmusjournal.org/node/151
Pinchoff J, Mahapatra B, Mishra R, Adedimeji A, Patel SK, Regules R (2024) Classifying climate vulnerability and inequalities in India, Mexico, and Nigeria: a latent class analysis approach. Environ Res Lett 19(3):034009. https://doi.org/10.1088/1748-9326/ad1fa6
Porecha M (2024, May 11) Study shows impact of climate hazards on women, children. The Hindu. https://www.thehindu.com/sci-tech/health/study-shows-impact-of-climate-hazards-on-women-children/article68164791.ece
Rao S (2020) A natural disaster and intimate partner violence: evidence over time. Soc Sci Med 247:112804. https://doi.org/10.1016/j.socscimed.2020.112804

Ravi A (2021) Global food systems are not thinking about women. A UN report calls for action. https://www.downtoearth.org.in/news/food/global-food-systems-are-not-thinking-about-women-a-un-report-calls-for-action-78246

Roy M, Venema HD (2002) Reducing risk and vulnerability to climate change in India: the capabilities approach. Gend Dev 10(2):78–83

Schwoebel M, Menon G (2004) Mainstreaming gender in disaster management support project

Singh C, Solomon D, Rao N (2021) How does climate change adaptation policy in India consider gender? An analysis of 28 state action plans. Climate Policy 21(7):958–975. https://doi.org/10.1080/14693062.2021.1953434

Sixth Assessment Report—IPCC (2023) https://www.ipcc.ch/assessment-report/ar6/

Sorensen C, Saunik S, Sehgal M, Tewary A, Govindan M, Lemery J, Balbus J (2018) Climate change and women's health: impacts and opportunities in India. https://agupubs.onlinelibrary.wiley.com/https://doi.org/10.1029/2018GH000163

UNDP (ed) (2011) Sustainability and equity: a better future for all. Palgrave Macmillan

What Is Climate Change? United Nations (n.d.) Retrieved 4 Mar 2024, from https://www.un.org/en/climatechange/what-is-climate-change

WHO (2023) Climate change. https://www.who.int/news-room/fact-sheets/detail/climate-change-and-health

World Meteorological Organisation (2024, February 16) The climate crisis is not gender neutral. World Meteorological Organization. https://wmo.int/media/news/climate-crisis-not-gender-neutral

Yadav SS, Lal R (2018) Vulnerability of women to climate change in arid and semi-arid regions: the case of India and South Asia. J Arid Environ 149:4–17. https://doi.org/10.1016/j.jaridenv.2017.08.001

# Chapter 8
# Theories Underpinning Studies on Extreme Weather Events and Spatio-temporal Disease Incidences in Marginal Spaces in the Global South

**Halleluah Chirisa and Abraham Matamanda**

**Abstract** Extreme weather events (EWEs) remain a threat to the world and the Global South, which remains behind in building resilience to climate change challenges. The study explores the impact of extreme weather events (EWEs) on disease incidence in marginalised areas of the Global South, highlighting the role of spatial temporal and power inequalities. It uses a desktop study methodology to conduct an integrative literature review, aiming to enhance the theoretical understanding of EWEs and disease dynamics. The study's aim is to improve our theoretical understanding of EWEs and disease dynamics in marginalised areas. By bridging gaps in knowledge, researchers can contribute to more effective policies, better disaster responses, and enhanced community resilience. The findings indicate that vulnerabilities in the Global South are exacerbated by historical colonialism and ongoing inequalities in race, ethnicity, and class. These marginalised communities lack power and are often excluded from state transformation programmes, making them more susceptible to EWE-induced diseases. The study emphasises the need for inclusive community health frameworks to improve resilience and suggests using Complexity Theory to address the multifaceted challenges of EWEs and disease incidence through a multi-sectoral approach. This theoretical advancement aims to inform better policies, disaster responses, and community resilience strategies.

We declare that, to the best of our knowledge, this article does not infringe upon anyone's copyright nor violate any proprietary rights and that any ideas, techniques, quotations, or any other material from the work of other people included in our article, published or otherwise, are fully acknowledged in accordance with the standard referencing practices.

H. Chirisa (✉)
Zimbabwe Ezekiel University (ZEGU), Bindura, Zimbabwe
e-mail: 2021688664@ufs4life.ac.za

Department of Geography, University of the Free State, Bloemfontein, South Africa

A. Matamanda
Department of Urban and Regional Planning, Bloemfontein, Republic of South Africa



W. Leal Filho and F. Wolf (eds.), *Climate Change and Health: Perspectives from Developing Countries*, Climate Change Management,
https://doi.org/10.1007/978-3-031-93177-2_8

**Keywords** Resilience · Power · Relegated · Community-health · State-nation · Inclusive · Extreme weather events · Disease incidence · Complexity theory · One health approach

## Introduction

Living conditions are a well-acknowledged determinant of health and health inequalities. Communities experience residential segregation and social marginalisation with disproportionate exposure and susceptibility to a combination of social determinants and risk factors associated with health outcomes (Caminada et al. 2021). Marginalised and impoverished communities are most susceptible to communicable diseases such as malaria, typhoid, and cholera in Southern Africa, owing to institutionalised and systematically enforced racial inequality (Faleye et al. 2023). Furthermore, poverty rates in South Africa are high amongst marginalised populations (Francis and Webster 2019). Therefore, this discourse brings to the fore the question of marginalisation and what it constitutes concerning EWEs, climate change, and emerging health challenges. Whereas national governments favour the rhetoric that EWEs result from climate change and directly lead to a development retrogression, contemporary geographers, political scientists, and sociologists explain the effect of EWEs as a multiplicity phenomenon. Particularly, through the lens of the Probabilistic Event Attribution (PEA) concept, the perspective that anthropogenic climate change has a heterogeneous effect on particular regions and nations is gaining traction (Lusk 2022). The challenges posed by EWEs differ according to space, thus developing into spatio-temporal disease incidences (Stott et al. 2016). Therefore, the purpose of this study is to unpack relevant theories as lenses through which such spatio-temporal disease incidences and marginality can be explored within the context of EWEs. Understanding different theories and using them as lenses to view the challenges of EWEs and disease incidences may assist governments in policy formulation and implementation. The significance of this study lies in accentuating background determinants of vulnerability in marginal spaces, which expose marginal populations to greater effects of climate change and EWEs than more privileged population groups.

Probabilistic Event Attribution shows that EWEs surpassed pre-industrial levels of anthropogenic global warming in 2017, contributing to an increase in public health challenges (Allen et al. 2018). As such, Springmann (2016) and Jones (2022) posit that climate change is the most significant global health threat of the twenty-first century that will directly and indirectly affect health. This will be a result of changes in temperatures, precipitation, and the occurrence of heatwaves, floods, and droughts. For example, in most heatwave-prone areas, there have been reports of diseases such as acute myocardial infarction, stroke, pulmonary embolism, and cardiomyopathy (Abrignani et al. 2022). Noteworthy is that EWEs such as heatwaves, floods, and droughts are a direct health threat to poor populations (Akhtar 2020). The discrepancies in the impact of EWEs on the wealth status of particular populations attest to the marginalisation of communities. In this regard, unless a thorough theoretical study is conducted, which leads to selecting a correct lens or approach, a wrongly chosen tool or lens will be disastrous, given the complex nature of the variables under study.

According to Varghese and Kumar (2022), "marginality" refers to the relative location of the communities in the geographical, political, social, or cultural spaces, and these spaces assume importance because the environment that shapes the actors is graded and endowed with differential resources. It is presumed that population groups that are resilient to the vagaries of EWEs and the resultant health challenges are affluent and adequately endowed and are, thus, not marginalised. Consequently, marginalisation may be depicted at individual, community, national, and regional levels. For example, at community levels, healthcare facilities may be overwhelmed by a surge in healthcare demand, leading to a deterioration of the health system to protect the population against current and future risks (Cianconi et al. 2020). Such a scenario resonates with the concept of territorial stigmatisation (Wacquant 2019: 24). Here, marginality, particularly in urban areas, results in some areas being relegated to the social space, bureaucratic, and symbolic power peripheries (Hincks and Powell 2022). Subsequently, assessing the conceptual undertones that guide studies on EWEs, public health challenges, and developments in the Global South is imperative.

## *Research Questions*

- What are the key theories that underpin the studies of marginality in the Global South?
- How do EWEs influence the spatio-temporal patterns of disease incidence in these marginal spaces?
- What are the specific vulnerabilities of marginal spaces in the Global South to EWEs and associated disease outbreaks?
- How do social, economic, and environmental factors in the Global South interact to influence the impacts of EWEs on disease incidences?
- What mitigation and adaptation strategies are most effective in managing the impacts of EWEs on spatio-temporal patterns of disease incidences in the Global South?

## Methodology

The study involved the following units of study:

- Theories
- EWEs
- Spatio-temporal Disease Incidences

- Marginal Spaces.

The desktop study used an integrative literature review, focusing on the four highlighted units. The search engine Google Scholar was used to gather literature on the four units. Search engines offer real-time access to up-to-date information, and subsequently, recent studies, news articles, and reports relevant to the study were sourced quickly. Desktop reviewing of the literature allowed for the systematic examination and analysis of relevant topics, identification of gaps, and positioning of the study within the existing body of knowledge.

The search engine enabled the sourcing of varied journal articles and books. The study focused on the Global South and gave examples of marginality within such context. The Global South was purposively selected due to the existence of 20 countries with the highest inequality amongst its populations globally, with South Africa scoring 63 on the Gini index in 2023—the highest of the 20 countries (Dyvik 2024).

Inclusion of the theories was based on the relevance of the theories to the research questions and the context of the study, Global South. Furthermore, the selection focused on well-established and widely accepted theories in the study of marginality. Theories that did not fit were excluded.

The common limitation of a theoretical paper is lack of empirical data to validate theory propositions. The aim of this paper was to analyse possible lenses to be used when studying the nexus between EWEs and disease incidences in marginal spaces. As such, the theoretical foundation of the study was well provided by this research paper.

## The Theories

The extreme weather–public health nexus is never a simple "black and white" narrative but a complex, multi-dimensional discourse. Hence, to do justice to the issue, various theoretical and conceptual aspects need to be addressed to simplify the intricacies. As such, an intersection of the following pillars of the argument—space, power dynamics, advanced marginality, necropolitics, and the proponents of related theoretical aspects—are discussed.

Space and power dynamics

- *Doreen B. Massey—The Space Concept*

Doreen Massey was born in Manchester in 1944 and grew up on the Wythenshawe council estate located in the southwest of the city. As a geographer, she notably contributed to the concept of space. In "*For Space*," three landmark contributions were championed: space is an interrelationship product; space is an intersectional phenomenon resulting from multiple factors; and space is dynamic as it is always under construction (Kitchin 2016: 815). This concurs with Anderson (2008), who posited that the fulcrum of the Space theory is that far from being inert and static, space becomes open to progressive politics, which can change how it is imagined and produced. Anderson's explanation brings into perspective the aspect of power. Massey (2009) observed that power and space are intertwined intimately, which is confirmed by the fact that some places, such as global cities, are associated with immense power. Nationally, capital cities within a country are regarded as citadels of power, and governmental administration is usually headquartered in such space. In the habitat hierarchy of urban spaces in the Global South, the development of primary cities vis-à-vis secondary cities connotes development inequalities promoting marginalisation and peripheralisation of urbanisation. In this regard, Massey (2009) posits that power geometrics highlight embedded inequalities and thus proposes the conceptualisation of space as a political task and a product of stakeholder relations. This implies that marginalised communities have limited political power and are, therefore, less resilient to the vagaries of extreme weather conditions and subsequent healthcare challenges.

- *Henri Lefebvre—The Social Production of Space Concept*

Henri Lefebvre was a French Marxist theorist who was also regarded as a philosopher, urban theorist, geographer, sociologist, political scientist, and historian. Lefebvre's main contribution, which is particularly relevant to this paper, is the concept of social production of space. Lefebvre (1974) argues that space cannot be understood as a simple collection of elements because a society's space is a product created through its spatial practice. As such, the Production of Space theory denotes three levels of social space: perceived social practices, conceived representations of space, and lived spaces of representation (Boanada-Fuchs and Boanada Fuchs 2018). Thus, a "dialectical relationship … exists within the triad of the perceived, the conceived, and the lived" (Lefebvre 1991, p. 39). For example, a conceived representation of space is abstract as it is rooted in the principles, beliefs, and visions held by practitioners, decision-makers, and others who are in a position to impose their notion of order on the concrete world and so create a practical impact on space within the social and political practice (Lefebvre 1991).

A representational space signifies a lived space whose subjects are inhabitants of particular areas, and the objects are social life, art, culture, images, symbols, systems of non-verbal symbols, and signs (Boanada-Fuchs and Boanada Fuchs 2018). Lefebvre's philosophical reasoning on space is assumed to have emerged from lived experiences in urban development, planning, and architecture (Stanek 2015). By and large, O'Connell (2015) observed that Lefebvre's theoretical account of space as a concrete abstraction bore the same paradoxical qualities of being a commodity and labour that it held for Marx.

In relation to this paper, the spatial forms of dwelling at the centre of Lefebvre's research were symptomatic of the post-war housing crisis in France and the privatisation of everyday life in the emerging consumer society (O'Connell 2014). Lefebvre offered a transdisciplinary reading of space as he shifted the dominant discussion from the ontology of the spatial to an epistemology of the urban framework, which captures the radical alterity of urbanisation (Elden 2004). This idea is supported by

O'Connell (2015), who observed that space is perceived as an abstraction of differentially experienced factors such as class, gender, age, and memory, calling for a need to embrace its diversity and effectively keep it open to all, with no marginalisation. The multiplicity of space is a significant factor in urban planning and development. Whereas inhabitants of a particular area may have perceived and conceived specific developmental trends within the region, it is the social responsibility of those in power to reduce marginalisation trends by conceiving better developmental trends to transform the lived space.

Advanced Marginality

- *Wacquant–Territorial Stigmatisation*

Wacquant (2008) observed that the political production of advanced marginality elucidates the nexus of class and race in the district dispossession or bas-quarters of the post-industrial metropolis in its phase of socio-spatial polarisation. Wacquant's observation captures the sudden implosion of the black American ghetto after the height of the civil rights movement and attributes it to the turnaround of local and federal policies after the mid-1970s. This policy turnaround accelerated the historical transition from the communal ghetto, confining all blacks in a reserved space that both entrapped and protected them, to the hyper-ghetto. In this regard, a hyper-ghetto is regarded as a territory of desolation that now contains only the unstable fractions of the African American working class. They are exposed to all manners of insecurity by unravelling a web of parallel institutions, which characterises the ghetto in its fully-fledged form (Flint 2018). Therefore, Hincks and Powel (2022) argued that existing accounts of territorial stigmatisation have tended to centre on urban locales of relegation, which can pragmatically be understood in Wacquant's schema as 'ghettos' and 'hyper ghettos'. Here these are referred to devices for the spatial confinement and seclusion of racialised groups. As such, territorial stigmatisation is perceived as a product of the relegation of some social classes as propelled by class differences. In this context, relegation is a critical process in contemporary advanced urban marginality involving banishing marginalised urban populations to particular socio-spatial formations (Flint 2018).

Wacquant (2011) argued that urban relegation obeys different logic as it is determined by ethnicity, modulated by class position, and aggravated by the state. Policies may have the effect of further relegating marginalised societies into even worse circumstances. Sennett (2006: 83) argued that contemporary social welfare systems may have the undesirable effect of generating a "specter of uselessness" and "permutations of dishonor" among the urbanites. Furthermore, Flint (2018) argued that the mentalities of individuals subject to welfare interventions are transformed by their routine engagement with hostility, interpersonal threats, and, in some cases, the coercive arm of the state. These interactions may alienate the people among themselves and among outsiders, resulting in further marginalisation of such societies during socio-economic hardships and challenges (Hancock and Mooney 2013).

- *Achille Mbembe—Necropolitics*

Necropolitics accounts for the state's ultimate sovereign power to control, regulate, and legitimise mortality. As elucidated by Mbembe, the concept of necropolitics builds on Foucault's notion of bio-power. In this regard, bio-power is an expansionary model of power whereby the expansion and efficiency of life, not its deduction, becomes the primary function of power, and deduction is merely an instrument in the service of expansion (Cisney and Morar 2015). As such, through the concept of necropolitics, Mbembe and Meintjes (2003) called attention to when authorities go beyond regulating life to regulating death itself. This has been notable in colonies, where in extra-legal spaces, the violence of the state is rationalised to be operating in the service of civilisation (Mbembe 2019). There, as in frontier regions, the lives of usually racialised communities are regarded as disposable; nobody feels any obligation to respond to, bear the responsibility of, or seek justice for legitimised deaths (Mbembe 2019).

Necropolitics still characterises much frontier expansion in Latin America, Indigenous Afro-Brazilian quilombolas and other local communities. In these areas, communities steadfastly give way to landscape transformation and large-scale resource extraction as forest dwellers are represented as lazy and lacking in industriousness or entrepreneurial drive (Ribeiro 2017). The significance of necropolitics in marginalisation studies is the notion that sovereign states, through their leaders, can decide which groups of people should live or be sacrificed for the greater good. This is explained by Gržinić (2021: 14), who points out, "What had been there as an exclusive right to protect nation-state borders was reframed by necropolitics as the right to decide who should live and who must die".

## *Disease Incidences and Household and Community Health*

- Sanders (1964)—*Measuring Community Health Levels*

Sanders (1964) argued that the assumption that a high morbidity rate in a community connotes deficient health care would seem plausible. Sanders (1964) listed obesity as the most prevalent condition, followed by mental disorders, where the development of tranquilizers in the 1950s had little impact on psychotic conditions. Otto et al. (2014) observed that even to date, with all the relevant medication and therapies available, mental illness treatments remain palliative rather than preventive or curative. Furthermore, Stott et al. (2016) observed that mental health challenges, as a rule, are not an immediate cause of death; however, preventive and control measures prolong life, including those of psychotic patients.

Mendin et al. (2023) observed that the third highest cause of morbidity is heart disease, and in combination with associated vascular and renal diseases, accounts for over two-thirds of all deaths globally. An increased survival rate, without cure of many chronic diseases, raises age-specific prevalence rates. Sanders (1964) proposed a modified life table to measure the comparative adequacy of health services for different population groups. Furthermore, Sanders (1964) suggested that the index of community preventive health services can be used as a comparative index. The index and the age-sex ratios of known and unknown diseases can be developed, such that the quality of health care for different population groups can be assessed by determining the proportion of known pathological conditions per ratio.

A disease-specific life table comparison was also proposed. These techniques include the modified life table analysis that can be used for specific diseases (Sanders 1964). Such comparisons between different communities includes the number of individuals with a particular disease (Sanders 1964). A community with more emerging diseases could necessitate better health services than one with less diseases.

## *Extreme Weather Events*

- Hulme (2014)—*Attributing Extreme Weather Extremes to Climate Change*

Hulme (2014) observed that humans and the societies in which they live have always sought to establish causes for their daily weather. The extreme weather blame tests scientific understanding and the veracity of climate simulation modelling in many ways (Hulme 2014). One approach to explain extreme weather phenomena is general physical reasoning. With rising greenhouse gas concentrations, thermodynamic arguments suggest that more intense precipitation events are expected in an atmosphere with more water vapour (O'Gorman and Schneider 2009). The danger of general physical reasoning is that it might be too specific to be of value or may be wrong (Hulme 2014). General physical reasoning can only be used for broad statements, such as when extreme weather is consistent with what is known about the human-enhanced greenhouse effect (Hulme 2014).

Hulme (2014) observed that a second approach uses statistical analyses of meteorological time series data to determine whether a particular weather (climate) extreme falls outside the range of what an average climate might have delivered. Such an analysis estimates the likelihood of a specific meteorological extreme occurring given no external human action (Hulme 2014). All extreme event analyses of meteorological time series are, at best, able to detect outliers but can never be decisive about possible causes. This approach can attribute a trend in extreme weather to human influence but not a specific weather event, and the extreme weather blame remains undetermined (Hulme 2014).

Hulme (2014) observes that the third approach uses multiple model simulations of climate systems. Initially, external bias is not included in the model. Thereafter, external bias, which is perceived to cause EWE, is included in the model. The new science of extreme weather attribution is developing in the context of politically charged debates about the causes and consequences of climate change and about appropriate responses (Hulme 2014).

- *Rudolf Virchow—One Health Approach*

According to Erkyihun and Alemayehu (2022: 03), "The One Health approach concept was launched in the nineteenth century by Rudolf Virchow (1821–1902) after he incorporated veterinary medicine into human health and asserted the absence of a dividing line between them" and has since been embraced worldwide as a public health strategy. The One Health approach may be described as "any research, policy development and implementation, capacity and infrastructure development, and system building or strengthening activities relevant to the health and well-being of humans, animals, plants, or ecosystems" (Mumford et al. 2023: 02). In this regard, the One Health approach is a practical approach to addressing complex health issues (Mackenzie et al. 2014). Noteworthy is that the approach is based on one central multisectoral assumption. This implies that the approach involves various stakeholders who deliberate on health delivery equilibrium, focusing on the environment, animals, and human beings. All levels of decision-making concerning health are included, that is, at local, national, and international levels. For example, Erkyihun and Alemayehu (2022) argue that because of the world's interconnectedness, the threat of a given zoonotic disease somewhere is a threat everywhere, and this requires inter-sectoral integration and collaboration to decrease the burden. Worldwide disease threats of such nature may only be resolved through the application of the One Health approach at international levels. Figure 8.1 depicts the central tenets of the One Health approach.

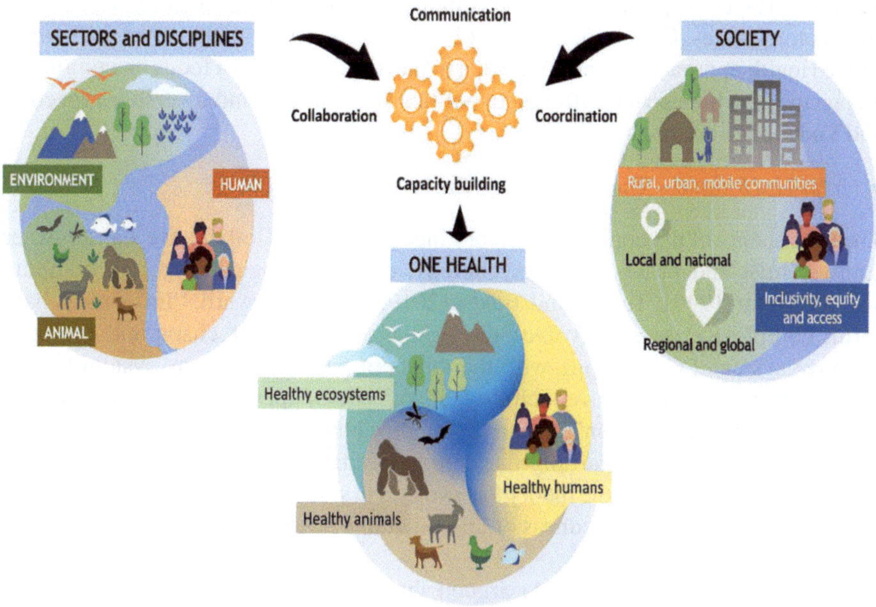


**Fig. 8.1** The one health approach (adapted from WHO 2021) https://www.who.int/multi-media/details/one-health-definition

The operational strategy of the One Health approach is centred on the "Four C's": communication, coordination, capacity building, and collaboration. Through such a structured strategy, stakeholders at all levels intersect and pool resources to realise the One Health approach. The Quadripartite Agreement exemplifies this. On March 17, 2022, the FAO, WOAH, WHO, and the UNEP jointly spearheaded the One Health Joint Plan of Action (OHJPA) during the COVID-19 pandemic (Erkyihun and Alemayehu 2022). The OHJPA was developed to improve the world's ability to "prevent, predict, detect, and respond to health threats and improve the health of humans, animals, and the environment while contributing to sustainable development" (FAO, UNEP, et al. 2022). From this perspective, the One Health approach has evolved holistically by moving away from an anthropocentric scope to a balanced eco-centric approach. Thus, the One Health approach can be applied in solving public health challenges in the wake of global health threats exacerbated by EWE (Jamal et al. 2023). For example, Wolmuth-Gordon and Mutebi (2023) argue that climate change events like floods and heatwaves are altering the distribution, transmission, and severity of zoonotic pathogens and their vectors; the One Health approach can have a significant impact through surveillance, prevention, and outbreak preparedness for zoonoses and any other disease. Applying the One Health approach to marginalised communities is exemplified in the Maasai communities. Gwakisa et al. (2023) explained that in order to stimulate collaboration and ownership of One Health interventions, a complete picture of the ecosystem in which the Maasai people live is essential. The concept of co-creation is indispensable for One Health innovations in such communities. Cunningham et al. (2017: 06) recommended that "at the heart of a One Health approach, in and for Africa in particular, must be a focus on community engagement, where social difference, politics and interests and diverse perspectives and knowledge are put center-stage".

- *John H. Holland—The Complexity Theory*

The proponents of the complexity theory include John H. Holland. He was an American scientist and professor of psychology, electrical engineering, and computer science (Agar 1999). Turner and Baker (2019: 12) put forward, "In its most basic form, complexity theory involves the primary concepts of chaos and [complex adaptive systems] CAS, along with the tenets of path dependence, system history, non-linearity, emergence, irreducibility, adaptiveness, operating between order and chaos, and self-organization". Figure 8.2 highlights the interconnectedness of these features and pathways regarding the complexity theory. This justifies the view of Talukder et al. (2024) that the theory has evolved to study contemporary problems and investigate the interconnectedness of unpredictable phenomena through a non-linear perspective.

Schneider and Somers (2006) argued that the basic tenets of the Complexity theory are non-linear dynamics, chaos theory, and adaptation/evolution. Larson (2016) asserted that other principles that define complexity theory include emergence, self-organisation, feedback, and chaos. From such tenets, it is noted that the theory may be described as consisting of non-linear systems, rendering future

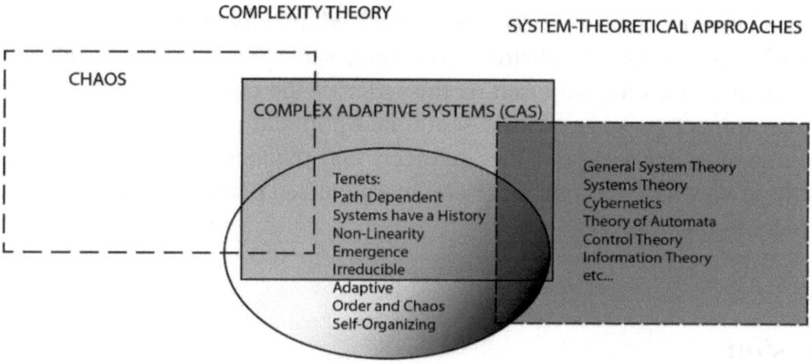


**Fig. 8.2** The complexity theory (adapted from Turner and Baker 2019: 12)

states unpredictable as systems transition from simple to complex (Turner and Baker 2019). Hence, de Mattos et al. (2012) summarised the complexity theory as a lens to investigate the properties and behaviour of the dynamics of non-linear systems. This resonates with He et al.'s (2011) assertion that the complexity theory highlights spatial self-organisation, non-linearity, plurality of equilibria, and the importance of co-evolutionary relationships. De Lombaerde et al. (2022) contended that the theory consists of heterogeneous elements that interrelate with one another and with their surroundings, and these interactions drive a system toward new emerging states as it coevolves within its environment (Strathern and McGlade 2014).

The complexity theory proffers an apt prognosis on the interrelationships between EWEs and spatio-temporal disease incidences in the Global South. Such incidences are described by their space, intensity, and frequency heterogeneity, confirming the phenomena of non-linearity, unpredictability, and complexity. In explaining the complexity of the health status of marginalised communities in urban areas, Chaudhry (2024: 01) points out, "The urban poor are disproportionately vulnerable to both the direct and indirect effects of climate change via several factors such as the precarious nature of their livelihoods, poor housing conditions, inadequate means to mitigate the impact of climate change, and limited access to healthcare sector resources". In employing the complexity theory, it is argued that such complex issues are better understood using multiple disciplines and perspectives rather than focusing on one.

## Results

The study on the theories underpinning EWEs and spatiotemporal disease incidences in marginal spaces revealed that the construction of space and the dominant ideas in society have a bearing on how community health responds to disease incidence. Stanek (2015) argued that Lefebvre's theory of space can be used to critique dominant architectural paradigms such as structuralism and functionalism. Space can be a commodity for consumption, appropriated and transformed. Lefebvre developed the spatial triad of how space is perceived, conceived, and lived. This shows that space, as designed, is a symbolic abstract construct, such that the views and beliefs of the dominant classes are imposed as the order of the concrete world. This creates a practical impact on space within social and political practices. A world order is created that suits the views of the architects and planners without considering the effects these views have on the poor and marginalised people, as the construction of space is centred on power.

## Discussion

The study highlights the interconnectedness of space and power, making marginalised communities vulnerable to diseases due to their lack of control. Communities in marginalised spaces are then vulnerable to disease resulting from EWEs because they lack control, such as the lack of power to influence planning and construction of infrastructure and the supply of water and electricity, which in most cases will not be available in marginalised communities. Incidences such as colonialism had dire consequences for developing countries where populations were relegated to disease infested areas. Mkhize (2018) observed that communities in marginalised spaces lack power, as their inequalities are based on an unbalanced historical chapter. Similarly, Kauma and Swart (2022) observed that poverty, hunger, and death are artificial in developing countries.

The study revealed that the product of space is a political task as class struggles continue and inequalities result in marginalisation. The study showed that an advanced nexus between marginality, class, race, and in some instances ethnicity exposes people to all kinds of insecurities. Mphambukeli (2018) observed that EWEs have a direct impact on the health of marginalised people, as in South Africa, where floods have led to typhoid and cholera in marginal areas. It further revealed that advanced marginality involves the punitive management of poverty through social deregulation and social protection cutbacks. In this case, the lack of development and the marginalisation of some communities in post-colonial Africa is making governments regulators of death through climate change-induced disease incidence.

The study also notes the connection between climate change, EWEs, and greenhouse gas emissions, though this reasoning is often vague. Meteorological statistics are typically available only to influential spaces, leaving marginalised areas without crucial information.

## Conclusion, Lessons, and Further Studies

World populations require climate justice that focuses on the impacts of climate change, addressed with the SDGs and other frameworks. This study concludes that the world should take a microscopic look into the effects of space and power on the marginalised people who bear the impact of climate change-induced disease. The study concludes that marginal areas barely contribute to global pollution and still face the effects of climate change, as they lack the power and space to negotiate their position and support development. The study further concludes that in developing countries where the vagaries of climate change are devastating and in marginal communities with disease prevalence, climate justice must be served, and the remnants of colonialism must be removed. Post-colonial states remain unreformed, moving on the same trajectory of marginalisation and exposing communities to the vagaries of climate change with no prioritisation of community health. It has been learned that a model of community health can be used to understand disease incidence, which can be adopted for climate disease incidence. The complexity theory applies a multidimensional approach involving all sectors in solving challenges imposed by EWEs in marginal spaces. A lesson has been drawn on how power and space are intertwined, as most communities that lack power are socially and politically excluded.

The study found that there are indicators of community health that can be adopted by developing countries to reduce diseases in marginal communities along with early diagnosis. The study revealed the proposed modified life table analysis method to measure the comparative adequacy of health services. This can be a step in the right direction for developing countries and marginal areas in developing countries.

## References


Abrignani MG, Lombardo A, Braschi A, Renda N, Abrignani V (2022) Climatic influences on cardiovascular diseases. World J Cardiol 14(3):152

Agar M (1999) Complexity theory: an exploration and overview based on John Holland's Work. Field Methods 11(2):99–120

Akhtar R (2020) Introduction: extreme weather events and human health: a global perspective. In: Extreme weather events and human health: international case studies, pp 3–11

Allen M, Dube OP, Solecki W, Aragón-Durand F, Cramer W, Humphreys S, Kainuma M (2018) Special report: global warming of 1.5 C. Intergovernmental Panel Climate Change (IPCC) 677:393

Anderson B (2008) For space (2005): Doreen Massey. Key Texts Hum Geogr 8:227–235

Boanada-Fuchs A, Boanada Fuchs V (2018) Towards a taxonomic understanding of informality. Int Dev Plann Rev 40(4):397–420

Bourdieu P (1991) Social space and symbolic space [1994]. Contemporary sociological theory

Caminada S, Turatto F, Iorio S, Paglione L, Errigo M, Mazzalai E, Jaljaa A, Giannini D, Tofani M, Michelazzo MB, Landi A (2021) Urban health and social marginality: perceived health status and interaction with healthcare professionals of a hard-to-reach community living in a suburban area of Rome (Italy). Int J Environ Res Public Health 18(16):8804

Carroll JM, Muller C, Grodsky E, Warren JR (2017) Tracking health inequalities from high school to midlife. Soc Forces 96(2):591–628

Caterino B (2022) The failure of the neo-liberal approach to poverty: the Rochester Monroe anti-poverty initiative. Springer Nature

Chaudhry D (2024) Climate change and health of the urban poor: the role of environmental justice. J Clim Change Health 100277

Chiva R, Ghauri P, Alegre J (2014) Organizational learning, innovation, and internationalization: a complex system model. Br J Manag 25:687–705

Cianconi P, Betrò S, Janiri L (2020) The impact of climate change on mental health: a systematic descriptive review—frontiers in psychiatry 11:74

Cisney VW, Morar N (2015) Biopower: foucault and beyond. University of Chicago Press

Cornell J (2021) Dynamics of identity and space in higher education: an institutional ethnographic case study of a transforming university

Cunningham AA, Scoones I, Wood JLN (2017) One health for a changing world: new perspectives from Africa. Phil Trans R Soc B 372:20160162. https://doi.org/10.1098/rstb.2016.0162

Curtis S, Fair A, Wistow J, Val DV, Oven K (2017) Impact of extreme weather events and climate change for health and social care systems. Environ Health 16:23–32

David Raj A, Padmapriya R, David Raj A (2024) Climate crisis impact on ecosystem services and human well-being. In: Climate crisis, social responses and sustainability: socio-ecological study on global perspectives. Springer Nature Switzerland, Cham, pp 3–36

De Lombaerde E, Vangansbeke P, Lenoir J, Van Meerbeek K, Lembrechts J, Rodríguez-Sánchez F, ... De Frenne P (2022) Maintaining forest cover to enhance temperature buffering under future climate change. Sci Total Environ 810:151338

deMattos P, Miller D, Park E (2012) Decision-making in trauma centers from the standpoint of complex adaptive systems. Manag Decis 50:1549–1569

Dyvik E (2024) Gini index—countries with the biggest inequality in income distribution 2023

Elden S (2004) Henri Lefebvre: theory and the possible

Erkyihun GA, Alemayehu MB (2022) One health approach for the control of zoonotic diseases. Zoonoses 2(37):1–11

Faleye O, Akande T, Moyo I (eds) (2023) Public Health in Post-colonial Africa: the social and political determinants of health. Taylor & Francis

FAO, UNEP, WOAH, & WHO (2022) One health joint plan of action (2022–2026): working together for the health of humans, animals, plants, and the environment. https://www.who.int/publications-detail-redirect/9789240059139

Flint J (2018) Encounters with the centaur state: advanced urban marginality and the practices and ethics of welfare sanctions regimes. Urb Stud 56(1):249–265

Francis D, Webster E (2019) Poverty and inequality in South Africa: critical reflections. Dev South Afr 36(6):788–802

Fuchs C (2019) Henri Lefebvre's theory of the production of space and the critical theory of communication. Commun Theory 29(2):129–150

Gržinić M (2021) Necropolitics by Achille Mbembe: extended essay on the book. Filozofski vestnik 42(1)

Gwakisa P, George J, Sindato C, Ngonyoka A, Nnko H, Justine Assenga J, Kimera S, Nessele MO (2023) Pillars for successful operationalization of one health as an ecosystem approach: experience from a human-animal interface in the Maasai steppe in Tanzania. One Health Outlook 5(11)

Hancock L, Mooney G (2013) Beyond the penal state: advanced marginality, social policy, and anti-welfarism. In: Squires P, Lea J (eds) Criminalisation and advanced marginality: critically exploring the work of Loic Wacquant. Policy Press, Bristol, pp 107–128

Harvey D (1990) Between space and time: reflections on the geographical imagination1. Ann Assoc Am Geogr 80(3):418–434

He Z, Rayman-Bacchus L, Wu Y (2011) Self-organization of industrial clustering in a transition economy: a proposed framework and case study evidence from China. Res Policy 40:1280–1294

Hincks S, Powell R (2022) Territorial stigmatization beyond the city: habitus, affordances, and landscapes of industrial ruination. Environ Plan A 54(7):1391–1410

https://www.statista.com/statistics/264627/ranking-of-the-20-countries-with-the-biggest-inequality-in-income-distribution/#:~:text=South%20Africa%20had%20the%20highest,neighbor%20Namibia%20followed%20in%20second. Accessed on 4 July 2024

Hulme M (2014) Attributing weather extremes to 'climate change': a review. Prog Phys Geogr 38(4):499–511

Jamal N (2023) One health and the opportunity for paradigm shifts through a new WHO pandemic agreement. Development 66(3):199–206

Jones A (2022) The health impacts of climate change: why climate action is essential to protect health. Orthopedics Trauma 36(5):248–255

Kalandides A (2020) Doreen Massey's 'a global sense of place ' is revisited. In: The Routledge handbook of place. Routledge, pp 32–41

Karndacharuk A, Vasisht P, Prasad M (2015) Shared space evaluation: O'Connell Street, Auckland. In: Proceedings of the Australasian Transport Research Forum

Kauma B, Swart S (2022) Hunger and power: politics, food (in) security and the development of small grains in Zimbabwe, 2000–2010. Historia 67(1):144–176

Kitchin R (2016) From a single line of code to an entire city: reframing the conceptual terrain of code/space. In: Code and the city. Routledge, pp 15–26

Larson JM (2016) Cheating because they can: the role of networks in informal governance. https://sciencespo.hal.science/hal-01420520/document

Lefebvre H (1991 [1974]) The production of space [translated by D. Nicholson Smith]. Blackwell, Oxford

Lefebvre H (2014) The production of space (1991). In: The people, place, and space reader. Routledge, pp 289–293

Lusk G (2022) Looking forward and backward at extreme event attribution in climate policy. Ethics Policy Environ 25(1):37–51

Mackenzie JS, McKinnon M, Jeggo M (2014) One health: from concept to practice. In: Yamada A, Kahn LH, Kaplan B, Monath TP, Woodall J, Conti L (eds) Confronting emerging zoonoses: the one health

Massey D (2009) Concepts of space and power in theory and political practice. Documents D'anàlisi Geogràfica 55:15–26

Mbembe A (2019) Bodies as borders. From the European South 4(1):5–18

Mbembé JA, Meintjes L (2003) Necropolitics. Publ Cult 15(1):11–40

Mendin SF, Krause JA, Gweh A, Baysah M, Nyumah J, Gaye C J, Hirschhorn LR (2023) Measuring health system responsiveness in a national community health worker primary care programme in rural Liberia. Int J Qual Health Care 35(2):mzad027

Mhlanga R, Gneiting U, Agarwal N (2018) Walking the talk: assessing companies' progress from SDG rhetoric to action

Mkhize NB (2018) Addressing the spatial inequality of economic infrastructure through spatial planning: a case of the proposed Edendale Town Center in Msunduzi municipality (Doctoral dissertation)

Mphambukeli TN, Nel V (2018) Migration, marginalization, and oppression in Mangaung, South Africa. Crisis, Identity, and Migration in Post-Colonial Southern Africa, pp 141–163

Mumford EL, Martinez DJ, Tyance-Hassell K, Cook A, Hansen GR, Labonté R, Mazet JAK, Mumford EC, Rizzo DM, Togami E, Vreedzaam A, Parrish-Sprowl J (2023) Evolution and expansion of the one health approach to promote sustainable and resilient health and well-being: a call to action. Front Public Health 10(1056459). https://doi.org/10.3389/fpubh.2022.1056459

Nolan C, Overpeck JT, Allen JR, Anderson PM, Betancourt JL, Binney HA, Brewer S, Bush MB, Chase BM, Cheddadi R, Djamali M (2018) Past and future global transformation of terrestrial ecosystems under climate change. Science 361(6405):920–923

O'Connell T, Rasanathan K, Chopra M (2014) What does universal health coverage mean? Lancet 383(9913):277–279

O'Connell D, Walker B, Abel N, Grigg N (2015) The resilience, adaptation and transformation assessment framework: from theory to application. CSIRO, Camberra

O'Gorman PA, Schneider T (2009) The physical basis for increases in precipitation extremes in simulations of 21st-century climate change. Proc Natl Acad Sci 106(35):14773–14777

Otto CA (2014) CEO optimism and incentive compensation. J Financ Econ 114(2):366–404

Paradigm (2014) Springer, Tokyo, pp 163–189

Ribeiro D (2017) Os índios e a civilização. Global Editora e Distribuidora Ltd
Sanders JV (1964) Colour of precious opal. Nature 204(4964):1151–1153
Sarriot EG, Sacks ER, Larson A (2016) Self-organization and emergence in global health. In: Insights from practice, blind spots and possible contributions from complexity science
Schneider M, Somers M (2006) Organizations as complex adaptive systems: implications of complexity theory for leadership research. Leadersh Q 17(4):351–365
Sennett R (2006) The culture of the new capitalism. Yale University Press, New Haven, CT
Springmann M, Godfray HCJ, Rayner M, Scarborough P (2016) Analysis and valuation of the health and climate change cobenefits of dietary change. Proc Natl Acad Sci 113(15):4146–4151
Stanek U (2015) Mobilities of architecture in the global cold war: from socialist Poland to Kuwait and back. Int J Islamic Archit 4(2):365–398
Stott PA, Christidis N, Otto FE, Sun Y, Vanderlinden JP, van Oldenborgh GJ, Zwiers FW et al (2016) Attribution of extreme weather and climate-related events. Wiley Interdiscipl Rev Clim Change 7(1):23–41
Strathern M, McGlade J (2014) The social face of complexity science: a festschrift for Professor Peter M. Allen. Emergent Publications, Litchfield
Talukder B, Schubert JE, Tofighi M, Likongwe PJ, Choi EY, Mphepo GY, Orbinski J (2024) Complex adaptive systems-based framework for modeling the health impacts of climate change. J Clim Change Health 15:100292
Turner JR, Baker RM (2019) Complexity theory: an overview with potential applications for the social sciences. Systems 7(4):1–22
Turner LR, Alderman K, Connell D, Tong S (2013) Motivators and barriers to incorporating climate change-related health risks in environmental health impact assessment. Int J Environ Res Public Health 10(3):1139–1151
Varghese C, Kumar SS (2022) Marginality: a critical review of the concept. Rev Dev Change 27(1):23–41
Wacquant L (2005) The two faces of the ghetto. Actes De La Recherche En Sciences Sociales 160(5):4–21
Wacquant L (2008) Relocating gentrification: the working class, science and the state in recent urban research. Int J Urban Reg Res 32(1):198–205
Wacquant L (2011) From 'public criminology to the reflexive sociology of criminological production and consumption: a review of public criminology? By Ian Loader and Richard Sparks (London: Routledge, 2010). Br J Criminol 51(2):438–448
Wacquant L (2013) The wedding of workfare and prison fare in the 21st century: responses to critics and commentators. In: Squires P, Lea J (eds) Criminalisation and advanced marginality: critically exploring the work of Loic Wacquant. Policy Press, Bristol, pp 243–258
Wacquant LJ (2019) Making class: the middle class (ES) in social theory and social structure. In: Bringing class back in. Routledge, pp. 39–64
Wacquant LJ (1997) Three pernicious premises in the study of the American ghetto. Int J Urban Reg Res 21(2):341–353
Walker M (2020) The well-being of South African university students from low-income households. Oxf Dev Stud 48(1):56–69
Weld ED, Waitt C, Barnes K, Garcia Bournissen F (2022) Twice neglected? Neglected diseases in neglected populations. Br J Clin Pharmacol 88(2):367–373
WHO (2021) Tripartite and UNEP support OHHLEP's definition of "One Health." Available from: https://www.who.int/news/item/01-12-2021-tripartite-andunep-support-ohhlep-s-definition-of-one-health
Wolmuth-Gordon H, Mutebi N (2023) Public health and climate change: a one health approach. UK Parliament POST Note 701

# Chapter 9
# Impacts of Climate Change on Pathogens and One Health

**Martin H. Richter, Jens-André Hammerl, Annette Johne, and Kaya C. Stollberg**

**Abstract** *Background.* This chapter discusses important changes of environmental factors attributed to climate change. Special attention is given to climate change impact on the One Health concept and its relation to human pathogen occurrence especially in developing countries and the overall implications on human health as a result. Furthermore, the chapter discusses impacts on food borne pathogens, how conditions favourable for pathogen survival in food may alter access and availability to safe food and food resources. *Methods.* Climate change-related scientific data often relies on prediction models in order to gain a better understanding of potential future impacts. However, prediction models are always an extrapolation of currently available data sets. This letter approaches the topic with available scientific evidence from previously published scientific data and data from the current studies of the authors, directly supporting the claim that changes in climate directly affect human pathogens. *Results and Conclusions.* Climate change is playing a profound role in the spread of infectious diseases in developing countries, with significant implications for public health. In order to address the health impacts of climate change in developing countries, it is crucial for governments and organisations to prioritise adaptation and resilience measures.

**Keywords** Human pathogens · Public health · Neglected tropical diseases · Food borne diseases · Countermeasures · Developing regions

## Introduction

The concept of One Health is described by the WHO One Health High Expert Level Panel as an integrated, unifying approach that aims to sustainably balance and optimise the health of people, animals, and ecosystems (One Health High-Level Expert, Panel et al. 2022).

M. H. Richter (✉) · J.-A. Hammerl · A. Johne · K. C. Stollberg
Department Biological Safety, German Federal Institute for Risk Assessment, Berlin, Germany
e-mail: Martin.Richter@bfr.bund.de



W. Leal Filho and F. Wolf (eds.), *Climate Change and Health: Perspectives from Developing Countries*, Climate Change Management,
https://doi.org/10.1007/978-3-031-93177-2_9

It recognises that the health of humans, domestic and wild animals, plants, and the wider environment (including ecosystems) are closely linked and interdependent (One Health High-Level Expert, Panel et al. 2022). Temperature, precipitation, humidity, vector abundance, and soil properties are important environmental factors that can alone negatively influence public health status for vulnerable populations. In addition, these factors can impact the spread and survival of zoonotic pathogens, as the survival of these pathogens is augmented by changes in temperature and humidity, such as permanently elevated ambient temperatures, increasing precipitation, or water scarcity in some regions.

The expected increase in high consequence pathogen occurrence and in foodborne infections presents a growing public health risk. As a result, climate change closely impacts the concept of One Health as an integrated, unifying approach and increases the challenges to sustain and balance the health of humans, animals, and the environment. Urgency is warranted in regions already struggling with continuous extreme weather events, such as already high mean yearly temperatures, challenging those regions with less room and time for countermeasures to take effect. Unfortunately, such regions are often part of or in their entirety developing countries that face multiple climate challenges, mainly of a socioeconomic nature. Collective international effort and support are urgently needed to develop and implement sustainable countermeasures as a holistic One Health approach.

Climate change may thus exert an increasing influence on more than half of all infectious diseases (Mora et al. 2022). This is not only true for already existing and endemic infectious agents; climatic changes also favour the establishment of novel infectious agents (emergence) as well as the return of pathogens suppressed in the past (re-emergence) (Thebo et al. 2017). For example, arthropod vectors as carriers for tropical diseases gain in abundance in temperate regions as ambient temperatures rise. As a consequence, the potential for spreading high consequence pathogens rises (Mojahed et al. 2022). Another example is the climate change-driven aspect of increased water scarcity, whereby agricultural operations may have to rely more frequently on treated wastewater for irrigation. This poses a number of risks to public health but also to food safety, including the contamination of irrigated produce by various types of pathogens (Campbell-Lendrum and Corvalán 2007; Thebo et al. 2017).

## Methodology

The authors present a scientific letter with the intent to provide an introduction on the complex matter of climate impact on ecosystems from a microbiological perspective, in which there exist potentially causing alterations or changes of ecosystem's microbiomes. Focus is laid on events that include climate related change in the spectrum of human and animal pathogens and climate related impact on the fitness of infectious agents affecting humans, especially in developing regions, either directly or indirectly via matrices such as food. Climate change-related scientific data often relies on prediction models in order to gain a better understanding of potential future impacts. However, prediction models are always an extrapolation of currently available data sets. This study approaches the topic with available scientific evidence from previously published scientific data and data from the authors' current studies to directly support the claim that changes in climate directly affect human pathogens with regards to the One Health concept, with special challenges for developing regions of the world. A non-systematic literature search was performed by applying the following search criteria: (i) pathogen survival [and] temperature; (ii) pathogen survival [and] temperature [and] humidity; (iii) pathogen prevalence surveillance. Results were screened for human pathogens, food borne pathogens. Moreover, publications were weighed to emphasise long term studies over single data point studies. Exclusion criteria were reviews and literature that applied prediction models in order to avoid extrapolation and condense search results to actual data point studies. "Region" was not applied as an exclusion criterion in order to account for publications on climate driven environmental changes in all regions of the world, with the intent to improve comparability of data between third and first world countries (Fig. 9.1).

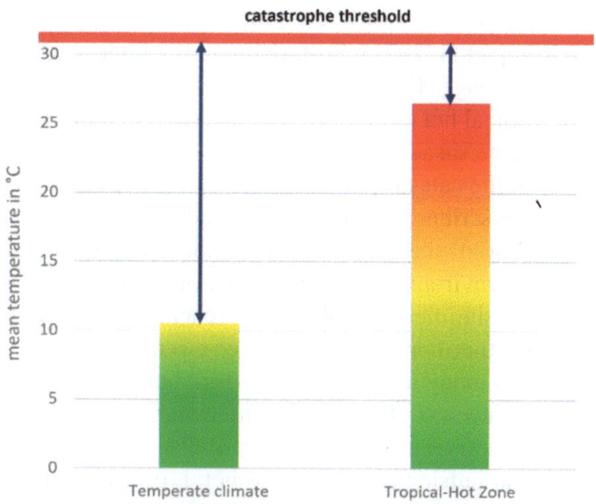


**Fig. 9.1** Illustration of relationship between mean temperature and time for countermeasures. *y*-axis; current yearly mean temperatures for temperate and tropical hot climate zones. Blue arrows, room/Δt; relative time available for countermeasures regarding human and environmental health until catastrophe threshold (red horizontal line) reach. If mean temperatures surpass catastrophe threshold, countermeasures remain largely ineffective. (Illustration Martin H. Richter©)

## *Overall Impacts of Climate Change in Developing Countries and Implications for Infectious Diseases*

Climate change can have a significant impact on the health of individuals worldwide but especially on those living in developing countries, leading to a range of adverse effects (Abbass et al. 2022; Abdallah et al. 2014; Campbell-Lendrum and Corvalán 2007). As global temperatures continue to rise and extreme weather events become more frequent, vulnerable populations in these countries are increasingly at risk (Butsch et al. 2023; Ebi et al. 2021).

One of the most direct ways in which climate change affects health in developing countries is through the increase of heat-related illnesses, accompanied by the lack of adequate access to health care and emergency care. Since the beginning of temperature recording, temperatures reach record highs for longer periods of time. Individuals, particularly those living in urban areas with limited access to air conditioning or cooling facilities, are at a higher risk of heat exhaustion and heatstroke (Butsch et al. 2023; Ebi et al. 2021). This can be particularly dangerous for children and the elderly, as well as individuals with pre-existing health conditions (Butsch et al. 2023). In many developing countries, the rising temperatures in already tropical hot climate zones leave much less time to develop and apply countermeasures (Kemp et al. 2022). Still, rising temperatures as a result of climate change is a global challenge. Figure 9.1 illustrates the room for possible countermeasure actions in tropical hot climate zones compared to temperate climate zones, emphasising a large time disadvantage for many developing countries in tropical hot climate zones until the catastrophe threshold is reached (Kemp et al. 2022). The catastrophe threshold describes a temperature-based limit. The closer mean temperatures reach this threshold, the less effective countermeasures become—e.g., for healthy living, healthy environmental conditions, or for providing sustainable and safe food sources until a significant breakdown is reached if temperatures surpass this threshold (Fig. 9.1). Furthermore, climate change has also been linked to the spread of infectious diseases in many regions of the world, but with more severe impact in developing countries (Anwar et al. 2019). It is noteworthy that temperatures become more conducive to the survival and proliferation of disease-carrying vectors, but also to the pathogen survival outside of a host in general, impacting pathogen virulence and higher pathogen tenacity (Beermann et al. 2023).

Therefore, climate change is playing a profound role in the spread of infectious diseases in developing countries, with significant implications for public health. The relationship between climate change and infectious diseases is influenced by various factors such as temperature, rainfall patterns, and population-driven changes in ecosystems such as habitation. For example, climate change is altering the geographic range and seasonality of disease-carrying vectors such as mosquitoes, ticks, and sandflies, as well as larger vectors such as rodents (Beermann et al. 2023). Rising temperatures and changing precipitation patterns are mainly responsible for creating more favourable conditions for these vectors to thrive and spread diseases like leptospirosis, malaria, dengue fever, chikungunya, and disease from Zika virus infections. In many developing countries, where vector-borne diseases are already a significant public health concern, the expansion of these diseases poses a major threat to the population.

## Impacts of Extreme Weather Events on Pathogen Occurrence

Climate change may also lead to an increase in extreme weather events such as flooding and storms worldwide but with potentially higher impacts in already economically struggling regions, as the recovery phase after such events has been shown to be longer compared to more developed regions in the world (Ebi et al. 2021). The elongation of the recovery phase in developing countries is not only based on economic differences and lack of disaster response infrastructure but also on the fact that severe weather events tend to take place more often in regions where many developing countries are located. As severe weather events rise so does exposure time to the pathogens associated with such events. Lack of access to proper sanitation and vector control and the expedited spread of disease-carrying vectors in turn leads to higher risk of infection and disease. Flooding, for example, often leads to the contamination of water sources, disrupting sanitation systems and increasing the risk of waterborne, rodent borne, and other vector borne diseases like cholera, typhoid fever, leptospirosis, or hantavirus related diseases, but also mosquito borne diseases such as Dengue fever. Hence, inadequate access to clean water and poor sanitation infrastructure in many developing countries further exacerbate pathogen survival, pathogen propagation, and consequently the spread of infectious diseases (Baldursson and Karanis 2011; Ferreira et al. 2021; Karanis et al. 2007; Rose and Jimenez-Cisneros 2017).

## Neglected Tropical Diseases and Resurgence of Diseases

Climate change significantly influences the distribution and prevalence of neglected tropical diseases (NTDs), which disproportionately affect populations in developing countries (Ochola et al. 2021). NTDs include a plethora of diseases caused by viruses, bacteria, and parasites. These diseases, including viral hemorrhagic fevers (e.g., Dengue Fever), bacterial diseases (e.g., leptospirosis, leprosy) and parasitic diseases (e.g., trypanosomiasis, schistosomiasis, lymphatic filariasis, soil-transmitted helminthiasis) and many more, are highly sensitive to environmental conditions (Booth 2018; Tidman et al. 2021; Uwishema et al. 2023). As the term NTD suggests, interest in combatting such diseases has always been low, even before climate change related problems became of interest to society. It is not only disinterest in general that leads to neglect, but also the economic situation in developing regions that prevents adequate countermeasures due to high developing costs but neglectable economic turnaround in profit-oriented health care settings. This includes the funding of research projects, development of drug or vaccine candidates for treatment and prevention, and vector control measures. As climate change progresses, we observe several key effects:

(i) Increased transmission: NTDs may experience heightened transmission due to changing environmental factors. Rising temperatures, altered rainfall patterns, and shifts in habitats can impact disease vectors and their life cycles, potentially leading to more widespread infections (Tidman et al. 2021; Uwishema et al. 2023).
(ii) Resurgence of Infectious Diseases: Climate change can trigger the resurgence of certain infectious diseases that were previously under control. For instance, as warmer temperatures allow disease-carrying vectors to expand their range to higher altitudes and latitudes, populations once shielded from diseases like malaria may now face new vulnerabilities (Mojahed et al. 2022; Wu et al. 2016).
(iii) Challenges for Disease Control Programmes: The evolving landscape of NTDs poses significant challenges for disease control efforts. Public health programmes must adapt to these changes, considering the shifting patterns of disease transmission and vulnerability (Baker et al. 2022; Ochola et al. 2021). This adaption is already challenging in well developed economies, as building such prevention strategies requires significant amounts of resources that are rarely made available, based on prediction models without the problem actually occurring. Therefore, prevention strategies are even more difficult to implement in structurally weaker regions. Hence, a better understanding of potential climate-related long term health effects and sufficient resources in order to develop and implement effective countermeasures are needed (Baker et al. 2022). One desired solution is the bundling of expertise from different fields of research. In the beginning, an important aspect of research projects is to recognise knowledge gaps in the respective individual line of work but that may have already been addressed by a sometimes-unrelated discipline. The One Health concept in particular offers a broader view and encourages us to bring together knowledge to seal these gaps in an efficient and economical way.

In summary, climate change has far-reaching implications for NTDs, necessitating comprehensive strategies to mitigate its impact on vulnerable populations. Efforts to address these challenges are crucial for maintaining effective disease control and safeguarding public health in developing countries but can only be sustainable if it is an international effort.

## *Foodborne Diseases*

Changes in temperature and humidity levels significantly influence the growth and transmission of foodborne pathogens, elevating the risk of diseases caused by, e.g., *Salmonella* spp. and *Escherichia (E.) coli* infections (EFSA/ECDC 2021). This impact is particularly concerning in developing countries where food safety standards may be lacking and refrigeration access is limited (Grace 2015).

Climate change directly affects various habitats, which can be altered by weather events such as prolonged droughts, rising temperatures, and heavy rainfall. These changes also impact the foods derived from these habitats and the microorganisms associated with them. For instance, foods may become more heavily contaminated with pathogens or harbor germs that were not previously present in concerning amounts (Dietrich et al. 2023).

Foodborne pathogens affected by climate change, besides *Salmonella* spp. and *E. coli*, include *Campylobacter* spp., *Vibrio* spp., and protozoan parasites (such as *Cryptosporidia* spp.). They thrive in higher ambient temperatures. The optimal temperature for Salmonella growth falls between 35 and 37 °C; below 15 °C, Salmonella growth is significantly reduced. Consequently, Salmonella multiplies faster at higher temperatures. The correlation between outdoor temperatures and Salmonella outbreaks has been well-documented, with studies reporting an increase in salmonellosis and other bacterial enteric diseases as temperatures rise (Akil et al. 2014; Dietrich et al. 2023).

According to Zhang et al., a 1 °C increase in the mean weekly maximum temperature leads to an expected 8.8% rise in weekly cases. Similarly, a 1 °C increase in the mean weekly minimum temperature results in a 5.8% increase in weekly cases (Zhang et al. 2010). Also, as ambient temperatures increase, the risk of cold chain disruption becomes higher. Further, refrigeration at high ambient temperatures is energy intensive and impacts the cooling performance of refrigeration devices. At the same time, the high demand for energy needed for cooling may not always be met, especially in developing countries. These factors have a significant impact on the microbiological status of food (Dietrich et al. 2023). The increase of severe weather events can also impact food safety, as severe flooding, for instance, can compromise wastewater treatment effectivity due to run off. This can lead to contamination of human, animal, and farm environments by human pathogens (Butsch et al. 2023).

Another example are Vibrio bacteria, which are strictly water borne pathogens. Dependent on concentration, these bacteria—as environmental pathogens—will often be present in seafood where they are able to further propagate, potentially leading to harmful concentrations within the food source (FAO/WHO 2020; Farmer et al. 2015). In the case of Vibrio infections, humans come in contact with the pathogen during food preparation or consumption of raw or improperly heated seafood (BfR 2022c; Dietrich et al. 2023). Infection can also occur by direct contact of compromised skin with bodies of water or food containing the bacteria, potentially leading to life threatening wound infections (Dupke et al. 2023). Toxicogenic vibrio bacteria are increasingly detected in consistently warm waters, posing challenges for regions already experiencing warm water temperatures, which is often the case in areas where developing countries are located (Baker-Austin et al. 2018; Vezzulli et al. 2020). Yet, as temperatures are on the rise in more temperate regions of the world as well, this phenomenon has also been registered more frequently now in usually colder sea waters such as the Atlantic coast in north-west Europe, the North Sea, and the Baltic Sea (Baker-Austin et al. 2018).

In seafood, the occurrence of Vibrio species pathogenic to humans also directly correlates with water temperatures (BfR 2022c; CDC 2019). As described, water temperatures continue to rise ubiquitously. Vibrio contamination amplifies exponentially in steady warm waters while pathogenic vibrio concentration remains low in bodies of water with fluctuating (cold–warm–cold) temperatures (Billaud et al. 2022). However, with ambient water temperatures on the rise also in temperate regions of the world, prediction analyses suggest increasing concentrations of Vibrio bacteria as cold-water seasons are becoming shorter (Dietrich et al. 2023).

The climate change-associated rise in ambient temperature also influences the virulence and tenacity of food associated parasites such as *Cryptosporidia* spp. (Dietrich et al. 2023). Cryptosporidiosis is one of the most common diarrhoeal diseases associated with the ingestion of contaminated water or food contaminated with the pathogen. Cryptosporidia can remain infectious for a long period of time (Gerace et al. 2019; Shirley et al. 2012). Their high stability against environmental influences, in particular, the long survivability outside of a host, suggests that these parasites could appear more frequently as food related pathogens in the future (Dietrich et al. 2023; Shirley et al. 2012). Extreme weather, such as heavy rainfall and flooding as a result of climate change, increases the risk of infectious oocysts entering bodies of water, as well as the risk of contamination of plant-based foods in particular (Ahmed et al. 2016).

Additionally, climate-related events indirectly impact public health and food safety. For instance, conversely to heavy rainfall and flooding, water scarcity depletes potable water resources and drives the more frequent use of treated wastewater as an only alternative for irrigation. This increases the potential for contaminating the irrigated produce if the wastewater isn't treated accordingly before reuse in irrigation (BfR 2020, 2022a, b). This is especially concerning for foods that are usually consumed raw, such as fruits and vegetables, where the edible part of the product comes in direct contact with the irrigation water.

## Conclusions

Climate change is playing a profound role in the spread of infectious diseases in developing countries, with significant implications for public health. In order to address the health impacts of climate change in developing countries, it is crucial for governments and organisations to prioritise adaptation and resilience measures. This includes investing in public health infrastructure, improving access to healthcare services, and implementing early warning and sentinel systems as well as diagnostic schemes. Additionally, efforts to mitigate climate change driven effects, such as reducing greenhouse gas emissions and promoting sustainable development practices, can help to protect the health and wellbeing of individuals in these vulnerable populations. If the current rapid increase in ambient temperature can be slowed, valuable time can be gained in order to develop, test, and implement direct counter- and safety measures regarding the containment of the spread of especially high consequence diseases. Similarly, indirect countermeasures such as effective vector control programmes and improved hygienic standards and sanitation will further support infection prevention. In summary, the following actions should be considered:

(i) Vector Control: Strengthening or implementing vector control measures to reduce human contact with the aim to prevent the spread of vector-borne diseases.
(ii) Disease Surveillance: Develop and implement or strengthen disease surveillance systems with the goal to monitor diseases as well as disease patterns in order to anticipate the spread of pathogens and to prevent outbreaks in response to climate-related factors.
(iii) Water and Sanitation: Clean potable water is adamant for human health. Hence, water and sanitation infrastructure must be established and protected to prevent waterborne diseases. Rigorously apply wastewater treatment measures if reused for irrigation.
(iv) Public Awareness: Raising awareness about the impact of climate change on human health, animal health, and environmental health. Understanding of links between climate change and infectious disease occurrence to reduce disease transmission.

In conclusion, worldwide efforts are needed to focus on the mitigation of climate change impacts on environmental, animal, and human health within the holistic One Health approach.

## References

Abbass K, Qasim MZ, Song H, Murshed M, Mahmood H, Younis I (2022) A review of the global climate change impacts, adaptation, and sustainable mitigation measures. Environ Sci Pollut Res Int 29(28):42539–42559. https://doi.org/10.1007/s11356-022-19718-6

Abdallah NA, Moses V, Prakash C (2014) The impact of possible climate changes on developing countries: the needs for plants tolerant to abiotic stresses. GM Crops & Food 5(2):77–80. https://doi.org/10.4161/gmcr.32208

Ahmed T, Scholz M, Al-Faraj F, Niaz W (2016) Water-related impacts of climate change on agriculture and subsequently on public health: a review for generalists with particular reference to Pakistan. Int J Environ Res Public Health 13(11). https://doi.org/10.3390/ijerph13111051

Akil L, Ahmad HA, Reddy RS (2014) Effects of climate change on Salmonella infections. Foodborne Pathog Dis 11(12):974–980. https://doi.org/10.1089/fpd.2014.1802

Anwar A, Anwar S, Ayub M, Nawaz F, Hyder S, Khan N, Malik I (2019) Climate change and infectious diseases: evidence from highly vulnerable countries. Iran J Public Health 48(12):2187–2195

Baker RE, Mahmud AS, Miller IF, Rajeev M, Rasambainarivo F, Rice BL, Takahashi S, Tatem AJ, Wagner CE, Wang L-F, Wesolowski A, Metcalf CJE (2022) Infectious disease in an era of global change. Nat Rev Microbiol 20(4):193–205. https://doi.org/10.1038/s41579-021-00639-z

Baker-Austin C, Oliver JD, Alam M, Ali A, Waldor MK, Qadri F, Martinez-Urtaza J (2018) Vibrio spp. infections. Nat Rev Dis Primers 4(1):8. https://doi.org/10.1038/s41572-018-0005-8

Baldursson S, Karanis P (2011) Waterborne transmission of protozoan parasites: review of worldwide outbreaks—an update 2004–2010. Water Res 45(20):6603–6614. https://doi.org/10.1016/j.watres.2011.10.013

Beermann S, Dobler G, Faber M, Frank C, Habedank B, Hagedorn P, Kampen H, Kuhn C, Nygren T, Schmidt-Chanasit J, Schmolz E, Stark K, Ulrich RG, Weiss S, Wilking H (2023) Impact of climate change on vector- and rodent-borne infectious diseases. Advance Online Publication. https://doi.org/10.25646/11401

BfR (2020) Aufbereitete Abwässer: Bakterielle Krankheitserreger auf frischem Obst und Gemüse vermeiden (021/2020). https://doi.org/10.17590/20200421-115529

BfR (2022a) Aufbereitete Abwässer: Protozoen auf pflanzlichen Lebensmitteln vermeiden (21/2022). Risikobewertung, Bundesinstitut für. https://doi.org/10.17590/20220727-111138

BfR (2022b) Aufbereitete Abwässer: Virale Krankheitserreger auf pflanzlichen Lebensmitteln vermeiden (019/2022). Risikobewertung, Bundesinstitut für. https://doi.org/10.17590/20220708-125409

BfR (2022c) Bakterielle Lebensmittelinfektionen durch Vibrionen: Gesundheitliche Bewertung zum Vorkommen von Vibrio spp. (Nicht-Cholera-Vibrionen) in Lebensmitteln (011/2022). Risiobewertung, Bundesinstitut für. https://doi.org/10.17590/20220413-083752

Billaud M, Seneca F, Tambutté E, Czerucka D (2022) An increase of seawater temperature upregulates the expression of vibrio parahaemolyticus virulence factors implicated in adhesion and biofilm formation. Front Microbiol 13:Article 840628. https://doi.org/10.3389/fmicb.2022.840628

Booth M (2018) Climate change and the neglected tropical diseases. Adv Parasitol 100:39–126. https://doi.org/10.1016/bs.apar.2018.02.001

Butsch C, Beckers L-M, Nilson E, Frassl M, Brennholt N, Kwiatkowski R, Söder M (2023) Health impacts of extreme weather events—cascading risks in a changing climate. J Health Monit 8(Suppl 4):33–56. https://doi.org/10.25646/11652

Campbell-Lendrum D, Corvalán C (2007) Climate change and developing-country cities: implications for environmental health and equity. J Urb Health: Bullet N Y Acad Med 84(3 Suppl):i109–i117. https://doi.org/10.1007/s11524-007-9170-x

CDC (2019) Vibrio and food

Dietrich J, Hammerl J-A, Johne A, Kappenstein O, Loeffler C, Nöckler K, Rosner B, Spielmeyer A, Szabo I, Richter MH (2023) Impact of climate change on foodborne infections and intoxications. J Health Monit 8(Suppl 3):78–92. https://doi.org/10.25646/11403

Dupke S, Buchholz U, Fastner J, Förster C, Frank C, Lewin A, Rickerts V, Selinka H-C (2023) Impact of climate change on waterborne infections and intoxications. J Health Monit 8(Suppl 3):62–77. https://doi.org/10.25646/11402

Ebi KL, Vanos J, Baldwin JW, Bell JE, Hondula DM, Errett NA, Hayes K, Reid CE, Saha S, Spector J, Berry P (2021) Extreme weather and climate change: population health and health system implications. Annu Rev Public Health 42:293–315. https://doi.org/10.1146/annurev-publhealth-012420-105026

EFSA/ECDC (2021) The European Union One Health 2020 Zoonoses Report (EFSA Journal 1831-4732). European Food Safety Authority, European Centre for Disease Prevention. https://doi.org/10.2903/j.efsa.2021.6971

FAO/WHO (2020) Risk assessment tools for Vibrio parahaemolyticus and Vibrio vulnificus associated with seafood. Microbiological risk assessment series; 20. Food and Agriculture Organization of the United Nations

Farmer JJ, III, Brenner FW, Cameron DN, Birkhead KM, Janda JM (2015) Vibrio: Bergey's manual of systematics of archaea and bacteria. In association with Bergey's Manual Trust. Wiley, pp 1–79

Ferreira DC, Graziele I, Marques RC, Gonçalves J (2021) Investment in drinking water and sanitation infrastructure and its impact on waterborne diseases dissemination: the Brazilian case. Sci Total Environ 779:146279. https://doi.org/10.1016/j.scitotenv.2021.146279

Gerace E, Lo Presti VM, Biondo C (2019) Cryptosporidium infection: epidemiology, pathogenesis, and differential diagnosis. Eur J Microbiol Immunol 9(4):119–123. https://doi.org/10.1556/1886.2019.00019

Grace D (2015) Food safety in low and middle income countries. Int J Environ Res Public Health 12(9):10490–10507. https://doi.org/10.3390/ijerph120910490

Karanis P, Kourenti C, Smith H (2007) Waterborne transmission of protozoan parasites: a worldwide review of outbreaks and lessons learnt. J Water Health 5(1):1–38. https://doi.org/10.2166/wh.2006.002

Kemp L, Xu C, Depledge J, Ebi KL, Gibbins G, Kohler TA, Rockström J, Scheffer M, Schellnhuber HJ, Steffen W, Lenton TM (2022) Climate endgame: exploring catastrophic climate change scenarios. Proc Natl Acad Sci USA 119(34):e2108146119. https://doi.org/10.1073/pnas.2108146119

Mojahed N, Mohammadkhani MA, Mohamadkhani A (2022) Climate crises and developing vector-borne diseases: a narrative review. Iranian J Publ Health. Advance Online Publication. https://doi.org/10.18502/ijph.v51i12.11457

Mora C, McKenzie T, Gaw IM, Dean JM, von Hammerstein H, Knudson TA, Setter RO, Smith CZ, Webster KM, Patz JA, Franklin EC (2022) Over half of known human pathogenic diseases can be aggravated by climate change. Nat Clim Chang 12(9):869–875. https://doi.org/10.1038/s41558-022-01426-1

Ochola EA, Karanja DMS, Elliott SJ (2021) The impact of Neglected Tropical Diseases (NTDs) on health and wellbeing in sub-Saharan Africa (SSA): a case study of Kenya. PLoS Negl Trop Dis 15(2):e0009131. https://doi.org/10.1371/journal.pntd.0009131

One Health High-Level Expert, Panel, Adisasmito WB, Almuhairi S, Behravesh CB, Bilivogui P, Bukachi SA, Casas N, Cediel Becerra N, Charron DF, Chaudhary A, Ciacci Zanella JR, Cunningham AA, Dar O, Debnath N, Dungu B, Farag E, Gao GF, Hayman DTS, Khaitsa M, Zhou L et al (2022) One health: a new definition for a sustainable and healthy future. PLOS Pathogens 18(6):e1010537. https://doi.org/10.1371/journal.ppat.1010537

Rose JB, Jimenez-Cisneros B (eds) (2017) Water and sanitation for the 21st century: health and microbiological aspects of excreta and wastewater management (global water pathogen project). In: Fayer R, Jakubowski W (eds), Part 3: specific excreted pathogens: environmental and epidemiology aspects—Section 9.3: Protists. E. Lansing, MI, UNESCO

Shirley D-AT, Moonah SN, Kotloff KL (2012) Burden of disease from cryptosporidiosis. Curr Opin Infect Dis 25(5):555–563. https://doi.org/10.1097/QCO.0b013e328357e569

Thebo AL, Drechsel P, Lambin EF, Nelson KL (2017) A global, spatially-explicit assessment of irrigated croplands influenced by urban wastewater flows. Environ Res Lett 12(7)

Tidman R, Abela-Ridder B, de Castañeda RR (2021) The impact of climate change on neglected tropical diseases: a systematic review. Trans R Soc Trop Med Hyg 115(2):147–168. https://doi.org/10.1093/trstmh/traa192

Uwishema O, Masunga DS, Naisikye KM, Bhanji FG, Rapheal AJ, Mbwana R, Nazir A, Wellington J (2023) Impacts of environmental and climatic changes on future infectious diseases. Int J Surg (Lond, Eng) 109(2):167–170. https://doi.org/10.1097/JS9.0000000000000160

Vezzulli L, Baker-Austin C, Kirschner A, Pruzzo C, Martinez-Urtaza J (2020) Global emergence of environmental non-O1/O139 Vibrio cholerae infections linked with climate change: a neglected research field? Environ Microbiol 22(10):4342–4355. https://doi.org/10.1111/1462-2920.15040

Wu X, Lu Y, Zhou S, Chen L, Xu B (2016) Impact of climate change on human infectious diseases: empirical evidence and human adaptation. Environ Int 86:14–23. https://doi.org/10.1016/j.envint.2015.09.007

Zhang Y, Bi P, Hiller JE (2010) Climate variations and Salmonella infection in Australian subtropical and tropical regions. Sci Total Environ 408(3):524–530

# Chapter 10
# Building a Climate Resilient and Low Carbon Health System in Nepal

**Mandira Lamichhane Dhimal, Bishal Dahal Khatri, Sailaja Ghimire, Bhim Prasad Sapkota, Erin A. Mordecai, Meghnath Dhimal, and Dinesh Bhandari**

**Abstract** Nepal is progressing towards a climate resilient and low carbon health system, encountering various opportunities and challenges. This chapter aims to assess the progress of the Nepalese health system by utilising the 10 building blocks outlined in the World Health Organization's framework for climate resilient and low carbon health systems. We present findings from a review of policy documents and research articles on climate change and health in Nepal. Our analysis revealed that the Nepalese government has demonstrated its commitments by formulating multiple policies to address climate mitigation, adaptation, and reduction of greenhouse gas (GHG) emissions. The government has been conducting periodic capacity-building activities, funding local evidence generation and adaptation implementation, integrating climate change into health programmes, and developing a framework for reducing GHG emissions. Despite these advancements, Nepal faces numerous challenges. The country's administrative division into three tiers reportedly results in insufficient communication and collaboration, which is a major obstacle. Additional challenges include the shortage of skilled human resources, limited and inconsistent data, limited access to advanced technology, and insufficient focus on evidence-based research. Nepal is likely to benefit from increased collaboration across various sectors and concerted efforts from the government and private agencies in its aspiration to develop a climate-resilient and low-carbon health system.

M. L. Dhimal · B. D. Khatri · S. Ghimire · E. A. Mordecai
Planetary Health Research Centre, Kathmandu, Nepal

B. P. Sapkota
Ministry of Health and Population, Kathmandu, Nepal

E. A. Mordecai
Biology Department, Stanford University, Stanford, USA

M. Dhimal (✉)
Nepal Health Research Council, Kathmandu, Nepal
e-mail: meghdhimal@nhrc.gov.np; meghdhimal@gmail.com

D. Bhandari
School of Nursing and Midwifery, Monash University, Clayton, VIC, Australia

 
W. Leal Filho and F. Wolf (eds.), *Climate Change and Health: Perspectives from Developing Countries*, Climate Change Management,
https://doi.org/10.1007/978-3-031-93177-2_10

**Keywords** Climate change · Health · Nepal · Low carbon · GHG emission · Resilient health system

## Introduction

Climate change poses a significant threat to global health, with particularly severe impacts in South Asia, including Nepal (World Health Organization 2023a). Global temperatures have risen by an average of 0.06 °C per year since 1850, leading to widespread environmental changes and health risks (IPCC 2022; World Health Organization 2023a). In South Asia, climate change is intensifying heatwaves, cyclones, droughts, and floods, potentially affecting living conditions for up to 800 million people, including some of the world's poorest and most vulnerable populations (The World bank 2024). Nepal, in particular, faces severe climate change impacts. The country is warming faster than the global average and ranks among the top 10 most affected countries in terms of long term climate related morbidity and mortality (South Asia Region Climate Knowledge Hub 2023). Projections indicate that Nepal may experience a mean temperature increase of 0.9–1.1 °C in the medium term (2016–2045), and 1.3–2.8 °C in the long term (2036–2065) when compared to the reference period of (1981–2010), with precipitation expected to increase by 2–6% and 8–12% in the medium- and long-term, respectively (ADB 2021; MoFE 2021c). These climate changes threaten key sectors in Nepal, including forestry, agriculture, and public health. The most prevalent climate-sensitive diseases in Nepal include vector-borne diseases (VBD), diarrhoeal diseases, cardiovascular diseases, cholera, malnutrition, mental health problems, and respiratory diseases (Dhimal et al. 2014, 2015a, 2021; MoFE 2021f; Shrestha and Nepal 2020). Climate change is estimated to double the number of people affected by river flooding and could reduce the annual Gross Domestic Product (GDP) by US$574 million by 2030. Despite contributing negligible greenhouse gas emissions, Nepal is committed to addressing climate change in line with the Paris Agreement. The country has been actively involved in international climate initiatives, including the Male Declaration 2017, and consistent participation in the Conference of Parties. Over the past decade, the Nepal's health sector has made significant progress towards protecting the health of its population from climate change impacts. In light of significant climate-driven health risks and its demonstrated commitment to climate action, it is critical for Nepal to develop a climate-resilient and low-carbon health system, defined by the World Health Organization (2023b) as "health systems capable of anticipating, responding to, coping with, recovering from, and adapting to climate-related shocks and stress, while minimizing greenhouse gas emissions and other negative environmental impacts to deliver quality care and protect the health and well-being of present and future generations".

**Rationale and objectives**

Low-resource countries like Nepal continue to face multiple acute challenges, including poor health service delivery, a shortage of skilled personnel, inadequate health sector financing, and insufficient diagnostic facilities. These immediate concerns often overshadow the integration of long-term climate change considerations into public health policies and decisions. Identifying the barriers to and opportunities for progress towards a climate-resilient and low-carbon health system is crucial for guiding resource allocation and policy decisions that support the implementation of long-term climate commitments. This chapter aims to examine the progress, barriers, and facilitators in the context of Nepal.

## Methodology

A scoping review was conducted using the Arksey and O'Malley (2005) framework, adhering to the five sequential stages: identifying the research question, identifying relevant studies, study selection, data extraction, and summarising and reporting the findings. The details on each stage are discussed in brief in the following section.

**Research Question**

The overall research question for our scoping exercise was: What are the challenges in building a climate climate-resilient and low-carbon health system in Nepal?

### *Identification of Relevant Studies*

**Search Strategy**

- We searched PubMed and Google Scholar databases and conducted direct Google searches.
- Searches used combinations of the keywords "Climate change", "Nepal", "Health Impacts", "Finance", "Leadership", "Human Resources", "GHG Emission", "Climate change", "Climate-resilient technologies", "Ministry of Forest and Environment", "Infrastructure", "Research", "Challenges".
- Keywords were searched by combining two or more than two words using Boolean operators like AND and OR. For instance:
  "Climate change AND Nepal", "Climate change AND Finance", "Climate change OR human resources OR Nepal AND challenges".

**Inclusion and Exclusion Criteria**

- This study contains reports and policy documents from governmental and non-governmental organisations. Only articles available from PubMed and Google Scholar were included. Documents written in Nepali and English language were given consideration, while documents in other languages were excluded.
- Articles that needed permission to read were excluded from the study, as were opinion pieces, online presentations and webinars, commentary studies, news reports, and editorials.

### *Study Selection*

Initially, the title and abstract were screened for relevance, then a full-text review was conducted.

### *Extracting and Categorising the Data*

Data were extracted using the data extraction table (Annex I) and were then categorised as per the WHO framework for climate-resilient and low-carbon health system (World Health Organization 2023a). The categorisation was conducted on ten major themes, which were leadership and governance, climate-smart health workforce, assessment of climate and health risks and GHG emissions, integrated risk monitoring and GHG tracking, health and climate research, climate-resilient health infrastructure technologies and supply chain, management of environmental determinants, climate-informed health programmes, climate-related emergency preparedness, and climate and health financing. Finally, thematic analysis was executed to ensure the consistency across the findings and to identify similarities and differences.

We also conducted a bibliometric analysis (which is a method to quantitatively analyse academic literature published by year, author, and keywords in a particular topic) of the articles included in the final synthesis, using VOS viewer software to map key concepts related to the research question and graphically illustrate the findings in chronological order by publication year.

## Results

The search retrieved 258 articles, of which 69 were discarded as duplicates. The remaining 189 studies were screened for title and abstract, with 121 ruled out from full text screening. Thereafter, 68 articles were full text screened and 40 included for data extraction and results synthesis. The other 28 articles, comprising opinion, editorial, news report, and commentary, were excluded from the final data analysis.

**Synthesis of the literature contributing to the development of climate-resilient and low-carbon health system in Nepal.**

The findings are summarised below based on the 10 components of the WHO operational framework that are required to build the climate-resilient and low-carbon health system of a country.

1. **Climate-transformative leadership and governance**

In the early 1990s, there was an absence of policy and political commitment toward climate change in Nepal. However, with the ratification of the United Nations Framework Convention on Climate Change (UNFCCC) in the 1990s, policy and political commitment related to climate change have been initiated in Nepal, such as the Nepal National Adaptation Programme of Action (NAPA) (MoE 2010), Local Adaptation Plan for Action (LAPA) (GoN 2011b), technology needs assessment of climate change mitigation (MoFE 2021e) and adaptation (MoFE 2021d), Climate Change Policy 2011 (GoN 2011a), Climate Change Policy 2019 (MoFE 2019b), National Health Policy 2014 (MOHP 2014), National Health Policy 2019 (MOHP 2019), Health National Adaptation Plan (H-NAP) (2017–2021) (MoHP 2018), Health National Adaptation Plan (2023–2030) (MOHP 2023) and National Adaptation Plan (NAP) 2021–2050 (GoN 2021a). The National Health Policy advocates for integrated preparedness and response measures to combat the threat of vector-borne and zoonotic diseases and other climate-sensitive health problems (MoHP 2019), while the climate change policy safeguards the country's socio-economic status by developing a climate resilient health system (MoFE 2019b). Furthermore, the Health National Adaptation Plan (H-NAP) focuses on adaptation measures to mitigate the impact of climate change, while the National Adaptation Plan (NAP) provides the context of the country's climate change situation and ways to maintain its climate resilient health system through various climate change mitigation and adaptation strategies. In 2018, the Climate Change Management Division (CCMD) was established under the Ministry of Forest and Environment, which makes plans, policies, and programmes related to climate change and is a focal point for the UNFCCC (MoFE 2024). Based on these policies, numerous mitigation and adaptation activities related to climate change have been carried out (CIAT et al. 2017; Climate and Development Knowledge Network (CDKN) 2020; MoALD 2019; MoFE 2021b, c). Furthermore, Nepal's government has started integrating climate change and health; for instance, they have started reporting climate-sensitive diseases from Early Warning and Reporting Systems (EWARS) (EDCD 2016), which is a good step forward, as this leads to prompt reporting and data-driven decision making.

Despite having policies and plans related to climate change, there is still a gap in implementing policies. Furthermore, collaboration and communication between various sectors are lacking, resulting in poor implementation of the policies

(Ranabhat et al. 2018). Nepal has developed its Nationally Determined Contributors (NDC), but health is not part of it; instead, it has only focused on emissions scenarios (GoN 2020). Nepal, like other countries, lacks a guideline or roadmap for reducing GHG emissions from health sectors; leadership and governance could play a significant role in these sectors.

2. **Climate-smart health workforce**

The health workforce is a crucial element in building a climate-resilient health system. Therefore, sensitising, increasing awareness, and providing education is important to mitigate the health impacts. A study conducted by Dhimal (2008) highlighted the lack of trained human resources, especially in the field of climate change. Recently, however, numerous capacity-building sessions related to climate change have been developed (HI-AWARE 2017) along with the training modules (WHO 2018) for government officials. The National Health Training Centre (NHTC) is also conducting training programmes related to climate change at all three levels of government (NHTC 2017). Likewise, the Nepal Health Research Council (NHRC) has conducted training related to climate change and has disseminated knowledge regarding climate change and health (NHRC 2018). On the other hand, obstacles remain in building a climate-smart health workforce due to geographical vulnerabilities (Bogren et al. 2020), political instability (Rai et al. 2024), institutional gaps (Rai et al. 2024), migration to other countries for better opportunities, and a high turnover rate (ILO 2017). Furthermore, there is an imbalance in the number of health workers present in urban and rural areas (Cao et al. 2021).

3. **Assessments of climate and health risks and GHG emissions**

Using climate and health vulnerability and adaptation (V&A) assessment guidance published by WHO, Nepal carried out a V&A assessment of the health sector in 2014 and 2022. Similarly, some indicators based on climate exposure, adaptive capacity, and sensitivity indicators for different sectors are used in the V&A assessment (Regmi et al. 2023). Using the findings of the V&A assessment, it was possible to design the Health National Adaptation Plan (H-NAP) (2017–2021) (MoHP 2018) and its update for 2023–2030 (MOHP 2023). From Nepal's V&A assessment for climate-sensitive diseases and health risks, it is clear that vulnerable groups such as women, children, and disabled people are at greater risk of contracting vector-borne diseases, food and water-borne diseases, respiratory diseases, malnutrition, and mental health issues (MOHP 2022). Challenges persist in inconsistency and misalignment across ministries in implementing adaptation options, and even if the implementation process begins, it is usually incomplete (MOHP 2022). Nepal has also designated a focal point (Multisectoral Coordination Section of Health Coordination Division) under the Management Division for climate change and health-related activities that collaborate and coordinate with other sectors such as health care waste management and environmental sections. Climate-related variables (temperature, precipitation, and humidity) are recorded by meteorological stations (MoEWI 2024), while health-related variables are recorded by the Epidemiology and Disease

Control Division (EDCD) (GoN 2024). Recently, EDCD has started weekly reporting of climate-sensitive diseases from all sentinel sites (GoN 2024). Thus, collaboration and coordination between these two government bodies are key to assessing the climate and health risks.

With the world population now over 7 billion, Nepal's contribution to climate change is diminutive—only less than 0.027% (GoN 2012)—while the Nepalese people represent about 4.27% of the global population (NSO 2021). This contrast underscores the reality that although Nepal is home to a significantly larger portion of the world population, it produces an almost unseen level of global emissions. From the base year of GHG inventory in 1994/95, Nepal's total net emission (emission minus removal) was at its peak with 39,265 $CO_2$ eq, which plunged to 13,447 $CO_2$ eq in 2000/01, and again significantly rose in the year 2011 (28,166.06 $CO_2$ eq). This latter rise may be due to data gaps, inconsistencies, and uncertainties in the GHG inventory (MoFE 2021c). Challenges, such as limited data on emissions (MoFE 2021e) and limited focus on GHG emissions from the livestock and health sector in national policies (MoFE 2019a), hinder complete accounting for GHG emissions and building a climate-resilient low-carbon health system. At the moment, various targets and plans are being set to mitigate the GHG emission in Nepal (USAID 2019).

4. **Integrated risk monitoring, early warning, and GHG emissions tracking**

Keeping track of climate-related variables is essential to predict environmental risks and emergencies while notifying the public to prevent numerous health impacts. These climatic variables are recorded by the hydrometeorological station (DHM 2024), which issues timely alerts to the public. Early Warning and Reporting (EWAR) guidelines have been developed, but the integration of climate-related variables remains a massive challenge to overcome (Dhimal et al. 2019). Meanwhile, surveillance related to climate-sensitive diseases such as malaria, dengue, and kala-azar is being done through the EWARS system, with routine and sentinel reporting (EDCD 2016).

To track GHG emissions, Nepal's Second Nationally Determined Contribution (NDC) has quantified targets for the years 2025 and 2030 in various energy sectors (GoN 2020). GHG tracking comes with many hurdles, as there are data gaps and challenges as well as challenges and limitations in forest regulations, which lead to forest fire, logging, and other hazards that make it complicated to track GHG emissions (MoFE 2023).

5. **Health and climate research**

The primary focus of research is to provide information to fill knowledge gaps on the linkages between climate hazards and health outcomes and to help policymakers address these risks through policy. Several research projects have been conducted recently in the field of climate change and health in Nepal. For instance, there are articles related to climate change and vector-borne diseases, water-related diseases, malnutrition, mental health, and non-communicable diseases (NCD) (Bhandari et al.

2020a, b; Dhimal et al. 2015b, 2018, 2021; Sen et al. 2017). We conducted a bibliometric analysis utilising the keywords "Climate change" "Nepal" AND "Health" to understand the nature of research conducted on climate change and health.

From an analysis of the recent literature (Fig. 10.1), it is clear that previous climate and health research was more associated with demographic and physical natural disasters, whereas recently, the research has shifted more towards vectors, global health problems, and statistical modelling. Although important studies have been conducted, the volume of peer-reviewed research is still limited. Even the Nepal National Adaptation Plan (NAPA) and National Adaptation Plan (NAP) have prioritised research related to climate and health (GoN 2021a; MoE 2010), but domestic funding allocation for climate change and health research is very limited or non-existent. Despite the research articles related to climate change and health that do exist, the literature does not fully capture the complex nature of climate change and health, in part due to data covering all dimensions of climate change (NSO 2024). This presents a ripe opportunity for collaborations across sectors and internationally. Moreover, research on the effectiveness of the adaptation and mitigation measures is also limited and much-needed (Karki et al. 2022).

6. **Climate-resilient and low-carbon infrastructure, technologies, and supply chains**

Climate-resilient healthcare infrastructure is critical for responding to climate-driven changes in agriculture, water, transport, and extreme weather events. To assess

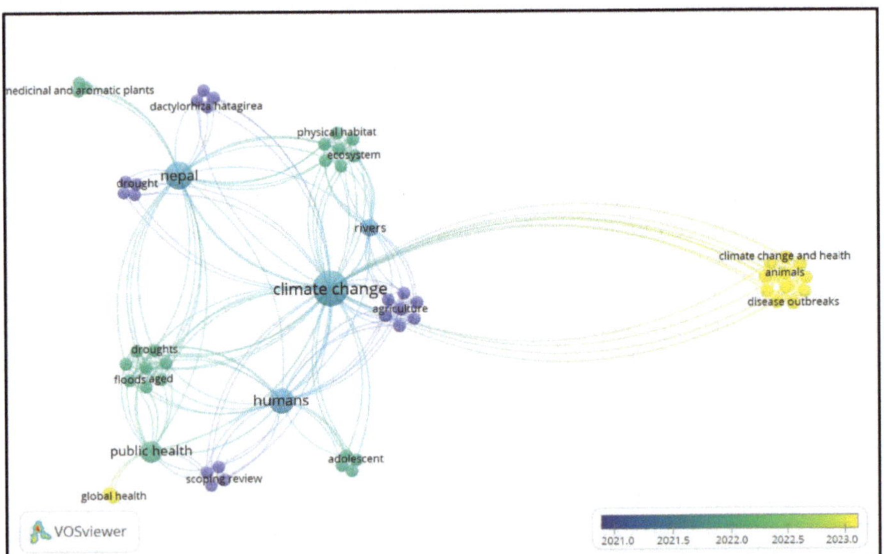


**Fig. 10.1** Bibliometric analysis using VOS viewer to elucidate keywords related to climate change and health research in Nepal. The size of the circle represents the number of occurrences of the keywords. The color indicates the date of publication, scaling from blue in 2021 to yellow in 2023

this capacity, Nepal conducted a technology needs assessment for both adaptation and mitigation; three priority sectors emerged as significant needs for adaptation, including agriculture, water, and public health (MoFE 2021d). In terms of agriculture, climate-smart agriculture such as crop diversification, water management, soil conservation, and energy efficiency are being identified and prioritised. Still, challenges continue to limit the ability to scale up climate-smart technologies, including inadequate institutional capacity, lack of conceptual clarity, insufficient charging infrastructure for electric vehicles (MoFE 2021b), and limited evidence-based knowledge and data on technology performance (MoALD 2019; Paudel et al. 2017). Similarly, in public health, activities like surveillance of drinking water quality are carried out, while for vector control Nepal government is planning to introduce Bacillus thuringiensis israelenis (Bti) for mosquito control. Finally, in the water sector, technologies such as water sprinklers, flood forecasting, and warning systems have been applied (MoFE 2021d). On the other hand, mitigation prioritised mainly three sectors—forestry, agriculture, and energy—with less of a focus on health (MoFE 2021e).

7. **Management of environmental determinants of health**

Climate change affects various dimensions of the environment like land, air, and water, which leads to numerous public health problems; therefore, managing these environmental determinants of health and their responses to climate is crucial. Recently, there have been developments in the field of waste management, and guidelines related to proper waste disposal practices are also available (MOHP 2014). Similarly, water quality surveillance and monitoring guidelines have also been formulated (ADPC 2021, 2023a, b). The integration of climate and agriculture is key to enhancing productivity, resilience, and reducing GHG emissions while increasing food security. Climate Smart Agriculture (CSA) practices such as crop diversification, water management, soil management, and energy efficiency have resulted in the reduction of GHG emissions, improved water management, and enhanced crop productivity. At the same time, challenges remain, including management of agricultural land and productivity, provision of irrigation, soil health, sustainable use of fertilisers and pesticides, gaps in food demand and supply, unstable market structure, and lack of financial access. Climate change, variability, and extreme weather events magnify such challenges by reducing crop yields and livestock productivity. Numerous factors such as land degradation, lack of funding, limited resource ownership, lack of coordination among policymakers, dependency on climate-sensitive agriculture for the production of crops, and geographical and ecological diversity deteriorate agricultural practices. In response, various mitigation strategies have been employed, for instance, promoting the conservation of soil organic matter, introducing integrated pest management and micro-irrigation systems, and adaptation strategies that include ensuring water efficiency and availability, crop diversification, and optimising soil moisture (CIAT et al. 2017).

8. **Climate-informed health programmes**

When building health programmes, the myriad impacts of climate change should be considered. Acknowledging this, the Ministry of Health and Population (MOHP) has started integrating climate change into health programmes (GoN 2012). Climate change has been considered in the curriculum of schools and universities as well (MoHP 2018). The Nepal government has initiated several health programmes such as awareness campaigns, vector control activities, community engagement, and training to enhance the general public and health workers' knowledge of climate change and health.

9. **Climate-related emergency preparedness and management**

Nepal is a country where diverse types of disasters occur, from flood to drought, and where climate change plays a critical role in those disasters. As highlighted in the study (ADRC 2020), Nepal had insufficient emergency preparedness planning and a lack of political willingness for Disaster Risk Reduction (DRR). Currently, however, Nepal has implemented the Sendai Framework for DRR (2015–2030), which underscored that the national policy is in sync with the international policy in developing better coordination, with a commitment to building a safer, more resilient country. Likewise, investment and delineation (Bhandari et al. 2020a, b) of authority in DRR also contributes to creating a climate-resilient nation (MOHA 2023). Meanwhile, challenges such as capacity building, resource constraints, coordination and collaboration, and translating policies into practical actions still prevail (MOHA 2023).

10. **Sustainable climate and health financing**

The Nepal government is receiving funds for climate and health from diverse sources such as the Global Environment Facility (GEF), Green Climate Fund (GCF), Adaptation Fund, and other multilateral development banks (ADPC 2023a). Meanwhile, Nepal is losing almost 1.5–2% of its annual GDP in climate-induced disasters, with costs potentially rising to 2–3% by mid-century (Ministry of Finance 2017).

Currently, Nepal has started integrating the climate budget into national planning to keep track of climate expenditures (Ministry of Finance 2017; MoALD 2019). There is a slight increase in climate funds from 5.36% (NRs: 27,749,423) in 2013/14 to 5.9% (NRs: 61,851,376) in 2016/2017 (Ministry of Finance 2017). The NAP also estimated the budget (i.e., USD 47.4 Billion), including health-related project costs of 4.75 billion, and the timeframe (until 2050) for implementing these adaptation programmes. Efforts were made to ensure that the programme is aligned with national policy while addressing the climate vulnerabilities and sectoral risks.

Even though there is funding, a financial gap persists, as there are insufficient domestic resources (ADPC 2023a; Ministry of Finance 2017). Financial challenges remain, such as lack of clarity on how to access financial resources and technical expertise, insufficient engagement, a lack of a conducive environment in the private sector, and most importantly, inadequate local resources (ADPC 2023a; Ministry of Finance 2017). Climate change mitigation and adaptation strategies engage diverse stakeholders, government bodies, private organisations, and Civil Society Organisations (CSOs), which are essential for achieving sustainable climate health financing (Ministry of Finance 2017; MoFE 2021a).

**Summary of the identified barriers to implementing a climate-resilient and low-carbon health system in Nepal**

Despite substantial progress, Nepal still faces many barriers to building a climate-resilient health system (Fig. 10.2). Meeting the technological challenges offers great potential for progress on climate resilience. Creating socially and culturally appropriate technologies represents a potential path to effectively interact with communities, thereby increasing the adoption of technology. The more familiar climate technologies are, the likelier the innovation can increase overall preparedness for climate change and events. New technologies can help promote progress, and investing in funding mechanisms that support the latest advancements could make a big difference. Strengthening resilience requires the development of institutional and human resource capacities. Better regulation and monitoring of wood logging and building up a record of climate-related variable data will allow decision-makers to make more informed decisions about how best to manage their resources. Systems are often improved by identifying lacking institutional and policy reforms to address these gaps. Avoiding the disadvantages of drawing lines on a map and improving policy delivery through improved implementation strategies will establish stronger, more resilient infrastructure. Developing a co-learning process based on evidence and enhancing collaboration and coordination, which upscales across sectors, paves the way to integrated climate adaptation approaches. If we can leverage these opportunities, then the health system that emerges from the ashes of this crisis will be much better equipped to face the challenges climate change throws at it. In summary, the translation of climate change and health-related policies and plans into practice is challenging due to the limited interministerial coordination and collaboration.

## Discussion

Our analysis revealed that a fundamental barrier to the development of a climate-resilient and low-carbon health system in Nepal is the insufficient communication and collaboration between various government departments, echoing findings from other studies (Ministry of Health and Medical Services 2020; Prescott et al. 2023; Schrage and Collins Adeyanju 2024). This lack of coordination hinders the implementation of policies, a challenge also noted by Pattanshetty et al. (2023). The potential reasons likely include a shortage of skilled human resources (Abdullahi and Othman 2020), insufficient data, inadequate funding, and other societal, political, cultural, and environmental factors.

The importance of health workers trained in building a climate-resilient health system is emphasised, aligning with various studies (Barber et al. 2024; Lokotola and Lokotola 2023; Ministry of Health and Medical Services 2020; Mosadeghrad et al.

**Fig. 10.2** Challenges in building a climate-resilient health system

2023; Pattanshetty et al. 2023; Schrage and Collins Adeyanju 2024). While Nepal has undertaken capacity-building activities (HI-AWARE 2017; NHRC 2018; NHTC 2017; WHO 2018), challenges persist, including geographical disparities (urban–rural; mountain, hill, and terai), technological constraints, and financial barriers. A long-term, comprehensive training package on climate change and health is essential to build a climate-smart health workforce.

This study emphasises the need for robust data collection and research on climate, health, and GHG emissions. However, sustaining financial and human resources for quality research remains a challenge. Prioritising climate and health risk modelling could significantly aid in preparedness (Gonçalves et al. 2022; Hewitt et al. 2021; Li et al. 2021), but this is hindered by the lack of skilled health workers and inadequate digital infrastructure and data. Nepal's government has developed a framework for reducing GHG emissions (GoN 2021b); however, it does not currently account for emissions from the health sector, which is a significant impediment to attaining a climate-resilient and low-carbon health system. The development of climate-resilient health infrastructure that incorporates renewable energy, waste disposal systems, ventilation systems, earthquake resistance building, and sustainable design is crucial (Gerwig 2022). However, Nepal faces unique challenges due to its resource limitations and geographical terrain. Similarly, protecting environmental determinants of health (air, water, land, forest, and agriculture) is vital for achieving a climate-resilient health system, but requires overcoming geographical and resource constraints (CIAT et al. 2017; Karki et al. 2022). In addition, integrating climate change into health programmes and emergency preparedness is a significant step toward building a climate-resilient health system as shown by a large body of literature (Mosadeghrad et al. 2023; Pattanshetty et al. 2023). Funding emerges as a critical factor in building and sustaining climate-resilient and low-carbon health systems, consistent with findings from various studies (Barber et al. 2024; Lokotola and Lokotola 2023; Mosadeghrad et al. 2023; Pattanshetty et al. 2023). Several entities such as multilateral and regional development banks, green climate funds, loss and damage funds, and various international organisations could be the potential funders. While Nepal's government receives funding from diverse sources, insufficient utilisation of local resources poses a challenge to achieving sustainable climate and health financing.

## Limitations

The study limitations include potential gaps in the literature reviewed and the scarcity of research specifically addressing climate-resilient and low-carbon health systems in the Nepalese context. Further research in this area is strongly encouraged to inform policy and practice.

## Conclusion

In conclusion, achieving a sustainable climate-resilient and low-carbon health system requires an interdisciplinary approach. Despite the numerous challenges identified, there is potential for progress through enhanced collaboration, coordination, and allied efforts from various entities. By addressing these key areas—improving interdepartmental communication, strengthening the health workforce, enhancing data collection and research capabilities, developing resilient infrastructure, protecting environmental determinants of health, and optimising funding strategies – Nepal can make significant strides towards a health system that is both resilient to climate change and contributes to climate change mitigation.

## Annex

### Data Extraction Table

| Article information | Data to be extracted |
|---|---|
| Study information | |
| | Title |
| | Authors |
| | Journal (year, volume) |
| | Aims/objective |

(continued)

(continued)

| Article information | Data to be extracted |
|---|---|
| | Population studied |
| | Concept |
| | Context |
| Information on climate change and health | |
| Challenges | What are the challenges in building climate resilient health system in Nepal in various components of the WHO operational framework |
| Leadership and Governance | What policy considerations were reported<br>What policies are there to address climate change and health issues in Nepal |
| Health workforce | What activities are carried out to develop human resources in climate change and health<br>Who are involved in carrying out those activities |
| Assessment of climate and health risk | How assessment of climate and health risk is conducted in Nepal<br>What is the situation of GHG emission in various sectors and who tracks GHG emission |
| Integrated risk monitoring and early warning | What climatic and health indicators were reported |
| Climate and health research | What is the context of climate and health research in Nepal |
| Climate-resilient infrastructure and technologies | What technologies are being applied to mitigate climate change |
| Management of environmental determinants | What environmental determinants are crucial in mitigating the impact of climate change on health |
| Climate-informed health programs | What programmes are there to lessen the impact of climate change on health in Nepal |
| Climate-related emergency preparedness and management | What kind of emergency preparedness plans are being deployed |
| Sustainable climate health financing | What are the sources of funding in Nepal for climate change and health<br>How much climate fund is available in Nepal |

## References

Abdullahi M, Othman N (2020) Bridging the gap between policy intent and implementation. J Sci Technol Innov Policy. https://jostip.utm.my/index.php/jostip/article/view/49/49

ADB (2021) Climate risk country profile: Nepal. World Bank Group, vol 3, Issue 4. www.worldbank.org

ADPC (2021) Water sector policies and guidelines of Nepal. https://www.adpc.net/igo/category/ID1811/doc/2022-otj7NA-ADPC-Water_Sector_Policies_and_Guidelines_of_Nepal.pdf

ADPC (2023a) Accessing climate finance in Nepal: issues and options. https://www.adpc.net/igo/category/ID1861/doc/2023-uAQc62-ADPC-Accessing_Climate_Finance_in_Nepal_ForWeb.pdf

ADPC (2023b) Assessment of water sector policies and guidelines of Nepal: identifying gaps and addressing needs. https://www.adpc.net/igo/category/ID1905/doc/2024-fJAm15-ADPC-Water_Sector_Policies_Nepal_WBCAR_Water_1.pdf

ADRC (2020) Nepal: a brief country profile on disaster risk reduction and management. Ministry of Home Affairs, Government of Nepal. https://dpnet.org.np/uploads/files/Nepal_CR2019B2021-08-1909-30-45.pdf

Arksey H, O'Malley L (2005) Scoping studies: towards a methodological framework. Int J Soc Res Methodol 8(1):19–32. https://doi.org/10.1080/1364557032000119616

Barber B, Rainham DG, Tyedmers P, Vandertuin T, Ritcey G, Christie SD (2024) Taking action towards climate-resilient, low-carbon, health systems: perspectives from Canadian health leaders and healthcare professionals. https://doi.org/10.1177/08404704241252032

Bhandari D, Bi P, Sherchand JB, Dhimal M, Hanson-Easey S (2020a) Assessing the effect of climate factors on childhood diarrhoea burden in Kathmandu, Nepal. Int J Hygiene Environ Health 223(1):199–206. https://doi.org/10.1016/J.IJHEH.2019.09.002

Bhandari D, Neupane S, Hayes P, Regmi B, Marker P (2020) Disaster risk reduction and management in Nepal: delineation of roles and responsibilities. Oxford Policy Management, May 29. https://www.preventionweb.net/files/72985_delineationofresponsibilityfordisas[1].pdf

Bogren M, Erlandsson K, Johansson A, Kalid M, Abdi Igal A, Mohamed J, Said F, Pedersen C, Byrskog U, Osman F (2020) Health workforce perspectives of barriers inhibiting the provision of quality care in Nepal and Somalia—a qualitative study. Sex Reprod Healthc 23:100481. https://doi.org/10.1016/J.SRHC.2019.100481

Cao WR, Shakya P, Karmacharya B, Xu DR, Hao YT, Lai YS (2021) Equity of geographical access to public health facilities in Nepal. BMJ Glob Health 6(10):6786. https://doi.org/10.1136/BMJGH-2021-006786

CIAT, World Bank, CCAFS, & LI-BIRD (2017) Climate-smart agriculture in Nepal. CSA country profiles for Asia series. International Center for Tropical Agriculture (CIAT); The World Bank; CGIAR Research Program on Climate Change, Agriculture and Food Security (CCAFS); Local Initiatives for Biodiver. CSA Country Profiles for Asia Series, 26

Climate and Development Knowledge Network (CDKN) (2020) Climate-smart agriculture takes off, thanks to 'women-friendly' tools and a gender-smart approach. Forest and rangeland soils of the United States under changing conditions: a comprehensive science synthesis, September ix–xv. https://doi.org/10.4135/9781446269084.n83

Dhimal M (2008) Climate change and health: research challenges in vulnerable mountainous countries like Nepal. J Nepal Health Res Council 7(2). https://doi.org/10.3126/jnhrc.v7i2.3025

Dhimal M, O'Hara RB, Karki R, Thakur GD, Kuch U, Ahrens B (2014) Spatio-temporal distribution of malaria and its association with climatic factors and vector-control interventions in two high-risk districts of Nepal. Malaria J 13(1). https://doi.org/10.1186/1475-2875-13-457

Dhimal M, Gautam I, Joshi HD, O'Hara RB, Ahrens B, Kuch U (2015a) Risk factors for the presence of chikungunya and dengue vectors (*Aedes aegypti* and *Aedes albopictus*), their altitudinal distribution and climatic determinants of their abundance in Central Nepal. PLoS Neglect Trop Dis 9(3). https://doi.org/10.1371/journal.pntd.0003545

Dhimal M, Gautam I, Joshi HD, O'Hara RB, Ahrens B, Kuch U (2015b) Risk factors for the presence of Chikungunya and dengue vectors (*Aedes aegypti* and *Aedes albopictus*), their altitudinal distribution and climatic determinants of their abundance in central Nepal. PLoS Neglect Trop Dis 9(3). https://doi.org/10.1371/JOURNAL.PNTD.0003545

Dhimal M, Dahal S, Dhimal ML, Mishra SR, Karki KB, Aryal KK, Haque U, Kabir MI, Guin P, Butt AM, Harapan H, Liu QY, Chu C, Montag D, Groneberg DA, Pandey BD, Kuch U, Müller R (2018) Threats of Zika virus transmission for Asia and its Hindu-Kush Himalayan region. Infect Dis Poverty 7(1). https://doi.org/10.1186/S40249-018-0426-3

Dhimal M, Pradhan B, Dahal S, Sah A, Gautam I, Dhimal ML, Lal BK, Anjani M, Jha K (2019) Review of existing diseases surveillance system in Nepal from climate change perspective study team members. https://nhrc.gov.np/wp-content/uploads/2021/02/Final-review-climate-change-and-survailalcne-in-Nepal.pdf

Dhimal M, Bhandari D, Dhimal ML, Kafle N, Pyakurel P, Mahotra N, Akhtar S, Ismail T, Dhiman RC, Groneberg DA, Shrestha UB, Müller R (2021) Impact of climate change on health and well-being of people in Hindu Kush Himalayan Region: a narrative review. Front Physiol 12:651189. https://doi.org/10.3389/FPHYS.2021.651189/BIBTEX

DHM (2024) DHM. https://www.dhm.gov.np/

EDCD (2016) Reporting system and reporting system (EWARS) guidelines. http://www.edcd.gov.np/public/uploads/attachments/1473142047.pdf

Gerwig K (2022) Climate change and healthcare: a complicated relationship. Front Health Serv Manage 39(1):4–10. https://doi.org/10.1097/HAP.0000000000000148

GoN (2011a) Climate change policy, 2011 (Issue January, pp 13–24). Government of Nepal. https://doi.org/10.5822/978-1-61091-178-8_2

GoN (2011b) National framework on local adaptation plans for action. Government of Nepal, Ministry of Science Technology and Environment. www.moenv.gov.np

GoN (2012) Nationally determined contributions. APN Sci Bull 2(1). https://doi.org/10.30852/sb.2012.38

GoN (2020) Second nationally determined contribution (NDC) Nepal. Government of Nepal. https://climate.mohp.gov.np/attachments/article/167/Second-Nationally-Determined-Contribution-(NDC)-2020.pdf

GoN (2021a) National adaptation plan (2021–2050). https://faolex.fao.org/docs/pdf/nep215543.pdf

GoN (2021b) Nepal's long-term strategy for net-zero emissions. https://unfccc.int/sites/default/files/resource/NepalLTLEDS.pdf

GoN (2024) EDCD. Resources. Epidemiology and Disease Control Division. https://edcd.gov.np/resources/tools

Gonçalves C, Honrado JP, Cerejeira J, Sousa R, Fernandes PM, Vaz AS, Alves M, Araújo M, Carvalho-Santos C, Fonseca A, Fraga H, Gonçalves JF, Lomba A, Pinto E, Vicente JR, Santos JA (2022) On the development of a regional climate change adaptation plan: integrating model-assisted projections and stakeholders' perceptions. Sci Total Environ 805:150320. https://doi.org/10.1016/J.SCITOTENV.2021.150320

Hewitt CD, Guglielmo F, Joussaume S, Bessembinder J, Christel I, Doblas-Reyes FJ, Djurdjevic V, Garrett N, Kjellström E, Krzic A, Costa MM, St. Clair AL (2021) Recommendations for future research priorities for climate modeling and climate services. Bull Am Meteor Soc 102(3):E578–E588. https://doi.org/10.1175/BAMS-D-20-0103.1

HI-AWARE (2017) Certificate programme in climate change (Issue level 6). https://hi-aware.org/

ILO (2017) Migration of health worker from Nepal. International Labor Organization, vol 01. https://www.ilo.org/sites/default/files/wcmsp5/groups/public/@asia/@ro-bangkok/@ilo-kathmandu/documents/publication/wcms_546805.pdf

IPCC (2022) Climate change 2022: impacts, adaptation and vulnerability, Working Group II contribution to the sixth assessment report of the intergovernamental panel on climate change. In: Working Group II Contribution to the Sixth Assessment Report of the Intergovernamental Panel on Climate Change (Issue August). https://doi.org/10.1017/9781009325844.Front

Karki G, Bhatta B, Devkota NR, Acharya RP, Kunwar RM (2022) Climate change adaptation (CCA) research in Nepal: implications for the advancement of adaptation planning. Mitig Adapt Strat Glob Change 27(3):1–13. https://doi.org/10.1007/S11027-021-09991-0/FIGURES/4

Li F, Liu J, Chen Z, Huang J, Liu C, Qu Z (2021) Navigating to urban environmental health: professionalized and personalized healthy living assistant based on intelligent health risk management. Urb Clim 40:101020. https://doi.org/10.1016/J.UCLIM.2021.101020

Lokotola CL, Lokotola C (2023) Towards a climate-resilient primary health care service. South African Family Practice 65(4). https://doi.org/10.4102/safp

Ministry of Finance (2017) Harnessing green climate fund resources for Nepal. Government of Nepal, Ministry of Finance (Issue July)
Ministry of Health and Medical Services (2020) Guidelines for climate-resilient and environmentally sustainable health care facilities in FIJI
MoALD (2019) Integrating climate change adaptation into agriculture sector planning of Nepal
MoE (2010) National adaptation program of action. Ministry of Environment. https://www.adaptation-undp.org/sites/default/files/downloads/nepal_napa.pdf
MoEWI (2024) Meteorological observations. Ministry of Energy, Water Resources, and Irrigation. https://www.dhm.gov.np/
MoFE (2019a) Barrier analysis and technology action plan for mitigation technologies. J Environ Sci 5
MoFE (2019b) National climate change policy, 2076, p 26. Government of Nepal. https://www.icimod.org/wp-content/uploads/2021/07/National-Climate-Change-Policy_english_2019_compressed.pdf
MoFE (2021a) Assessment of climate financing allocation: unpacking eighty per cent allocation to the local level
MoFE (2021b) Assessment of electric cooking targets for Nepal's 2020 nationally determined contributions (NDC) (Issue March). www.mofe.gov.np
MoFE (2021c) Nepal's third national communication to the United Nations Framework Convention on Climate Change (UNFCCC). Government of Nepal Ministry of Forests and Environment
MoFE (2021d) Technology needs assessment for climate change adaptation (Issue August).
MoFE (2021e) Technology needs assessment for climate change mitigation. In: Technology needs assessment for climate change mitigation. Ministry of Forest and Environment, vol 5, Issue October
MoFE (2021f) Vulnerability and risk assessment and identifying adaptation options in health sector in Nepal. https://www.mofe.gov.np/downloadsdetail/3/2018/98853603/
MoFE (2023) Nepal's existing GHG inventory arrangement and way forward (Issue August). Ministry of Forest and Environment. https://climate-transparency-platform.org/sites/default/files/2023-08/PPT_MoFE/Nepal.pdf
MoFE (2024) Climate change management division. Ministry of Forest and Environment. https://www.mofe.gov.np/divisions/climate4954change4954management4954division-8457
MOHA (2023) Mid-term review implementation of the Sendai framework for disaster risk reduction
MoHP (2018) Climate change health adaptation strategies and action plans of Nepal (2017–2021). https://www4.unfccc.int/sites/NAPC/Documents/Parties/Nepal/HNAP.pdf
MoHP (2019) National health policy of Nepal, 2076. https://drive.google.com/file/d/1jZ6KrLmjJNjMvbemJ6dSox-30PJndfAT/view
MOHP (2014) Health care waste management guideline health care waste management guideline. https://climate.mohp.gov.np/downloads/Health_Care_Waste_Management_Guideline_2071.pdf
MOHP (2022) Vulnerability and adaptation assessment of climate sensitive diseases and health risks in Nepal. https://mohp.gov.np/uploads/Resources/VA/Assessment/Report-MoHP-2022.pdf
MOHP (2023) Health national adaptation plan (2023–2030). https://climate.mohp.gov.np/attachments/article/178/H-NAP-English_2024_FINAL.pdf
Mosadeghrad AM, Isfahani P, Eslambolchi L, Zahmatkesh M, Afshari M (2023) Strategies to strengthen a climate-resilient health system: a scoping review. Glob Health 19(1):1–11. https://doi.org/10.1186/S12992-023-00965-2/FIGURES/6
NHRC (2018) Annual report, vol 75, Issue July 2017. https://www.prudential.com.vn/export/sites/prudential-vn/.galleries/our_company/pdf/bap-cao-tai-chinh-2018.pdf
NHTC (2017) Major activities. National Health Training Center. http://nhtc.gov.np/
NSO (2021) National population and housing census 2021. National Statistic Office. https://censusnepal.cbs.gov.np/results/downloads/national

NSO (2024) National climate change survey 2022. National Statistic Office. https://api.giwms.gov.np/storage/36/posts/1715605547_98.pdf?fbclid=IwZXh0bgNhZW0CMTEAAR0E3BvqhXoRWqJRVc_I8Shwu20R5387ihi-1RMJNwljqfJ9eVJ_XHce_M0_aem_AS2nPO2DBqExSvBv4t3tpqoK4-5sYsNveJTYxwJ7zRbDC9eyAS2IiYtHS8dAQrQ4pTfgc-ctqc_LxbkZZiT4eg

Pattanshetty S, Inamdar A, Bhatt K, Paths HB (2023) Approaches to building climate-resilient health systems. Orfonline.Org. https://www.orfonline.org/wp-content/uploads/2023/11/ORF_Monograph_Converging-Paths_n.pdf#page=78

Paudel B, Khanal RC, Bhatta K, Chaudhary P et al (2017) Climate-smart agriculture in Nepal. CSA Country Profiles for Asia Series, vol 26

Prescott S, Mahmood J, Saraiva A, Mutanga O, Studer M, Bjorgaard J, Al Nahian M (2023) Public health impact and health system preparedness within a changing climate in Bangladesh: a scoping review. Challenges 14(1):4. https://doi.org/10.3390/CHALLE14010004

Rai P, O'Connor DA, Ackerman IN, Buchbinder R (2024) Healthcare system sustainability challenges in Nepal and opportunities offered by alternative healthcare delivery arrangements. J Nepal Med Assoc 62(273):347–349. https://doi.org/10.31729/JNMA.8584

Ranabhat S, Ghate R, Bhatta LD, Agrawal NK, Tankha S (2018) Policy coherence and interplay between climate change adaptation policies and the forestry sector in Nepal. Environ Manage 61(6):968–980. https://doi.org/10.1007/s00267-018-1027-4

Regmi BR, Sapkota R, Paudyal A, Gautam DK, Thapa R, Joshi R, Shah SGCG, Mishra B (2023) Co-development of vulnerability and risk assessment framework and methodology for Nepal. Environ Monit Assess 195(6). https://doi.org/10.1007/s10661-023-11330-6

Schrage P, Collins Adeyanju G (2024) Building climate resilient health systems: panacea for mitigating climate change effects in Abdullahi M, Othman N (2020) Bridging the gap between policy intent and implementation. J Sci Technol Innov Policy. https://jostip.utm.my/index.php/jostip/article/view/49/49

Sen B, Dhimal M, Latheef AT, Ghosh U (2017) Climate change: health effects and response in South Asia. BMJ (Clin Res Ed) 359:j5117. https://doi.org/10.1136/bmj.j5117

Shrestha A, Nepal S (2020) Climate change and its impacts on HKH The Hindu Kush Himalaya. ICIMOD. https://www.icimod.org/wp-content/uploads/2020/11/Day2_Present_IITM_CORDEX_ArunSantosh0201013_SN_V2.pdf

South Asia Region Climate Knowledge Hub (2023) Climate change in Nepal—SAR-CLIMATE. http://sar-climate.adpc.net/index.php/countries/nepal/#1660371418344-67e96782-e3648930-8e80

The World Bank (2024) Climate and development in South Asia. World Bank Group. https://www.worldbank.org/en/region/sar/brief/integrating-climate-and-development-in-south-asia/integrating-climate-and-development-in-south-asia-region

USAID (2019) Greenhouse gas emissions in Nepal, Issue 2014, p 2. https://www.climatelinks.org/sites/default/files/asset/document/2019_USAID_Nepal_GHG_Emissions_Factsheet.pdf

WHO (2018) Nepal Climate change, health & WASH. WHO. https://www.chatpdf.com/c/oARmJy1F6Jq6Mtw8O8tkY

World Health Organization (2023a) Climate change. WHO. https://www.who.int/health-topics/climate-change#tab=tab_1

World Health Organization (2023b) Operational framework for building climate resilient and low carbon health systems, pp 1–56. https://www.who.int/globalchange/publications/building-climate-resilient-health-systems/en/%0Afile:///Users/rosalind/Dropbox/1Mac/Papers2/Reports/2017/WHO/2017WHO.pdf%0Apapers2://publication/uuid/FBDE32AB-02C3-419F-AA2B-78E4BE9C197D

# Chapter 11
# Strengthening the Resilience of the Turkish Healthcare Systems Through a Novel Climate and Health Curriculum

**Mine Durusu Tanriover, Begüm Güler Şentürk, Zeynep Komesli Ergin, Sıla Gürbüz, Kıvılcım Pınar Kocabıyık, Banu Binbaşaran Tüysüzoğlu, Melih Elcin, and Emine Didem Evci Kiraz**

**Abstract** *Background*. The 'Climate Clinic' project was launched to enhance healthcare system resiliency by prioritising the upskilling of the workforce. This study aimed to describe a process for developing a pre-graduate climate and health curriculum. *Methods*. The climate and health education literature was reviewed with a focus on competencies. Focus group meetings with national and international experts, as well as pre-graduate students, were conducted using preformed questions, followed by a needs analysis. We selected core competencies from the Global Consortium on Climate and Health Education (GCCHE) to guide curriculum development, ensuring alignment with recognised standards. A curriculum framework with defined learning objectives and outcomes was proposed, along with flexible curricular programmes adaptable to various pre-graduate settings. *Results*.

---

M. Durusu Tanriover (✉)
Faculty of Medicine, Hacettepe University, Ankara, Türkiye
e-mail: mdurusu@hacettepe.edu.tr

Vaccine Institute, Hacettepe University, Ankara, Türkiye

B. Güler Şentürk
Istanbul Goztepe Prof. Dr. Suleyman Yalcin City Hospital, Istanbul, Türkiye

Z. Komesli Ergin
Memorial Healthcare, Istanbul, Türkiye

S. Gürbüz
Faculty of Medicine, Karadeniz Technical University, Trabzon, Türkiye

K. P. Kocabıyık · B. Binbaşaran Tüysüzoğlu
Yuvam Dünya Association, Istanbul, Türkiye

M. Elcin
Springfield College School of Health Sciences, Springfield, MA, USA

E. D. Evci Kiraz
Faculty of Medicine, Adnan Menderes University, Aydin, Türkiye

 
W. Leal Filho and F. Wolf (eds.), *Climate Change and Health: Perspectives from Developing Countries*, Climate Change Management,
https://doi.org/10.1007/978-3-031-93177-2_11

Key needs identified from the focus group meetings include establishing a pool of instructors, providing content on health-climate interaction, tailoring programmes to Türkiye-specific risks, strategically positioning educational programmes for flexible basic competencies, offering specialisation and certification options for trainers, and adopting a participatory policy at the leadership level involving students. *Conclusions*. This study is the first to propose a methodology, systematically conduct a needs analysis, and outline a roadmap for integrating a climate and health curriculum into pre-graduate programmes in Türkiye.

**Keywords** Climate change · Healthcare · Resiliency · Education · Training needs

## Introduction

Climate change is significantly impacting human health and healthcare systems worldwide (Romanello et al. 2021). Rising temperatures, air pollution, diminishing water resources, and extreme weather events contribute to disasters, accelerate the spread of infectious diseases, disrupt food sources, and exacerbate chronic health issues (Ebi et al. 2021). Although communities are already experiencing these health effects, their frequency and severity are expected to increase in the coming years. The Intergovernmental Panel on Climate Change (IPCC) has documented that climate-related risks are emerging faster than anticipated and will become more severe (Masson-Delmotte et al. 2024). This report underscores how climate change exacerbates health inequalities, heightens the risk of emerging infectious diseases, and affects the prevalence of both infectious and non-infectious diseases. Additionally, the World Health Organization (WHO) projects that climate change could cause approximately 250,000 additional deaths per year globally between 2030 and 2050 due to direct and indirect causes (Organization WH 2018).

Türkiye is not exempt from these risks. A 2022 WHO risk profile for Türkiye highlights significant health impacts related to increased temperatures, heatwaves, drought, and heavy precipitation (Organization WH 2022). Notably, heat-related deaths among the elderly, which averaged 5.2 per 100,000 between 1961 and 1990, are projected to rise to 62 per 100,000 by 2080 under high-emission scenarios (Organization WH 2022). Moreover, climate change significantly affects the healthcare workforce and infrastructure. The Lancet Countdown study has shown that healthcare systems globally are not adequately prepared for climate change (Romanello et al. 2021). The ability of healthcare systems to protect patients and the promptness of healthcare professionals in responding to climate crises are critical determinants of the health risks posed by climate change (Council and Committee 2018).

Recognising these challenges, there is a growing emphasis on integrating climate change education into healthcare curricula. Health protection and improvement have finally started to gain attention as principal aims in the fight against the climate crisis. One of the most concrete examples of progress is the organisation of 'Health Day' for the first time at the 28th United Nations Climate Change Conference (COP28) discussions in 2023 (Liu 2023). Health Day focused on evidence-based research on the relationship between climate change and human health, understanding the needs of healthcare systems, and coping with the effects of climate change on human health (Liu 2023).

Globally, the need for education on climate change has gained significant momentum. The Global Consortium on Climate and Health Education (GCCHE), established in 2017, now represents over 300 healthcare professionals from 56 countries and reaches an estimated 175,000 students annually by supporting the development of curriculum programmes (Council and Committee 2018). As awareness of climate change grows, countries such as the United States, Canada, Australia, and the United Kingdom have integrated 'climate change and health' into their medical school curricula (Visser et al. 2023; Collaborative IU 2020; Maxwell and Blashki 2016; Ford 2009). The Planetary Health Report Card (PHRC), an international student-led initiative, expanded its reach in 2023, increasing the number of participating medical faculties to 96 by including institutions from Greece, India, New Zealand, South Africa, and Switzerland. In the same year, 13 medical faculties from Türkiye joined the PHRC project, and a pharmacy faculty is set to participate in the 2024 report (Irlam et al. 2024).

Despite these global efforts, significant gaps remain in many countries. A 2020 study evaluating the knowledge of Turkish healthcare professionals regarding the health impacts of climate change revealed that 79.1% of participants felt they lacked sufficient knowledge and skills to address these issues (Akay et al. 2020). The first climate change and health education module in Türkiye was developed in 2019 (Kiraz 2019). However, a 2022 study by Özmen and Evci Kiraz highlighted the persistent absence of climate-related content in medical education and underscored the need for comprehensive reform (Özmen and Kiraz 2022). In response, Türkiye organised its first interdisciplinary and inter-sectoral education on the relationship between climate and health in 2022, titled "Health and Climate Change Education," during the II International Health and Climate Change Congress (Evci Kiraz 2022).

The Yuvam Dünya Association is committed to reshaping our generation's approach to combatting the climate crisis and drive societal transformation across all sectors (https://yuvamdunya.org/). Central to its mission is altering the language used in climate crisis discourse and spearheading projects that promote cultural change and societal impact. The 'Climate Clinic' project was established on this foundation, serving as an umbrella initiative focused on enhancing the resilience of healthcare systems by educating healthcare professionals (The Climate Clinic 2024). By developing realistic and scalable curriculum models for continuous professional development both during and after formal education, the project aims to create a roadmap that enables healthcare workers across all disciplines to effectively address the health impacts of the climate crisis in their professional practice.

Building on these initiatives, the first aim of this study is to propose a methodology for developing a climate and health curriculum that can be integrated into the national core curricula. The second aim is to conduct a needs assessment analysis as a foundation for establishing programme requirements and outlining the curriculum.

## Methods

The Climate Clinic project began with a thorough evaluation of medicine and health sciences programmes at the pre-graduate level in Türkiye. The goal was to design a flexible climate and health curriculum that could be integrated into various programmes. After assembling a multidisciplinary team, a step-by-step methodology was established.

The first step involved conducting a narrative review to identify gaps in climate change education within existing medical and healthcare programmes. We used the keywords "climate change," "health," "curriculum," and "competency" in our search. Although this review was comprehensive, it was not a systematic review, which we acknowledge as a limitation of this study.

Simultaneously, we prioritised the identification of core competencies in climate change and health education. We chose to adopt the competencies developed by the Global Consortium on Climate and Health Education (GCCHE) as they have been widely recognised and recently updated (Global Consortium on Climate and Health Education 2023). This approach allowed us to focus on competencies rather than specific curriculum examples, recognising that curricula can vary widely based on local needs and requirements. By establishing a set of core competencies, we created a foundation for developing a curriculum that can be adapted to diverse contexts while adhering to recognised standards.

Following the literature review and competency selection, we conducted focus group meetings with national and international experts, as well as pre-graduate students. These discussions, guided by preformed questions, provided deeper insights into the specific needs and gaps in current curricula. The data from these focus groups, combined with our literature review findings, were analysed to define the key competencies required for healthcare professionals to address climate-related health challenges effectively.

Based on this needs analysis, we proposed a curriculum framework that included specific learning objectives and outcomes aligned with the defined competencies. Finally, we developed example curricula designed for flexible integration into various pre-graduate programmes. These examples were crafted to align with the defined competencies and address the identified needs, ensuring that the proposed curriculum can be tailored to different educational settings while maintaining high educational standards (Table 11.1).

## Results

a. **Analysis of the focus group meetings**

Focus group discussions were conducted online by task members using standardised questions. Two meetings were held with six national experts on August 23, 2023 and September 12, 2023 and with five international experts on September 28, 2023,

**Table 11.1** Monthly timeline for the tasks allocated

| | Month 1 | Month 2 | Month 3 | Month 4 | Month 5 | Month 6 | Month 7 | Month 8 | Month 9 |
|---|---|---|---|---|---|---|---|---|---|
| **Literature Review and Adding Articles to the Library** | ■ | | | | | | | | |
| **Compilation of Examples in Practice (National)** | | ■ | | | | | | | |
| **Compilation of National/International Reports** | | ■ | | | | | | | |
| **National Expert Interviews** | ■ | ■ | | | | | | | |
| **National Student Interviews** | | ■ | | | | | | | |
| **International Expert Interviews** | ■ | ■ | | | | | | | |
| **Preparation of Needs Analysis Report** | | | ■ | ■ | | | | | |
| **Determination of Program Objectives and Competencies** | | | | ■ | ■ | | | | |
| **Preparation of Sample Training Programs** | | | | | ■ | ■ | | | |
| **Implementation and Evaluation of Training Programs in Pilot Studies** | | | | | | ■ | ■ | | |
| **Curriculum Development** | | | | | | | ■ | ■ | ■ |
| **Implementation of the Curriculum in Health Sciences and Medical Faculties** | | | | | | | | ■ | ■ |
| **Integration of the Curriculum into the Core Education Programme** | | | | | | | | | ■ |

September 29, 2023, and October 9, 2023. Additionally, six students participated in a focus group meeting on November 1, 2023. The responses to the standardised questions from all focus group discussions have been summarised as follows:

**Question 1**: What are the deficiencies in climate change and health impact education?

Both national experts and students identified the lack of awareness at the educational level, the absence of a pool of competent and adequately informed educators, and the failure of educators to incorporate the impacts and information of climate change into their respective fields as the most significant deficiencies. At the educational programme level, it was emphasised that the seriousness of the issue and its impacts on health are not adequately reflected in existing programmes. There is a lack of learning topics emphasising the impacts of climate change on health in the national core curriculum, and the few courses offered in some faculties are purely theoretical and superficial, lacking practical application. Therefore, it was stated that pre-graduation education in the health field does not equip graduates with the skills to prepare them for real-life situations, and the existing courses are mostly included in public health courses, making it difficult to translate them into clinical practice and actionable health services.

**Question 2**: Are there any studies in your country/region to identify these deficiencies?

Students were more aware of nationally conducted studies and platforms such as the Science Virus platform, the Climate Ambassadors project, Planetary Health Report Card, Climate Clinic, European Medical Students' Association Türkiye Policy Reports, the World Medical Association declaration, Scientific and Technological Research Council of Türkiye, and Brain and Climate research.

**Question 3**: What basic knowledge, skills, and competencies do we need to train?

Students and national experts listed independent topics or competencies to answer this question: being able to clinically intervene in the effects of climate change on health, risk management, competence in disaster and emergency response, knowing the environmental impacts of products, establishing cause-and-effect relationships, crisis management skills, being focused on preventive medicine rather than pathology, effective communication and social skills, information literacy, being a climate advocate, and being an activist leader in society. International experts grouped these competencies into specific areas and emphasised the importance of putting them into practice.

**Question 4**: What content should we provide for this?

Students suggested presenting topics with scientific content that teach cause-and-effect connections and mechanisms. They also emphasised providing preventive health courses from the perspective of climate change, highlighting locally prominent health issues, using interesting examples such as microplastics passing through the placenta, and emphasising a comprehensive interdisciplinary approach that integrates different systems in climate change. National experts underlined the need for joint educational goals for all health faculties while recognising the importance of determining specialised knowledge, skills, and competencies according to disciplines. International experts stressed the importance of having a clear vision, generating ideas for change, advocating for climate change, and emphasising the importance of leadership skills.

**Question 5**: What methods will be necessary for this?

Students and national experts emphasised the use of various methods, including case studies, fieldwork, standardised patient interviews, simulations, role-playing, examining good practice examples from around the world, research projects, advocacy projects, elective courses, compulsory courses, and exam questions, online education models, problem-based learning methods, social responsibility projects, and community-based work. They all agreed on the importance of developing a standardised curriculum for teacher training in climate and health education. International experts emphasised the urgency of implementing changes and stressed the need for creating a unified "One Health" discipline.

**Question 6**: What challenges might be encountered in integrating the educational content into the curriculum?

Students and national experts highlighted challenges such as the perceived insignificance of climate and health issues, reluctance, conservatism, and lack of adaptation among administrators and faculty members, especially clinicians, the inability to allocate new course hours in an already busy curriculum, challenges in creating interdisciplinary programmes, and difficulty in finding common ground among different disciplines. International experts also saw leadership and educator shortages as well as everyone being too busy as significant issues. However, they emphasised that climate change is the most significant challenge to health and humanity's existence and should not be overshadowed by other difficulties.

**Question 7**: What solutions can be proposed to overcome these challenges?

Suggestions from students and national experts included soliciting faculty input and determining their baseline awareness levels, creating a pool of educators, conducting pilot programmes, and standardising educator training curricula, using interactive teaching materials and student-centred teaching approaches, engaging in competitions, peer education, assignments, and research in these areas, and incorporating climate and health education into the curriculum, especially in the first three years of study. International experts emphasised the importance of creating a united "One Health" discipline and the need for leadership in addressing the issue.

**Question 8**: Are there any "good practice" examples of climate change and health impact education in your country/region?

Students and national experts cited examples of topics being added to the curriculum at some universities, while at others, discussions are being held at the deanery and faculty levels. They mentioned local learning modules and initiatives like iklimin.org and iklimeuyum.org as examples. Internationally, examples such as the Health and Environment Council in Portugal, the Rising Leaders in Environmental Sustainability in Healthcare, and a 40-h champion programme for training healthcare workers in Switzerland were cited as good practices.

**Question 9**: Is general education on the health effects of climate change sufficient, or is specialisation necessary?

All participants in the focus group discussion believed that we need to train more teachers in planetary health, climate, and health topics, and specialisation is particularly important for educators. Therefore, there is a need to develop specialised training programmes, certification programmes, or modular training for clinicians. It was also emphasised that since we are facing a global crisis, the basic competencies of all disciplines should be part of the curriculum. However, it was acknowledged that over time, there will inevitably be a shift towards specialisation. Finding a balance and designing the content together with students and academicians were highlighted as crucial for success.

b. **Needs assessment**

After reviewing the literature on climate and health training and curricula in Türkiye and globally and extracting key points from the focus group discussions, a needs assessment was conducted. The needs for a climate and health curriculum in Türkiye were summarised into six main domains (Table 11.2).

c. **Competency target concepts**

To define a climate and health curriculum programme, several concepts were identified, around which competencies should be established. These are summarised in Table 11.3.

d. **Sample curricula to be integrated into pre-graduate programmes**

**Table 11.2** The results of the needs assessment

| Domain | Need |
|---|---|
| Creation of a competent pool of instructors | Establish a pool of instructors who are highly aware of the priority requirements, knowledgeable, and skilled in education |
| Two-way interaction | Provide content on the two-way interaction between health and climate, skills related to mitigation and adaptation strategies, climate communication, advocacy, leadership, and other action-oriented competencies |
| Adaptation of existing international curricula | Adjust the content of these programmes to address the prominent risks and situations specific to Turkey, rather than creating an entirely new curriculum from scratch |
| Flexibility in curriculum placement | Strategically position for each educational programme, offering at least basic competencies with flexibility |
| Specialisation and certification for instructor pool | Implement programmes that offer options such as specialisation, certification, or postgraduate education to create a pool of specialised instructors |
| Inclusive policy for student involvement | Adopting a participatory policy at the leadership level involving students is crucial for the success and impact of the outputs of the process |

**Table 11.3** Competency target concepts

| |
|---|
| Climate change is one of the greatest threats to human health in the twenty-first century, and urgent measures are needed to address this trend |
| Human health is significantly influenced by climate, which is a determinant of all life, such as air, water, and food |
| Climate change is primarily a result of human-induced factors |
| The effects of climate change vary depending on location and geography |
| Demographic structure, economic development, technology, and other activities affect climate and the environment |
| Climate change has broad and profound effects on human health |
| Mitigation and adaptation to climate change are necessary to reduce health burdens now and in the future |
| Both short-term and long-term health benefits can be obtained from mitigating climate change |
| Health systems should implement effective adaptation solutions at both individual and population levels |
| Planetary health, environmental health, climate justice, and ecological health are overlapping frameworks in all policies, including health and one health, intersecting with climate change and health |
| Climate change poses significant risks to health infrastructure and affects emergency and disaster planning |
| Global, regional, national, local, and environmental data should be used to guide health decisions to prevent, prepare for, and respond to climate-related events |
| Health professionals should consider ethical, professional, and legal obligations related to climate and health |

By synthesising the evidence gathered through our literature search, focus group meetings, and needs assessment, we defined sample curricula that incorporate a variety of pedagogical approaches. These approaches include case studies, interactive sessions, field visits, community projects, and simulations. Additionally, we aimed to ensure that these curricula align with the competencies outlined by the GCCHE. The resulting sample curriculum frameworks are presented here.

**Box 1. Curriculum for the Climate and Health course offered in Aydın Adnan Menderes University Faculty of Medicine (Curriculum for the Climate and Health course offered in Aydın Adnan Menderes University Faculty of Medicine n.d.) that is designed to be integrated into the course programs of the 1st, 2nd, 5th, and 6th grades of the Medical Faculty**

1st year:

Public Health and Prevention

Environment, Climate, and Health Relations

City Health

Global Risks and Planet Health

2nd year:

Determinants of Health

Sustainable Development Targets

5th year:

Environmental History and Epidemiology

Determinants of Health (Demographic, Environmental, Social, and Climate Determinants)

Health Impact Assessment

One Health Concept

City Health Measuring

Sectoral Health Approach

6th year:

Environment and Health Legislations

City Health Profile and Development Plan

Climate and Health Local Action Plan

**Box 2. 10-day programme inspired from the Academic Life in Emergency Medicine (Climate Change and Emergency Medicine. Available at: https://www.aliem.com/climate-change-emergency-medicine/)**

Overview of the Climate Crisis and Current Situation (1 Hour)

Climate and Health Impacts (2 h)

Vulnerable Populations and Environmental Justice (2 h)

Global Health and Climate Change (2 h)

COVID-19 and Climate Change (1 h)

Building Resilient Health Systems (1 h)

Sustainability in Health (2 h)

Climate Change Education (1 h)

Climate Change Policy (2 h)

Climate Change Advocacy (2 h)

**Box 3. Sample 16-Hour Medium-Length Programme that can be named 'Climate Change and Health: In-Depth Understanding'**

Module 1: Basic Concepts and Scientific Foundations of Climate Change

Objective: To ensure participants understand the basic concepts and scientific foundations of climate change.

Module 2: Direct and Indirect Effects of Climate Change on Health

Objective: To enable participants to learn in detail about the direct and indirect effects of climate change on health.

Module 3: Strategies and Practices for Combatting Climate Change in the Healthcare Sector

Objective: To enable healthcare professionals to learn strategies and practices they can use to combat climate change.

Module 4: Climate Change and Emergency Preparedness

Objective: To help healthcare professionals understand the factors that need to be considered regarding climate change when preparing for emergencies in the healthcare sector.

Module 5: Increasing Climate Change Awareness and Advocacy in the Community

Objective: To help participants develop skills to increase climate change awareness and advocacy in the community.

**Box 4. Sample 4-hour programme named 'Basic Climate Change Knowledge and Health Impacts'**

Lesson 1: What is Climate Change?

Objective: To help participants understand the concept of climate change.

Lesson 2: The Health Impacts of Climate Change

Objective: To help participants understand the direct and indirect effects of climate change on human health.

Lesson 3: The Role of Healthcare Workers in Combatting Climate Change

Objective: To help healthcare workers understand how they can contribute to combatting climate change.

Lesson 4: Coping Strategies for Climate Change

Objective: To help participants learn coping strategies for dealing with climate change on personal and professional levels.

## Discussion

In this study, we proposed a methodology for developing a framework for a climate and health curriculum to be integrated into pre-graduate programmes in health sciences, including, but not limited to, medicine. Our findings reveal specific needs that both align with and expand upon the experiences of other countries, offering a comprehensive understanding of the challenges and opportunities in this field.

One of the key findings of this study is the critical need to develop a pool of instructors who are not only highly aware of the priority requirements but also knowledgeable and skilled in education. This is essential for establishing a successful climate and health pre-graduate programme. However, Türkiye faces an additional challenge: building this expertise from a relatively small base of specialists in climate-health intersections, which underscores the need for extensive faculty training programmes. Specialisation and certification of instructors were identified as essential strategies to enhance faculty expertise and ensure that students receive high-quality education from well-qualified instructors. Similar approaches have been successfully implemented at institutions like the University of California, San Francisco, and the University of Washington, where faculty development programmes focus on specialised training (Haines et al. 2014).

Another significant need identified is the inclusion of content that addresses the bidirectional relationship between health and climate. This includes equipping students with skills related to mitigation and adaptation strategies, climate communication, advocacy, leadership, and other action-oriented competencies. This approach is reflected in the comprehensive curricula of institutions such as Harvard University's T.H. Chan School of Public Health and Columbia University's Mailman School of Public Health, which integrate theoretical knowledge with practical training through internships and fieldwork (Shaw et al. 2014). Adapting this model to Türkiye's specific regional needs, such as addressing local climate risks, allows for a curriculum that offers both global and local perspectives on climate-health interactions. Rather than creating a completely new curriculum from scratch, tailoring existing content to address Türkiye's specific challenges, such as extreme weather events and air pollution, presents a valuable opportunity (Ebi et al. 2021). This approach is consistent with strategies used by institutions like the University of California, which embedded climate-related modules within existing public health courses to maintain relevance without overloading students (Haines et al. 2014).

Implementing a health and climate curriculum poses challenges, particularly within the already compact nature of existing curricula. Strategically positioning each educational programme to cover at least the fundamental competencies with flexibility allows for adaptation to changing needs and the integration of new knowledge and skills as they emerge. There are several examples of best practices where international curricula have been successfully adapted to local contexts. However, the challenge of sustainably integrating a new, interdisciplinary curriculum into already full programmes remains. Ensuring the scalability of the curriculum is also crucial, alongside addressing potential resistance and skepticism from various stakeholders. Given the rapidly evolving nature of climate science, the curriculum content must be regularly updated to reflect the latest scientific evidence and emerging approaches, particularly in areas such as advocacy and behaviour change. Active learning methods, such as problem-based learning, simulations, and field projects, have proven effective in climate-health education. For instance, the London School of Hygiene & Tropical Medicine employs simulations of climate-related health emergencies to train students in crisis management and response, thereby enhancing critical thinking and problem-solving skills (Daalen et al. 2022). These methods not only deepen students' understanding of the subject matter but also prepare them to address real-world health challenges related to climate change.

Finally, adopting a student-centred, participative approach at the leadership level is essential for the curriculum's effectiveness and impact. Actively involving students in the development and evaluation of the curriculum fosters a sense of ownership and ensures that the curriculum meets their needs and expectations. Continuous feedback from students and professionals is crucial for adapting and improving the curriculum to meet evolving educational and professional requirements (Frumkin et al. 2008). This participative approach has been highlighted as a critical factor in the success of climate-health education programmes worldwide (Levy and Patz 2015).

Raising awareness among healthcare professionals and the public about the health impacts of climate change is also crucial. Healthcare personnel trained in this area can educate the public on these impacts and guide necessary precautions. Assessing the broader impact of integrating climate change education into the medical and healthcare community will be essential. Moreover, there is a need for continuously updated data and information on the health effects of climate change. Therefore, a curriculum that promotes research and innovation in addressing climate change within health sciences and medicine can better prepare healthcare professionals to provide effective and up-to-date services. To fine-tune and continuously improve these programmes, formative and summative assessment tools aligned with learning objectives are necessary. These tools, along with accreditation processes by auditing institutions, should be used to implement feedback mechanisms for continuous improvement. Evaluating the curriculum's effectiveness in enhancing climate literacy, advocacy, and preparedness will be key to building a resilient healthcare system (Molthan-Hill et al. 2022).

One of the strengths of this study is the integration of international knowledge and expertise to develop a curriculum framework adapted to local environments and needs. The proposed framework will first be applied and evaluated through pilot studies. Initially, a pool of trainers will be established through train-the-trainer sessions, followed by training sessions for various target groups to assess whether the competencies and programmes meet their goals. The final version will then be implemented in pre-graduate education in medicine and health sciences, ensuring that the curriculum can be successfully integrated into the Core Training Programme. Partnerships with healthcare institutions and community organisations will also be sought to provide practical learning opportunities.

A primary limitation of this study is the small sample size of academicians, predominantly from medical faculties, which may not fully represent all healthcare professionals across the country. Future research should focus on the implementation of the target competencies and core skills across different health sciences programmes.

## Conclusions

To our knowledge, this study is the first to propose a methodology, systematically conduct a needs analysis, and draw a roadmap for establishing a curriculum around climate and health to be integrated into pre-graduate programmes in Türkiye. The direct and indirect effects of climate change on individual, societal, and systemic health primarily compel the health sector to rapidly develop adaptation strategies. Empowering healthcare professionals is a crucial cornerstone in strengthening the resilience of healthcare systems.

## *References*


Akay DS, Akca G, Atik AD, Erkoç F (2020) Yaşam Bilimleri Profesyonellerinin İklim Değişikliğinin Sağlık Etkileri Konusunda Eğitim İhtiyaçları Var Mı? J Int Lingual Soc Educ Sci 6(1):141–151

Collaborative IU (2020) Global health education in medical schools (GHEMS): a national, collaborative study of medical curricula. BMC Med Ed 20

Council GA, Committee C (2018) Global Consortium on Climate and Health Education (GCCHE) Core climate & health competencies for health professionals. Columbia University

Ebi KL, Vanos J, Baldwin JW, Bell JE, Hondula DM, Errett NA et al (2021) Extreme weather and climate change: population health and health system implications. Annu Rev Public Health 42(1):293–315

Evci Kiraz ED ÇE. Sağlık Ve İklim Değişikliği Eğitimi. II Uluslararası Sağlık Ve İklim Değişikliği Kongresi Online 2022

Ford LB (2009) Climate change and health in Canada. McGill J Med: MJM 12(1):78

Frumkin H, Hess J, Luber G, Malilay J, McGeehin M (2008) Climate change: the public health response. Am J Public Health 98(3):435–445

Global Consortium on Climate and Health Education. Columbia Mailman School of Public Health (2023). Available from https://www.publichealth.columbia.edu/file/11940/download?token=ILZgbU2L

Haines A, Ebi KL, Smith KR, Woodward A (2014) Health risks of climate change: act now or pay later. The Lancet 384(9948):1073–1075

Irlam J, Reid S, Rother H-A (2024) Education about planetary health and sustainable healthcare in low-and middle-income countries: planetary health report card assessment of perceptions at university of cape town faculty of health sciences. Educ Health 37(1):8–19

Kiraz E (2019) İklim Değişikliği Eğitim Modülleri Serisi 14: İklim değişikliğinin insan sağlığına etkileri. İklim Değişikliği Alanında Ortak Çalışmaların Desteklenmesi Projesi, İklimİN, We Global

Levy BS, Patz JA (2015) Climate change, human rights, and social justice. Ann Glob Health 81(3):310–322

Liu Q (2023) The 28th United Nations climate change conference (COP28)—November 30 to 12 Dec 2023. China CDC Weekly 5(49):1093

Masson-Delmotte V, Zhai P, Pörtner H, Roberts D, Skea J, Shukla P et al (2018) Intergovernmental Panel on Climate Change (IPCC). Global warming of 1.5∘ C: an IPCC special report on the impacts of global warming of 1.5 °C above pre-industrial levels and related global greenhouse gas emission pathways, in the context of strengthening the global response to the threat of climate change. Sustainable development, and efforts to eradicate poverty. Available online https://www.ipcc.ch/sr15/. Accessed on 17 Jan 2024

Maxwell J, Blashki G (2016) Teaching about climate change in medical education: an opportunity. J Public Health Res 5(1)

Molthan-Hill P, Blaj-Ward L, Mbah MF, Ledley TS (2022) Climate change education at universities: relevance and strategies for every discipline. Springer, Handbook of climate change mitigation and adaptation, pp 3395–3457

Organization WH (2018) COP24 special report: health and climate change

Organization WH (2022) Health and climate change: country profile 2022: occupied Palestinian territory. World Health Organization

Özmen A, Kiraz EDE (2022) Sağlıkta iklim değişikliği müfredatı: Türkiye tıp fakülteleri içerik analizi. Sürekli Tıp Eğitimi Dergisi 31(5):388–400

Romanello M, McGushin A, Di Napoli C, Drummond P, Hughes N, Jamart L et al (2021) The 2021 report of the Lancet countdown on health and climate change: code red for a healthy future. The Lancet 398(10311):1619–1662

Shaw C, Hales S, Howden-Chapman P, Edwards R (2014) Health co-benefits of climate change mitigation policies in the transport sector. Nat Clim Chang 4(6):427–433

The Climate Clinic (İklim Kliniği) 2024. Available from: https://tr.linkedin.com/company/the-climate-clinic?trk=ppro_cprof

van Daalen KR, Romanello M, Rocklöv J, Semenza JC, Tonne C, Markandya A et al (2022) The 2022 Europe report of the Lancet countdown on health and climate change: towards a climate resilient future. The Lancet Public Health 7(11):e942–e965

Visser EH, Oosterveld B, Slootweg IA, Vos HM, Adriaanse MA, Schoones JW et al (2023) The development and characteristics of planetary health in medical education: a scoping review. Acad Med

Yuvam Dünya. Available from https://yuvamdunya.org/

# Chapter 12
# Knowledge, Perception, Attitude, and Practice of Climate Change Among Healthcare Students and Practitioners: A Scoping Review from a Developing Countries Perspective

**P. R. Shankar, A. Guragain, S. Ganesan, and N. Jha**

**Abstract** *Background.* Healthcare students (HCS) and healthcare practitioners (HCP) have an important role in addressing climate change (CC). This study aims to review original research on knowledge, perception, attitude and practice of CC among HCS and HCP. *Methods.* Two databases, PubMed and Google Scholar, were searched for original research articles using specific keywords. Studies were also obtained by examining the reference list of included articles. Only English language publications from 2008 to 2024 were included. Study quality was assessed using suitable checklists. *Results.* Sixty-six studies were included. Most were published during the period from 2017 to 2024, used quantitative methods, and were conducted in developing nations. The United States (US) published the highest number. Most studies were conducted among nursing and medical students, did not include an intervention, and were of good quality. Among HCS and faculty in developed nations, knowledge was good but practice required improvement. Among developing countries, knowledge ranged from poor to moderate and there were deficiencies in practice. *Conclusions.* Respondents wanted sessions on CC. More studies are required. Some studies

P. R. Shankar (✉)
IMU Centre for Education, IMU University, Kuala Lumpur, Malaysia
e-mail: ravi.dr.shankar@gmail.com

A. Guragain
Kathmandu, Nepal
e-mail: anusmritiii@gmail.com

S. Ganesan
IMU University, Kuala Lumpur, Malaysia
e-mail: shahanapriyatinii@gmail.com

N. Jha
KIST Medical College, Lalitpur, Nepal
e-mail: nishajha32@gmail.com

 
W. Leal Filho and F. Wolf (eds.), *Climate Change and Health: Perspectives from Developing Countries*, Climate Change Management,
https://doi.org/10.1007/978-3-031-93177-2_12

had methodological limitations. Studies from non- or under-represented regions and countries are required.

**Keywords** Developed countries · Climate change · Developing countries · Delivery of healthcare · Curriculum · Faculty · Perception

## Introduction

Climate change (CC) is an important threat facing humanity today (Costello et al. 2009), with significant health consequences due to its impact on access to food, water, air and clean environment (Intergovernmental panel on climate change 2014). CC can increase the risk of thermal injury and microbial proliferation and have deleterious consequences on nutrition and mental health (McMichael et al. 2006). Developing nations will be disproportionately affected by health effects due to heat, extreme weather events, air pollution, allergic diseases, infectious diseases, and malnutrition. Increasing temperatures is also likely to affect mental health (Padhy et al. 2015). Advocating for CC action is not easy due to the complexity of the problem and the wide range of stakeholders involved (Wabnitz et al. 2020). Health care practitioners (HCPs) can play an important role in educating the public about CC and planetary health (Maibach et al. 2015) and helping populations cope with the health effects of CC. Mountainous developing countries like Nepal may be more vulnerable.

Due to limited financial and health system resources, developing countries are at higher risk (Tong and Ebi 2019). Hence, this scoping review is done to obtain information on published original research on the knowledge, perception, attitude, and practice of climate change among healthcare students and practitioners with a focus on developing countries to plan educational and other interventions among this group.

## Methods

Arksey and O'Malley's methodological approach was used (Arksey and O' Malley 2005). The review was exempt from ethical review by the Institutional Review Board of KIST Medical College, Lalitpur, Nepal.

### *Databases and Search Strategy*

This review adhered to the Preferred Reporting Items for Systematic Reviews and Meta-Analyses (PRISMA) 2020 guidelines (Page et al. 2021) in conjunction with the PRISMA Extension for Scoping Reviews (PRISMA-ScR) (Tricco et al. 2018). The appendix describes the reporting of this review according to PRISMA-ScR. Between January 15, 2024, and June 21, 2024, two online databases—PubMed and Google Scholar—were searched for original research publications pertaining to the study topic. The key words used were 'Healthcare students', 'Healthcare professionals', 'Knowledge, attitude and practice', 'Perception', 'Climate change', 'One health' and 'Planetary health'. Further studies were obtained by examining the reference lists of articles obtained.

## *Peer-Reviewed Literature Screening*

The titles and abstracts were independently reviewed for eligibility. The inclusion and exclusion criteria used are shown below. As this scoping review focused on original publications, other types of publications were excluded.

Inclusion Criteria:

- Original Articles
- Published between 2008 and 2024
- English language articles
- Conducted among HCS and HCPs

Exclusion Criteria:

- Any article other than an original article (e.g. review, opinions, editorial, etc.)
- Articles published before 2008
- Non-English language articles
- Studies carried out in populations other than HCS and HCPs

## *Data Analysis*

A data extraction form was designed with title of article, year of publication, journal, research type, authors' country, number of authors, country studied, developing country or not, the quality checklist used and the quality of the study, respondent category, study type, outcomes measured, and interventions. The quality of the papers was independently assessed by two researchers using the AXIS Quality Instrument for quantitative research (Downes et al. 2016) and a quality checklist for qualitative studies (National Institute for Care and Health Excellence 2012). The average of the two scores was used. Discrepancies were discussed until a consensus was reached. Figure 12.1 shows the data charting process.

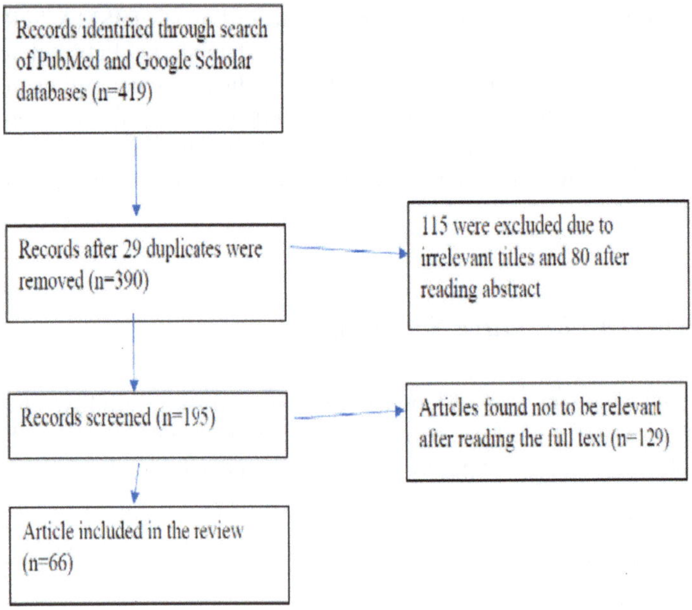


**Fig. 12.1** Charting process of the review

## Results and Discussion

Sixty-six studies on knowledge, perception, attitude, and practice of climate change and one health were included.

Table 12.1 shows the characteristics of the studies. Most were published during the last seven years. Most were quantitative studies done using questionnaires, and a large number were conducted in the US. Fifteen of the 27 countries studied (55.5%) were developing nations.

Knowledge was studied in 50 studies, attitudes in 47, and practice in 22 studies. Most studies examined multiple parameters.

### *The Quality of the Studies Included*

The quality of the studies was assessed using the instruments described previously. The maximum possible score for the AXIS checklist is 20 (Table 12.2).

Among the weaknesses noted were the use of non-standard questionnaires, low response rates, not mentioning the limitations of the study, and in some cases lack of information about ethical approval. Insufficient details were provided on methodology in some qualitative studies. The summary of the included studies is available at https://doi.org/10.6084/m9.figshare.25293343 (Shankar et al. 2024).

**Table 12.1** Characteristics of the studies included ($n = 66$)

| Characteristic | Number (Percentage) |
|---|---|
| *Years of publication* | |
| 2009–2016 | 18 (37.3) |
| 2017–2024 | 48 (72.7) |
| *Type of research* | |
| Quantitative | 48 (72.7) |
| Qualitative | 7 (10.6) |
| Mixed methods | 11 (16.7) |
| *Country of affiliation of authors* | |
| United States | 14 (21.2) |
| India | 8 (10.6) |
| Turkey | 4 (6.1) |
| Egypt | 9 (12.1) |
| Germany | 6 (9.1) |
| Nigeria | 2 (3) |
| China | 4 (6.1) |
| Sweden | 4 (6.1) |
| Spain | 3 (4.5) |
| Australia | 4 (6.1) |
| United Kingdom | 5 (7.6) |
| South Africa | 6 (9.1) |
| Nepal | 2 (3) |
| Canada | 2 (3) |
| New Zealand | 2 (3) |
| Kazakhstan | 1 (1.5) |
| Italy | 1 (1.5) |
| Iraq | 1 (1.5) |
| Palestine | 1 (1.5) |
| Saudi Arabia | 1 (1.5) |
| Philippines | 1 (1.5) |
| Kuwait | 1 (1.5) |
| France | 1 (1.5) |
| *Number of authors* | |
| One | 2 (3) |
| Two | 9 (13.6) |
| Three | 17 (25.7) |
| Four | 13 (19.7) |
| Five | 2 (3) |
| Six and more | 23 (34.8) |
| *Country where the study was done* | |

(continued)

**Table 12.1** (continued)

| Characteristic | Number (Percentage) |
|---|---|
| United States | 14 (21.2) |
| India | 8 (10.6) |
| China | 4 (6.1) |
| Egypt | 9 (12.1) |
| Turkey | 4 (6.1) |
| South Africa | 6 (9.1) |
| Germany | 6 (9.1) |
| Nigeria | 2 (3) |
| Sweden | 4 (6.1) |
| Australia | 4 (6.1) |
| United Kingdom | 5 (7.6) |
| Spain | 3 (4.5) |
| Nepal | 2 (3) |
| New Zealand | 2 (3) |
| Canada | 2 (3) |
| Kazakhstan | 1 (1.5) |
| Italy | 1 (1.5) |
| Iraq | 1 (1.5) |
| Palestine | 1 (1.5) |
| Ethiopia | 1 (1.5) |
| Saudi Arabia | 1 (1.5) |
| Philippines | 1 (1.5) |
| Jamaica | 1 (1.5) |
| Kuwait | 1 (1.5) |
| Chile | 1 (1.5) |
| Colombia | 1 (1.5) |
| Uruguay | 1 (1.5) |
| *Number of nations studied* | |
| One | 60 (90.9) |
| Multinational | 6 (9.1) |
| *Types of respondents* | |
| Nursing students | 17 (25.7) |
| Medical students | 18 (27.3) |
| Multiple professions students | 14 (21.2) |
| Health professionals | 22 (33.3) |
| Public health | 4 (6) |
| Faculty | 6 (9.1) |
| Nutrition professionals | 2 (3) |
| Pharmacy students | 2 (3) |
| *Intervention done along with the study* | |
| Yes | 5 (7.6) |
| No | 61(92.4) |

In certain cases, the total is more than 66 as multiple countries, and respondents were studied

**Table 12.2** Type of studies included in the review

| Score | Number of studies |
|---|---|
| *Quantitative* | |
| 18 | 3 |
| 17.5 | 1 |
| 17 | 2 |
| 16.5 | 5 |
| 16 | 9 |
| 15.5 | 7 |
| 15 | 7 |
| 14.5 | 4 |
| 14 | 4 |
| 13.5 | 6 |
| 13 | 2 |
| 12.5 | 2 |
| 12 | 1 |
| 11.5 | 4 |
| 11 | 3 |
| *Mixed-methods and qualitative* | |
| 2+ | 12 |
| 1+ | 5 |

The total number is greater than 66 as mixed methods studies had both a quantitative and qualitative score

## *Knowledge, Perception, Attitude, and Practice of Students and Faculty in Developed Nations of Climate Change and Planetary Health*

In the US, students were aware of the health impact of CC and wanted the topic to be incorporated into curricula (Breakey et al. 2023). The faculty surveyed, however, were not comfortable teaching the subject. A group of German researchers developed a questionnaire to study knowledge and attitudes toward CC (Bugaj et al. 2021). Future doctors were well-informed about the health consequences of CC but their perception about their responsibility to address CC was weaker. A multinational study conducted using the sustainability attitudes in nursing survey (SANS) questionnaire found students were aware of and had positive attitudes to introducing sustainability issues in the curriculum (Alvarez-Nieto et al. 2022). In Sweden, nursing students perceived their role as important to promote CC action and mentioned these issues should be better integrated into the curriculum (Anaker et al. 2021). In the US, clinical vignettes influenced fourth year medical students' opinions on CC (Prasad et al. 2011). Studies among veterinary students were less common. Veterinary students (US) believed that they should take a leadership role to tackle CC but reported a lack of educational opportunities within their programme (Pollard et al. 2021).

Medical, nursing and physician assistant students at Yale University (US) opined that the relationship between pollution, CC, and health should be addressed in their curricula (Ryan et al. 2020). Undergraduate nursing students in England and China had a positive attitude toward sustainability, but the Chinese students rated certain items higher in importance (Chen and Price 2020). Nearly 70% of nursing faculty in South Carolina (US) did not include health implications of sustainability in their teaching, while over 80% did not include the implications of CC (Amerson et al. 2022). In Montana (US), nursing faculty regarded flooding-related displacement as the most recognised CC impact, and other impacts mentioned were air quality related issues and extreme weather events (Streich 2014). In Ulm (Germany), a mandatory elective increased students' environmental knowledge and students were positive about the course (Muller et al. 2023). In Australia, pharmacy students lacked knowledge of environmentally sustainable pharmacy practice (ESPP), and only a few reported having curricular exposure to ESPP content (Chen et al. 2023). A study published in 2021 noted that only 56% of medical schools surveyed include One Health-related subject matter in their curricula. The challenges mentioned for incorporating One Health into medical school curricula include a packed curriculum and a lack of expertise in the subject (Docherty and Foley 2021). A selective course on 'The Impact of Climate Change on Human Health' was offered at Stanford University, US, resulting in a greater intention of students to change personal behaviors and to apply new knowledge in future clinical practice (Gomez et al. 2021). Bridges to Access (B2A): Health and Climate was a student-led interdisciplinary conference in the US with positive impacts on the participants' knowledge, confidence, and beliefs about climate change (Moon et al. 2022). At the University of New Mexico, US, an interprofessional education course on Climate Change and Public Health Preparedness supported students' professional interest and action competence (Rogers et al. 2023). Students' knowledge and perceived abilities improved as they progressed through the course.

Overall, respondents from developed nations had good knowledge, but their practice required improvement. They wanted CC and one health to be included in the curriculum.

## *Knowledge, Perception, Attitude and Practice of Students and Faculty in Developing Nations of Climate Change, One Health, and Planetary Health*

In the US, only 44 of the 90 accredited public health institutions were found to offer a climate change-related course to their students (Arora et al. 2023).

A study conducted in four universities found nursing students exhibited positive attitudes toward CC, and Saudi Arabian students exhibited the most positive attitudes (Cruz et al. 2018). In China, most medical students were aware of the health impacts of CC and embraced their role in addressing the same (Liao et al. 2019). Students of preventive medicine were more likely to perceive their responsibility, and respondents were in favour of incorporating CC into the curricula. In China, knowledge of the causes of CC was a significant predictor of increased awareness of the negative impacts of CC among medical and nursing students, but not among public health students (Yang et al. 2018). In Ethiopia, respondents had insufficient knowledge to understand the public health impacts of CC, and the use of electronic mass media was associated with greater knowledge (Nigatu et al. 2014). In India, poor CC knowledge was noted among medical students and postgraduates, and most did not have environment-friendly practices (Reddy et al. 2022).

In a Turkish Faculty of Medicine, students of lower grades and females had higher CC knowledge (Deveci et al. 2019). In Nigeria, all students were aware of CC and its effect on health and regarded the causes as man-made, but only 34% acted to mitigate the consequences (Nzeobi et al. 2020). In Egypt, most of the nursing students studied had satisfactory knowledge, negative attitudes, and inadequate CC practices (Abdel Nabi et al. 2023). The authors recommended implementing an CC educational programme. A health promotion programme regarding environmental literacy and CC health risk among nursing students improved their knowledge and practice scores (Lotfy Ahmed et al. 2023).

A mixed-methods study among Turkish nursing students found most knew the main cause of CC, and the score was higher among environmentally active students (Ergin et al. 2021). Turkish student nurses embraced their role in responding to CC but believed they were not adequately prepared (Tuna et al. 2022). In the Philippines, medical students had fair to good CC knowledge, and favourable attitudes (Domantay et al. 2021). Health science students in the Pokhara Valley, Nepal had unsatisfactory knowledge, and many were unaware of the possible impact of CC (Chalise et al. 2015). At a rural university in South Africa, students had low knowledge of CC, and the authors mentioned that if their knowledge can be strengthened, they can make a positive contribution to their communities' climate resilience (Mugambiwa and Dzomonda 2018). A study conducted among five medical universities in China showed that most students were aware of the health impacts of climate change, but they were not ready to respond to the same (Liao et al. 2019).

Medical and paramedical students in Tamil Nadu, India had moderate CC knowledge but poor attitude (Nijamudeen et al. 2020). In Egypt, an awareness programme improved total knowledge level, positive perception, and practices of nursing students regarding CC (Abdallah and Wagdy Farag 2022). At the University of Cape Town, South Africa, health science students had little understanding about CC, and their main source of information was the internet and social media; nearly half wanted the curriculum to address this topic (Irlam et al. 2023). In Nigeria, 73% of undergraduate medical students believed that they contributed to global warming, although only 9% of respondents were in favour of using public transport (Isah et al. 2011). In India, students mentioned human activities contributed to CC, and their information sources was newspapers and magazines (Pandve and Raut 2011). Also in India, interns had good knowledge of the consequences of CC but limited knowledge of health protection measures (Majra and Acharya 2009). In a medical college in Navi Mumbai (India), faculty were aware of global warming and mentioned it was due to man-made causes (Nalgundwar et al. 2014).

Most health science students in Udupi, India had a favourable attitude and believed CC will have a significant health impact with a positive correlation between knowledge and attitudes (Rynjah 2020). In Iran, students had a moderate level of environmental knowledge, and there was a significant difference between males and females (Hemayatkhah et al. 2018). Egyptian nursing students had poor CC knowledge and practice, and less than one-third agreed that these issues should be included in the curriculum (Mohammed et al. 2022). In Kathmandu, Nepal, public health students believed that CC has the greatest effect on deprived and vulnerable populations and that skin cancer, heat stroke, infections, malaria, and cardiovascular diseases could result from CC (Mahato et al. 2014).

In Egypt, among nursing school faculty, climate anxiety was strongly associated with environmental attitudes and job engagement. (Atta et al. 2024). At Ain Shams medical faculty, Egypt, medical students had good CC perception and its importance in the curriculum (Momen et al. 2024). A qualitative study among HCS in Kazakhstan highlighted interprofessional collaboration in addressing CC (Balay-Odao et al. 2024). In an Egyptian university, there were significant differences in knowledge, perception, attitudes, and practice among nursing students after an educational intervention (Mohammed et al. 2024).

There were more studies from developing countries. The knowledge ranged from poor to moderate. Some studies showed poor attitudes, and there were deficiencies in practice. Students also wanted CC to be included in the curriculum.

## *Knowledge, Perception, Attitude, and Practice of HCPs in Developed Nations on Climate Change, One Health and Planetary Health*

In Rome, Italy, correct answers were more likely to be provided by professionals and varied depending on the respondents' location (La Torre et al. 2023). Television was the most common information source. A multinational study among HCPs in both developed and developing countries found that respondents understood CC is happening, is caused by humans, viewed it as an important and growing cause of health harm, and felt a responsibility to educate the public and policymakers about the problem. (Kotcher et al. 2021). Many participants indicated that personal, professional, and societal barriers impede their involvement. In Canada family physicians opined a 3-h, electronic, continuing medical education (eCME) activity covering CC topics would be useful, allowing them to use the knowledge in their practice (Valois et al. 2016).

In the US, African American physicians mentioned that CC was directly relevant to patient care and that many of their own patients were moderately or severely affected (Sarfaty et al. 2014). American Thoracic Society members believed that CC is happening and is driven by human activity (Sarfaty et al. 2015). Physician organisations should play an important role in educating the public, patients, and policy makers. Similar results were noted in Wisconsin (US), but only about 17% were comfortable counselling their patients (Boland and Temte 2019). A study among Swedish nurses showed an incongruence between climate and environmental issues on one hand and the nurses daily work and the belief that public health work can mitigate CC on the other hand (Anaker et al. 2015). Nursing directors of US health departments viewed the environment as under threat and saw a role for nurses in addressing it (Polivka et al. 2012).

International members of the American Thoracic Society mentioned CC is happening, driven by man-made factors, and is relevant to patient care (Sarfaty et al. 2016). They noted an increase in chronic disease severity, allergies, and weather-related injuries. American Society of Anaesthesiologists members reported recycling practices in their operating rooms, but most did not have information about hospital-wide sustainability practices (Ard Jr et al. 2016). General practitioners in Baden-Wurttemberg, Germany knew about heat health impacts, especially on cardiovascular morbidity and volume imbalances (Herrmann and Sauerborn 2018). Among Basque health workers in Spain, moderate levels of environmental health (EH) knowledge and skills were reported, and nurses and nursing students reported higher EH skills than other professionals (Subiza-Perez et al. 2024). In Germany, approximately 70% of caregiving professionals surveyed considered impacts on the mental health of children and adolescents due to extreme weather events as important (Hieronimi et al. 2024).

In Germany, stakeholders of outpatient practices had a mixed degree of awareness, ranging from a passive role, handing over responsibility to others, and a low perceived self-efficacy to a proactive implementation of adaptation strategies (Litke et al. 2024). Australian psychologists were not prepared to identify and manage mental health presentations related to CC and were not engaged in CC mitigation (Stilita and Charlson, 2024). Most UK researchers strongly supported climate advocacy. However, a high workload, uncertainty about what actions to take, a perceived lack of agency or power, inflexible university processes, and pressure to travel were some of the barriers mentioned (Latter et al. 2024). In the UK, groups connected with epilepsy had a high level of concern about CC impacts and positive attitudes toward climate action (Blenkinsop et al. 2024). Most rehabilitation professionals within the US and elsewhere understood that CC impacts human health and thought actions should be taken; however, they also perceive barriers to their own involvement (Campbell et al, 2023).

In developed nations, HCPs were aware of CC and its impact on their patients' health. They wanted CMEs on CC and planetary health. There were more studies among HCPs in developed countries.

### *Knowledge, Perception, Attitude and Practice of HCPs in Developing Nations of Climate Change, One Health and Planetary Health*

Among health professionals working at the Chinese Centre for Disease Control and Prevention (CDC) a gap between perceptions and actions regarding CC was seen (Wei et al. 2014). Local guidelines and action to balance CC and local sustainable development were highlighted. Randomly selected nurses in central China were aware of CC but did not know that their work and CC mutually influenced each other (Xiao et al. 2016). Education on this topic should be strengthened in Chinese universities. In Ghana, a study among the public and health experts on the linkage between CC and health (Hussey and Arku, 2019) found disease was the top concern, but the public is less likely to link disease to CC when compared to experts. Among key stakeholders in Puducherry, India, a knowledge and education gap on the health impacts of CC and a need for multi-level intervention and awareness programmes was noted (Shrikhande et al. 2023). In South Africa, health workers—mainly doctors—had a reasonable knowledge of CC and its possible health impact and recommended improvements in environmental health education (Manga et al. 2022). In Turkey, nutrition professionals' CC knowledge and awareness levels were correlated and significantly affected each other (Ozel et al. 2022). In South Africa, HCPs had overwhelmingly positive attitudes and a high degree of interest in education, implementation, and taking more environmental sustainability responsibilities, but their knowledge was poor (Lister et al. 2022). These studies found reasonable levels of knowledge, but CMEs were not mentioned.

## Limitations

The authors searched using two commonly available databases. They examined the reference list of the studies identified to look for other studies. It is possible that not all studies were included. The analysis was done manually. Studies from before 2008 were excluded. Only English language studies were included. A few studies were conducted in several countries, and the countries were not completely listed.

## Recommendations

More global studies are required. The knowledge and perception of HCS and HCPs were studied at a basic level, and so studies assessing these at a higher level and how they will address CC health issues among the community and patients are required. The quality of the studies may need strengthening. Different studies had used different questionnaires, so developing standards for the methodology of these studies can be considered. CC consequences at a regional and local level can be explored. Most students wanted CC and planetary health to be included in the curriculum, and the incorporation of these topics requires careful planning. Studies from under and non-represented countries and regions are necessary. A systematic review of studies in this area can be considered.

## Conclusions

Studies showed poor to moderate knowledge, poor to good attitudes, and weak practices regarding CC and one health. The methodological aspects of the studies may need to be strengthened. Most respondents wanted these topics to be included in the curriculum (in case of HCS) and as CME (for HCPs). These studies can be used to develop educational interventions. Further studies among HCPs in developing nations should be a priority. The studies among HCS were predominantly from India, Turkey, China, Egypt, and Nigeria. Studies from other countries are required.

## Appendix

Preferred Reporting Items for Systematic reviews and Meta-Analyses extension for Scoping Reviews (PRISMA-ScR) Checklist.

| Section | Item | Prisma-ScR checklist item | Reported on page # |
|---|---|---|---|
| *Title* | | | |
| Title | 1 | Identify the report as a scoping review | 1 |
| *Abstract* | | | |
| Structured summary | 2 | Provide a structured summary that includes (as applicable): background, objectives, eligibility criteria, sources of evidence, charting methods, results, and conclusions that relate to the review questions and objectives | 1 |
| *Introduction* | | | |
| Rationale | 3 | Describe the rationale for the review in the context of what is already known. Explain why the review questions/objectives lend themselves to a scoping review approach | 1 and 2 |

(continued)

(continued)

| Section | Item | Prisma-ScR checklist item | Reported on page # |
|---|---|---|---|
| Objectives | 4 | Provide an explicit statement of the questions and objectives being addressed with reference to their key elements (e.g., population or participants, concepts, and context) or other relevant key elements used to conceptualise the review questions and/or objectives | 2 |
| *Methods* | | | |
| Protocol and registration | 5 | Indicate whether a review protocol exists; state if and where it can be accessed (e.g., a Web address); and if available, provide registration information, including the registration number | Not registered |
| Eligibility criteria | 6 | Specify characteristics of the sources of evidence used as eligibility criteria (e.g., years considered, language, and publication status), and provide a rationale | 2 and 3 |
| Information sources | 7 | Describe all information sources in the search (e.g., databases with dates of coverage and contact with authors to identify additional sources), as well as the date the most recent search was executed | 2 and 3 |
| Search | 8 | Present the full electronic search strategy for at least 1 database, including any limits used, such that it could be repeated | 2 |
| Selection of sources of evidence | 9 | State the process for selecting sources of evidence (i.e., screening and eligibility) included in the scoping review | 2 and 3 |
| Data charting process | 10 | Describe the methods of charting data from the included sources of evidence (e.g., calibrated forms or forms that have been tested by the team before their use, and whether data charting was done independently or in duplicate) and any processes for obtaining and confirming data from investigators | 3 |
| Data items | 11 | List and define all variables for which data were sought and any assumptions and simplifications made | 3 |
| Critical appraisal of individual sources of evidence | 12 | If done, provide a rationale for conducting a critical appraisal of included sources of evidence; describe the methods used and how this information was used in any data synthesis (if appropriate) | 3 and 4 |
| Synthesis of results | 13 | Describe the methods of handling and summarising the data that were charted | 3 |
| *Results* | | | |

(continued)

(continued)

| Section | Item | Prisma-ScR checklist item | Reported on page # |
|---|---|---|---|
| Selection of sources of evidence | 14 | Give numbers of sources of evidence screened, assessed for eligibility, and included in the review, with reasons for exclusions at each stage, ideally using a flow diagram | 3 |
| Characteristics of sources of evidence | 15 | For each source of evidence, present characteristics for which data were charted and provide the citations | 4 to 6 |
| Critical appraisal within sources of evidence | 16 | If done, present data on the critical appraisal of included sources of evidence (see item 12) | 3 and 4 |
| Results of individual sources of evidence | 17 | For each included source of evidence, present the relevant data that were charted that relate to the review questions and objectives | 4 to 6 |
| Synthesis of results | 18 | Summarise and/or present the charting results as they relate to the review questions and objectives | 3 and 4 |
| *Discussion* | | | |
| Summary of evidence | 19 | Summarise the main results (including an overview of concepts, themes, and types of evidence available), link to the review questions and objectives, and consider the relevance to key groups | 4 to 6 |
| Limitations | 20 | Discuss the limitations of the scoping review process | 6 |
| Conclusions | 21 | Provide a general interpretation of the results with respect to the review questions and objectives, as well as potential implications and/ or next steps | 7 |
| *Funding* | | | |
| Funding | 22 | Describe sources of funding for the included sources of evidence, as well as sources of funding for the scoping review. Describe the role of the funders of the scoping review | 7 |

## *References*

Abdallah ZA, Wagdy Farag AA (2022) Impact of awareness program regarding health consequences of climate change on knowledge, perception and daily life practices among nursing students. Egypt J Nurs Health Sci 3(1):367–390

Abdel Nabi E, Shafik SA, Saad AM (2023) Assessment of nursing students' awareness regarding climate change. Helwan Int J Nurs Res Pract 2(1):69–82

Álvarez-Nieto C, Richardson J, Navarro-Perán MÁ et al (2022) Nursing students' attitudes towards climate change and sustainability: a cross-sectional multisite study. Nurse Educ Today 108:105185

Amerson RM, Boice O, Mitchell H, Bible J (2022) Nursing faculty's perceptions of climate change and sustainability. Nurs Educ Perspect 43(5):277–282

Anaker A, Nilsson M, Holmner A, Elf M (2015) Nurses' perceptions of climate and environmental issues: a qualitative study. J Adv Nurs 71(8):1883–1891

Anåker A, Spante M, Elf M (2021) Nursing students' perception of climate change and sustainability actions–a mismatched discourse: a qualitative, descriptive exploratory study. Nurse Educ Today 105:105028

Ard JL Jr, Tobin K, Huncke T, Kline R, Ryan SM, Bell C (2016) A survey of the American Society of Anesthesiologists regarding environmental attitudes, knowledge, and organization. A&a Practice 6(7):208–216

Arksey H, O'Malley L (2005) Scoping studies: towards a methodological framework. Int J Soc Res Methodol 8(1):19–32

Arora M, Comrie AC, Ernst KE (2023) Assessing climate and health curriculum in graduate public health education in the United States. Front Public Health 11:1124379

Atta MHR, Zoromba MA, El-Gazar HE et al (2024) Climate anxiety, environmental attitude, and job engagement among nursing university colleagues: a multicenter descriptive study. BMC Nurs 23(1):133

Balay-Odao EM, Colet PC, Almazan JU et al (2024) Environmental sustainability in healthcare: a qualitative study of the perspectives of nursing, medical and public health students in Kazakhstan. Nurse Educ Pract 76:03917

Blenkinsop S, Wardrope A, Willis J, Sisodiya SM (2024) Climate change: attitudes and concerns of, and learnings from, people with neurological conditions, carers, and health care professionals. Epilepsia 65(1):95–106

Boland TM, Temte JL (2019) Family medicine patient and physician attitudes toward climate change and health in Wisconsin. Wilderness Environ Med 30(4):386–393

Breakey S, Starodub R, Nicholas PK, Wong J (2023) A cross-sectional study to assess faculty and student knowledge of climate change and health: Readiness for curricular integration. J Adv Nurs 79(12):4716–4731

Bugaj TJ, Heilborn M, Terhoeven V et al (2021) What do final year medical students in Germany know and think about climate change?—The climattitude study. Med Educ Online 26(1):1917037

Campbell E, Alexander M, Kotcher J, Maibach E (2023) Rehabilitation medicine professionals' views on climate change and health. J Int Soc Phys Rehabil Med 6(4):90–95

Chalise M, Pathak G, Parajuli S, Chalise AM, Thakur P, Chauhan HS (2015) Knowledge on climate change and its impact in human health among health sciences students of Pokhara Valley Nepal. Int J Health Sci Res. 6(1):362–370

Chen MJ, Price AM (2020) Comparing undergraduate student nurses' understanding of sustainability in two countries: a mixed method study. Nurse Educ Today 88:104363

Chen EY, Forrester C, McEvoy AM, Singleton J (2023) Pharmacy students' perceptions on environmental sustainability in pharmacy education and practice. Explor Res Clin Soc Pharm 12:100366

Costello A, Abbas M, Allen A et al (2009) Managing the health effects of climate change. Lancet 373:1693–1733

Cruz JP, Felicilda-Reynaldo RF, Alshammari F et al (2018) Factors influencing Arab nursing students' attitudes toward climate change and environmental sustainability and their inclusion in nursing curricula. Public Health Nurs 35(6):598–605

Deveci SE, Kurt O, Pirinçci E, Oğuzöncül AF (2019) Knowledge and awareness of a medical faculty students in Turkey about global warming, climate change and their consequences. J Hum Sci 16(2):679–689

Docherty L, Foley PL (2021) Survey of One Health programs in US medical schools and development of a novel one health elective for medical students. One Health 12:100231
Domantay JA, Leochico CF, Salvador PA et al (2021) Knowledge and attitudes of future physicians in the Cordillera region of the Philippines towards climate change: a pre-pandemic cross-sectional study. J Clim Chang Health 4:100063
Downes MJ, Brennan ML, Williams HC, Dean RS (2016) Development of a critical appraisal tool to assess the quality of cross-sectional studies (AXIS). BMJ open 6(12)
Ergin E, Altinel B, Aktas E (2021) A mixed method study on global warming, climate change and the role of public health nurses from the perspective of nursing students. Nurse Educ Today 107:105144
Gomez J, Goshua A, Pokrajac N et al (2021) Teaching medical students about the impacts of climate change on human health. J Clim Chang Health 3:100020
Hemayatkhah M, Rahmanian V, Mansoorian E (2018) Evaluating the level of environmental behaviors among students at Jahrom University of Medical Sciences, 2016–2017. J Environ Health Sustain Dev 3(3):567–577
Herrmann A, Sauerborn R (2018) General practitioners' perceptions of heat health impacts on the elderly in the face of climate change—a qualitative study in Baden-Württemberg, Germany. Int J Environ Res Public Health 15(5):843
Hieronimi A, O'Reilly F, Schneider M et al (2024) A Germany-wide survey of caregiving professionals on climate change and mental health of children and adolescents-factors influencing their relevance rating of extreme weather event associated mental health impairments. BMC Pub Health 24(1):120
Hussey LK, Arku G (2019) Conceptualizations of climate-related health risks among health experts and the public in Ghana. Social Sci Med. 223:40–50
IPCC 2014 (2014) 2014-Impacts, adaptation and vulnerability: regional aspects. In: Contribution of working group III to the fifth assessment report of the intergovernmental panel on climate change. Cambridge, United Kingdom and New York, NY, USA. Cambridge University Press
Irlam J, Razzak Z, Parker QA, Rother HA (2023) Student knowledge and perceptions of climate change and environmental sustainability at the faculty of health sciences, University of Cape Town, South Africa. Afr J Health Prof Educ 15(1):2–6
Isah YC, Idolor EE, Ofili AN, Isah EC (2011) Awareness and perception of global warming among undergraduate medical students in a Nigerian University. J Community Med Prim Health Care 23(1–2):32–40
Kotcher J, Maibach E, Miller J et al (2021) Views of health professionals on climate change and health: a multinational survey study. Lancet Planet Health 5(5):e316–e323
La Torre G, Sestili C, Cocchiara RA, Barbato D, Mannocci A, Del Cimmuto A (2023) Knowledge and perception about climate change among healthcare professionals and students: a cross-sectional study. South East Eur J Public Health 13
Latter B, Demski C, Capstick S (2024) Wanting to be part of change but feeling overworked and disempowered: researchers' perceptions of climate action in UK universities. PLOS Clim 3(1):e0000322
Liao W, Yang L, Zhong S et al (2019) Preparing the next generation of health professionals to tackle climate change: are China's medical students ready? Environ Res 168:270–277
Lister HE, Mostert K, Botha T et al (2022) South African healthcare professionals' knowledge, attitudes, and practices regarding environmental sustainability in healthcare: a mixed-methods study. Int J Environ Res Public Health 19(16):10121
Litke NA, Poß-Doering R, Fehrer V et al (2024) Building climate resilience: awareness of climate change adaptation in German outpatient medical practices. BMC Health Serv Res 24(1):184
Lotfy Ahmed A, Abd Elftah Mohamed N, Sobhy Hassan G (2023) Effect of health promotion program regarding environmental literacy and climate change health risks among newly nursing students. Egypt J Health Care 14(3):561–578
Mahato S, Singh R, Maharjan K (2014) Perception of public health students of Kathmandu, Nepal towards effect of climate change. Health Sci Res 1(4):88–92

Maibach EW, Kreslake JM, Roser-Renouf C et al (2015) Do Americans understand that global warming is harmful to human health? Evidence from a national survey. Ann Glob Health 81(3):396–409

Majra JP, Acharya D (2009) Protecting health from climate change: preparedness of medical interns. Indian J Community Med. 34(4):317

Manga A, Dartchiev D, Variava E (2022) Healthcare and climate change: a South African health professionals' perspective. Wits J Clin Med 4(3):173–177

McMichael AJ, Woodruff RE, Hales S (2006) Climate change and human health: present and future risks. Lancet 367:859Y869

Mohammed E, El-Mouty A, Ameen N (2022) Nursing students knowledge, attitude, and practice regarding health effect of climate change. Mansoura Nurs J 9(2):589–601

Mohammed EGF, Fahmy LM, Megahed FIAL (2024) Effect of educational program regarding climate change on nursing students' awareness, attitude and practices in Suez Canal University. Int Egypt J Nurs Sci Res 4(2):297–316

Momen M, Abdellatif AM, Yousef SS (2024) Perception of climate change at ain shams medical campus: a survey of Egyptian medical students. Egypt J Community Med 42(1):22–29

Moon AB, Xie W, Suh JH, Moritz BW (2022) Bridges to access 2022: the impact of an interdisciplinary, educational conference about climate-related health effects and climate disparities on pre-health and health professional student knowledge, beliefs, and confidence. J Clim Chang Health. 8:100170

Müller L, Kühl M, Kühl SJ (2023) Climate change and health: changes in student environmental knowledge and awareness due to the implementation of a mandatory elective at the medical faculty of Ulm? GMS J Med Educ 40(3)

Mugambiwa SS, Dzomonda O (2018) Climate change and vulnerability discourse by students at a South African University. Jàmbá. 10(1):1–6

Nalgundwar AN, Velankar DH, Joshi SM, Mankar MJ, Pinto VN (2014) A study of awareness and lifestyle related practices regarding global warming among the teaching staff of a medical college in Navi Mumbai. Med J Dr. DY Patil Univ 7(6):714–716

National Institute for Health and Care Excellence (2012) Appendix H quality appraisal checklist—qualitative studies. https://www.nice.org.uk/process/pmg4/chapter/appendix-h-quality-appraisal-checklist-qualitative-studies#footnote_17

Nigatu AS, Asamoah BO, Kloos H (2014) Knowledge and perceptions about the health impact of climate change among health sciences students in Ethiopia: a cross-sectional study. BMC Pub Health 14(1):1

Nijamudeen SS, Nisha B, Dutta R, Jain T (2020) Awareness of climate change and its impact on public health among medical and paramedical students in a teaching hospital of Tamil Nadu, Ann TropMed Public Health 23(S22):SP2323111

Nzeobi JR, Chineke HN, Ubajaka CF, Adogu PO (2020) Knowledge of health impact of climate change and practice of preventive measures among students of a Nigerian Tertiary Institution. Asian J Environ Ecol 13(4):77–87

Özel M, Bogueva D, Marinova D, Tekiner IH (2022) Climate change knowledge and awareness of nutrition professionals: a case study from Turkey. Sustainability 14(7):3774

Padhy SK, Sarkar S, Panigrahi M, Paul S (2015) Mental health effects of climate change. Indian J Occup Environ Med 19(1):3

Page MJ, McKenzie J, Bossuyt P et al (2021) The PRISMA 2020 statement: an updated guideline for reporting systematic reviews. Int J Surg 88:10590

Pandve HT, Raut A (2011) Assessment of awareness regarding climate change and its health hazards among the medical students. Indian J Occup Environ Med 15(1):42

Polivka BJ, Chaudry RV, Mac CJ (2012) Public health nurses' knowledge and attitudes regarding climate change. Environ Health Perspect 120(3):321–325

Pollard AE, Rowlison DL, Kohnen A et al (2021) Preparing veterinarians to address the health impacts of climate change: student perceptions, knowledge gaps, and opportunities. J Vet Med Educ 48(3):343–350

Prasad V, Thistlethwaite W, Dale W (2011) Effect of clinical vignettes on senior medical students' opinions of climate change. South Med J 104:401–404

Reddy GP, Rajamouli J, Arora KD, Jothula KY, Amaravadi S, Boda A (2022) Knowledge, perceptions and practices of medical students towards climate change and global warming: a cross sectional study. J Family Med Pri Care. 11(6):2557

Rogers HH, Tucker M, Couig MP, Svihla V (2023) Facilitating an interprofessional course on climate change and public health preparedness. Int J Environ Res Public Health 20(10):5885

Ryan EC, Dubrow R, Sherman JD (2020) Medical, nursing, and physician assistant student knowledge and attitudes toward climate change, pollution, and resource conservation in health care. BMC MedEduc 20:1–4

Rynjah B (2020) A descriptive study to assess the knowledge and attitude towards global warming and perceived impacts of regional climate change on health among the health science students of Udupi District, Karnataka. Manipal College of Nursing, Manipal theses and dissertations, 32. https://impressions.manipal.edu/mcon/32

Sarfaty M, Mitchell M, Bloodhart B, Maibach EW (2014) A survey of African American physicians on the health effects of climate change. Int J Environ Res Public Health 11(12):12473–12485

Sarfaty M, Bloodhart B, Ewart G et al (2015) American Thoracic Society member survey on climate change and health. Ann Am Thorac Soc 2(2):274–278

Sarfaty M, Kreslake J, Ewart G et al (2016) Survey of international members of the American Thoracic Society on climate change and health. Ann Am Thorac Soc 13(10):1808–1813

Shankar PR, Guragain A, Ganesan S, Jha N (2024) Knowledge of HPE students, faculty and practitioners on climate change. figshare. Dataset. https://doi.org/10.6084/m9.figshare.25293343.v1

Shrikhande SS, Merten S, Cambaco O et al (2023) Climate change and health?: Knowledge and perceptions among key stakeholders in Puducherry, India. Int J Environ Res Public Health 20(6):4703

Stilita G, Charlson F (2024) Keeping Sane in a changing climate: assessing psychologists' preparedness, exposure to climate-health impacts, willingness to act on climate change, and barriers to effective action. Int J Environ Res Public Health 21(2):218

Streich JL (2014) Nursing faculty's knowledge on health impacts due to climate change. Doctoral dissertation, Montana State University-Bozeman, College of Nursing. https://scholarworks.montana.edu/xmlui/handle/1/9140

Subiza-Pérez M, Vrotsou K, Esnal H et al (2024) Environmental health knowledge and competences in Basque health workers. A comparison of different professional profiles. Environ Res 243:117789

Tong S, Ebi K (2019) Preventing and mitigating health risks of climate change. Environ Res 174:9–13

Tricco AC, Lillie E, Zarin W et al (2018) PRISMA extension for scoping reviews (PRISMA-ScR): checklist and explanation. Ann Intern Med 169:467–473

Tuna HI, Tuna PT, Molu B, Keskin AY (2022) Determination of nursing students' awareness of the health effects of climate change. Int J Caring Sci 15(2):1149–1156

Valois P, Blouin P, Ouellet C, Renaud JS, Bélanger D, Gosselin P (2016) The health impacts of climate change: a continuing medical education needs assessment framework. J Contin Educ Health Prof 36(3):218–225

Wabnitz KJ, Gabrysch S, Guinto R et al (2020) A pledge for planetary health to unite health professionals in the Anthropocene. Lancet 396(10261):1471–1473

Wei J, Hansen A, Zhang Y, Li H, Liu Q, Sun Y, Bi P (2014) Perception, attitude and behavior in relation to climate change: a survey among CDC health professionals in Shanxi province. China Environ Res 134:301–308

Xiao J, Fan W, Deng Y, Li S, Yan P (2016) Nurses' knowledge and attitudes regarding potential impacts of climate change on public health in central of China. Int J Nurs Sci 3(2):158–161

Yang L, Liao W, Liu C, Zhang N, Zhong S, Huang C (2018) Associations between knowledge of the causes and perceived impacts of climate change: a cross-sectional survey of medical, public health and nursing students in universities in China. Int J Environ Res Public Health 5(12):2650

# Chapter 13
# Defining Research Priorities on Climate-Sensitive Infectious Disease Justice for South America

**Romina Rekers, Marcelo de Araujo, Timothy Daly, Pedro Fior Mota de Andrade, María Victoria Gerbaldo, Euzebiusz Jamrozik, Yasna Palmeiro Silva, Carlos Yabar, and Florencia Luna**

**Abstract** Setting research priorities helps to optimise responses to climate-sensitive infectious diseases. When it comes to infectious disease justice, it is essential to identify the just paths for building resilience. This is particularly challenging in the area of climate-sensitive infectious diseases, in which available adaptation paths differ depending on epidemic evolution and future mitigation scenarios. Moreover, due to the communicable nature of infectious diseases and poorly controlled borders, transformative changes are necessary for controlling climate-sensitive infectious diseases in regions such as South America (SA). Attributable to the particularities of this continent, some containment measures, such as lockdowns or school closures, can create unjust burdens, as the case of the COVID-19 pandemic has shown. This study shows the outcome of a philosophical conversation aimed to identify the regional

---

R. Rekers (✉)
University of Graz, Graz, Austria
e-mail: romina.rekers@uni-graz.at

M. de Araujo · P. F. M. de Andrade
State University of Rio de Janeiro, Rio de Janeiro, Brazil

T. Daly · F. Luna
Bioethics programme, FLACSO, Buenos Aires, Argentina

M. V. Gerbaldo
National University of Cordoba, Cordoba, Argentina

E. Jamrozik
Ethox Centre, Oxford, UK

Y. P. Silva
University College London, London, UK

C. Yabar
Universidad de San Martín de Porres, Lima, Peru

 
W. Leal Filho and F. Wolf (eds.), *Climate Change and Health: Perspectives from Developing Countries*, Climate Change Management,
https://doi.org/10.1007/978-3-031-93177-2_13

background conditions that make climate-sensitive infectious disease justice in SA necessary and possible.

**Keywords** Research priorities · Climate-sensitive infectious diseases · Justice · Climate adaptation · South America

## Introduction

The Global South will be most affected by anthropogenic climate change, for which it is, historically speaking, hardly responsible (Hickel 2020). SA countries have collectively contributed less than 10% of historical GHG emissions (Bárcena-Ibarra 2018), and the region's consumption-based $CO_2$ per capita emissions remain below the global average (Global Carbon Project 2021). At the same time, three SA countries (Brazil, Chile, and Colombia) are among the ten countries with the highest forest GHG removals (Harris et al. 2021). Paradoxically, deforestation in SA, attributed mainly to the expansion and intensification of agriculture and cattle ranching, continues to be one of the largest global sources of gross forest-related emissions (Harris et al. 2021). SA forests also play an important role in climate adaptation as they influence vector–pathogen-host interactions (Castellanos et al. 2022). Much of the forested area in SA protected territories belonging to indigenous communities who face violence from both state and non-state actors aiming to reduce their land control (Global Witness 2023).

Changes in climate influence the spread and timing of infectious disease (Romanello et al. 2022). While the epidemic burdens in SA tend to be high mainly due to VBDs (Mosegui et al. 2022), climate change increases the climatic suitability for infectious diseases (Rocklöv and Dubrow 2020) extending their geographical and seasonal range (López et al. 2018). The worst impacts affect countries with low adaptive capacity where original endemicity is high and control status is poor. The last dengue outbreak in Brazil, Argentina and other countries in the region is an example of regional public health impacts that are attributed partly to climate change (Dias and McCoy 2024). That is why the challenges posed by climate change cannot be treated in isolation from global health justice. The intersection between climate justice and health justice is especially relevant in regions such as SA, where climate impacts on health pose high risks due to overlapping vulnerabilities and the related exposure paths of different populations.

Setting research priorities in the area of climate-sensitive infectious diseases is required for developing evidence-based climate-health strategies for SA. It is essential for identifying the just paths for building resilience to climate-sensitive diseases and contributes to maximising the opportunities for successful adaptation in the face of three challenges. Firstly, controlling infectious diseases requires a demanding kind of cooperation due to their communicable nature and the porous borders for people and goods that are poorly controlled in SA. Secondly, some control measures, such as lockdowns or school closures, can lead to unjustified burdens, as the case of the COVID-19 pandemic has shown (Rekers and Luna 2023). This increases the necessity for developing less burdensome containment measures. Finally, in the area of infectious diseases, available adaptation paths change depending on epidemic evolution and under different future climate scenarios.

## Methodology

This study identifies the background conditions that should shape the research agenda on climate-sensitive infectious disease justice in SA. According to the view that justice is a problem-responsive value, identifying the background conditions that make justice necessary and possible is a preliminary task for developing justice conceptions (Hume 1978, 1998). In this sense, the conditions that make climate-sensitive infectious disease justice necessary and possible play a relevant role in the identification of the principles and rules that should guide the setting of research priorities. Reflective equilibrium is the methodology that we adopted for this task. Frequently, reflective equilibrium is a methodology that a philosopher uses alone at a desk. Recently, some philosophers have proposed to open the reflective equilibrium to incorporate public views (Wolff and De-Shalit 2023) and experts from different disciplines (Savulescu et al. 2021).

This article identifies the background conditions that define the role of climate-sensitive infectious disease justice in SA. We conducted a philosophical conversation inspired by open forms of reflective equlibrium including authors from different disciplines and regions that took place in the "Climate change and health in South America—defining research priorities" workshop (2023 Oxford Global Health & Bioethics Conference). The conversation is structured around three questions around climate-health adaptation: (1) What are the main justice challenges for adapting to climate-sensitive infectious diseases in SA? (2) What should be the research priorities for climate-sensitive infectious disease justice? (3) How can SA contribute to the global health justice challenge posed by climate-sensitive infectious diseases?

## Results and Discussion

### *Challenges for Climate-Sensitive Infectious Disease Justice in South America*

We identified the following background conditions that constrain adaptation, limit its transformational potential, and increase the probability of maladaptation to climate-sensitive infectious disease in SA.

I. **Lack of awareness about the relationship between climate change and health impacts**

The COVID-19 pandemic has worsened this problem. It affected education, which is crucial for climate-health adaptation capacity at the collective and individual levels. It is estimated that more than three million children in Latin America may permanently drop out of school because of school closures during the pandemic (UNICEF 2021). In addition, disinformation limits adaptation capacity and hinders the participation of vulnerable groups in climate-health policy. Misinformation and mistrust may lead to deadly impacts related to infectious diseases. Misinformation about unscientific COVID-19 treatments shared in social media, for example, reached a large part of the population with fatal results (Furlan and Caramelli 2021). Moreover, corruption in SA leads to mistrust (Parra Saiani et al. 2021) and predisposes populations to believe in misinformation and the spread of non-validated treatments (Borges do Nascimento et al. 2022).

II. **Climate change is not part of the agenda of regional integration schemes, nor is it a priority in national adaptation planning**

Although climate-sensitive infectious disease control requires regionally coordinated efforts, climate change is still absent from the agenda of the regional integration schemes—Andean Community, Mercosur, Central American Common Market—(Fernández et al. 2023). Examples of multinational cooperation to develop and implement adaptation strategies for climate-sensitive infectious diseases are still few, although they have proven effective, as in the case of a multi-national team from Ecuador and Peru working on climate-informed dengue surveillance (Quichi et al. 2016) and their successful binational collaboration which resulted in the local elimination of malaria (Krisher et al. 2016). At the national level, while VBDs are the most common hazard included in National Adaptation Plans (Paz-Soldán et al. 2023), only four countries—Argentina, Brazil, Chile, and Suriname—have developed their Health National Adaptation Plans (Rekers et al. 2025).

III. **Populations within the region are disproportionately exposed to social, environmental, and commercial determinants of poor health**

Due to different historical legacies, populations within the SA region are disproportionately exposed to social, structural, and environmental determinants of poor physical (de Andrade et al. 2015) and mental health (Buedo and Daly 2024; Daly and Buedo 2024). This is exacerbated by widespread socioeconomic inequality (de Andrade et al. 2015). Under these conditions, climate-sensitive infectious diseases like Chikungunya can lead to severe illness and long periods of deterioration that are neither tracked nor treated (Sakhamuri et al. 2022, Ledford 2024).

One increasingly studied aspect of non-medical determinants of health is the set of market-driven factors (Friel et al. 2023), which include unhealthy commodities and industrial practices that lead to different forms of pollution (Mialon 2020). With a growing urban population, SA faces a public health emergency caused by air pollution (Husaini et al. 2024).

IV. **Vulnerable health systems and underfunded scientific systems**

The COVID-19 pandemic has highlighted the structural constraints and low resilience of SA health systems for facing overlapping health and social emergencies (Halpern and Ranzani 2022). At the beginning of the COVID-19 emergency, for example, the increase in epidemic burdens affected the development and manufacturing of tests for many tropical diseases (de Souza et al. 2020). Moreover, the medium- and long-term COVID-19 pandemic impacts on the well-being of the SA population limit its capacity to adapt to climate-sensitive infectious diseases (Rekers and Luna 2023).

The resource allocation for science and development by national governments in SA is not proportional to the climate-health risks that science can help to reduce. For example, although Argentina spends just 0.5% of its GDP on research and development, which is below the average for MICs in the region such as Brazil, the government plans to cut investment in health and science (Nature Editorial 2023). This is part of an anti-science trend promoted by right-wing populist governments in the region (Rodrigues 2021), the same type of trend that was partly responsible for the devastating surge of COVID-19 cases in Brazil (Taylor 2021). From an optimistic side, the COVID-19 pandemic has shown that rapid and coordinated action for facing emergencies is feasible (de Araujo and Meyer 2022). In addition, there was an increase in the amount of collaborative research and research protocols not funded by industry (Chapman et al. 2022).

V. **Precarious livelihoods, poverty, and inequality increase the justificatory burden of non-pharmacological measures**

In SA, black and indigenous peoples, as well as women, are those who face the most significant climate risks due to their level of vulnerability and exposure paths related to their livelihoods. Indigenous populations face high climate-health risks since they live in geographically isolated areas and resource-dependent and/or impoverished communities (Cissé et al. 2022). These risks increase when geographical isolation hinders access to medical care during extreme weather events. At the same time, various economic and social factors contribute to women having less capacity to adapt (Castellanos et al. 2022). For instance, they are more vulnerable to low-agency migration (Silva Rodríguez de San Miguel et al 2021).

The COVID-19 pandemic has increased poverty and extreme poverty rates due to the lockdowns and high healthcare-related expenses (Benza and Kessler 2022). By the end of 2020, there were 22 million more people living in poverty than in 2019, with eight million more people living under conditions of extreme poverty (ECLAC 2021). Poor populations are highly vulnerable to the impacts of climate change, and they usually live in areas with high exposure to extreme climate events (Nagy et al. 2018). This exacerbates and triggers poverty cycles in urban and rural areas due to the loss of assets and livelihoods (Castellanos et al. 2022), which contributes to the spread of infectious diseases.

Inequality poses a special challenge for climate-sensitive infectious disease control. At the beginning of the COVID-19 vaccination programme, for example, most native communities in Amazonia were unable to be included because they were not registered with the country's unified health system (Dos Santos Costa et al.

2021), which resulted in only half of them being vaccinated (Benites et al. 2021). Similarly, in Argentina, the vaccination programme required electronic registration, thereby excluding those populations that live in informal settlements without internet access or electricity (Pereira 2021).

VI. **Mental health is under-integrated in climate-health national planning**

The prevalence of poor mental health among adults is particularly high in the region. For instance, the prevalence of depression is estimated at 5% worldwide, and yet a recent systematic review found a regional prevalence of 12% (Errazuriz et al. 2023). This suggests the need for a greater understanding of local and regional contextual factors that contribute to the genesis and maintenance of mental health conditions over the course of a lifetime (Buedo and Daly 2024). Climate change brings a new set of challenges for the ethical treatment of poor mental health (Buedo and Daly 2024), with widespread eco-anxiety being one major factor (Pearson 2024). Despite mental health aspects being recognised as major health challenges that are set to worsen with climate change (Paz-Soldán et al. 2023), national adaptation plans are not addressing them. Chile is the only country in the region that includes baseline information on mental health in its national adaptation plan, but it has not developed concrete proposals, indicators, or financing related to it.

VII. **Low agency climate displacement increase the risk of infectious diseases spreading and limit the feasibility of control measures**

Migration flows contribute to the spread of infectious diseases, especially due to the poor status of border controls. In SA, migration is mainly related to political turmoil. This has been the case for the 6.1 million people who have left Venezuela due to the political and social crisis (IOM 2022). Peru and Colombia are among the countries with higher migration rates due to climate impacts, while Brazil has a high rate of internal displacement due to climate disasters (Castellanos et al. 2022). The precarious legal situation and structural discrimination patterns faced by migrants put them at a higher climate-health risk. The Chilean 2021 COVID-19 vaccination plan that only included Chilean citizens is an example of the differentiated risk that migrants face regarding infectious diseases (Ministerio de Salud de Chile 2021).

VIII. **Financial constraints for climate-health adaptation**

In Central and SA, less than a quarter of climate financing is dedicated to adaptation, (Castellanos et al. 2022). Further, only 10% of adaptation funds support health adaptation measures (Hartinger et al. 2023). Considering the size of its GDP, Brazil is receiving disproportionately more international climate funds (Castellanos et al. 2022). Several factors contribute to the unequal distribution of financial resources, such as disparities in technical infrastructure, different levels of awareness about the necessity of climate action, and lack of knowledge about the available financial options (Hickmann et al. 2021). This unequal regional distribution of resources may frustrate efforts to control infectious diseases to the extent that it depends on regional coordination and shared information.

Some SA countries might indeed be able to deploy their own funds for climate-health adaptation. However, if some countries such as Venezuela and Brazil decide to use revenues generated by their national oil companies to promote climate adaptation at the national level (Petrobrás 2023, CNN Brasil 2023), this might result in a transnational and intergenerational shift of vulnerabilities and exposures that count as maladaptation.

While most SA countries face internal financial limitations that constrain the feasibility of climate-health adaptation policies, allocation of funds is among the unaddressed aspects in national climate communications (Paz-Soldán et al. 2023). In this context, governments are legally bound to establish institutional frameworks to secure permanent international financing for adaptation and cover losses and damages (CIDH & REDESCA 2021).

IX. **Political instability and corruption necessitate long-term strategies and engage grassroots organisations**

Political practices, local institutions, patterns of exclusion and epistemic injustice may hinder the climate-health policy from being deployed in favour of more vulnerable populations. Populism and short-termism constrain the law's capacity for driving the climate-health transition in SA. The climate policy of the former president of Brazil, Jair Bolsonaro, is arguably an example of this. In his four-year mandate as President of Brazil, logging in the Amazonia reached its highest level in two decades (Stabile et al. 2022). In 2018, following Trump's 2017 attempt, Bolsonaro threatened to withdraw from the Paris Agreement. On 4 July 2022, nevertheless, the Supreme Federal Court of Brazil recognised the Paris Agreement as a "human rights treaty". Regarding governance, climate events are associated with increases in crime and violence that governments cannot fully resolve before the end of their mandate, leading to further instability (Fernández et al. 2023).

X. **The climate-health agenda is sometimes perceived as reflecting the interests of the Global North and ignoring structural problems in the region**

In this scenario, the "debt-for-climate swaps" form a financial mechanism that could drive climate commitment in SA (Patel et al. 2022). This mechanism becomes even more compelling for creditors in the Global North, given the cross-border and interconnected nature of risks posed by infectious diseases (Suk et al. 2014), especially when the freed-up resources are pledged to control climate-sensitive infectious diseases.

## *Research Priorities on Climate-Sensitive Infectious Disease*

So far, most of the research in the field of climate-health has focused on understanding the impacts of climate change on population health and predicting the health burdens attributed to climate change, with less attention given to adaptation and mitigation strategies (Hartinger et al. 2023). This scenario offers a reason to give additional priority to research in climate-health mitigation and adaptation. Since climate mitigation is already receiving most of the international funding (Castellanos et al. 2022), climate-health adaptation should receive additional attention in the research agenda. Besides, only a small part of the literature (in English) provides explicit normative analyses of the intersection between ethics and health justice and climate-sensitive infectious diseases. Nevertheless, there is renewed and ethically informed research on climate-sensitive infectious disease in low-and middle-income countries (Markle et al. 2023).

In the area of adaptation to climate-sensitive infectious diseases, the IPCC 6AR identifies several research gaps. More research is needed, for example, on the link between rodent-borne disease outbreaks and climate conditions or on the climate impacts on water quality, which is relevant for the area of waterborne infectious diseases (Cissé et al. 2022). But what should be the justice criteria for identifying research priorities? Should we prioritise research in areas with low levels of evidence, or with low levels of agreement, or with inconsistent results? Should we prioritise research that benefits the more vulnerable groups? Should we prioritise research that benefits a large number of people? Should we consider the interests of future generations? Should the interests of future generations count as much as those of the present generation?

One way of identifying research priorities is to apply the just transition framework (Table 13.1). First, we should identify the aim of the climate-health transition in the area of climate-sensitive infectious diseases. Reducing the climate-health risk posed by climate-sensitive infectious diseases seems a relevant aim in this area. The climate risks posed by climate-sensitive infectious diseases can be reduced by reducing the hazard of disease spread, reducing different types of vulnerabilities (age, gender, mobility, etc.), and changing the exposure paths (outdoor employment, livelihood, etc.) (Cissé et al. 2022).

When it comes to infectious diseases as a climate health hazard, research is focused on the impacts of climate change on infectious diseases (Castellanos et al. 2022) and how extreme climate events and climate displacement increase the risk of epidemic outbreaks (McMichael 2015; Semenza et al 2022). Studies of how the geographical and seasonal spread of climate-sensitive infectious diseases is projected under different future mitigation and adaptation scenarios should be developed. The fact that adaptation is poorly represented in present modelling frameworks counts as an additional reason for giving priority to adaptation to climate-sensitive infectious disease scenarios (Van Maanen et al. 2023).

**Table 13.1** Identifying research priorities for a just transition

| Aims | General enablers | Justice enablers |
|---|---|---|
| Reducing the risk posed by CSID | Governance | Recognitional justice |
| CSID hazards | Knowledge systems | Procedural justice |
| Vulnerabilities | | Distributive justice |
| Exposure paths | | Intergenerational justice |

Assessing vulnerability to climate-sensitive infectious diseases should also be a research priority. According to the WHO Health and Climate Change Global Survey in SA, only Brazil has reported to have completed a climate change and health vulnerability and adaptation assessment (WHO 2021).

The SA climate-health research agenda should also prioritise studies on how to reduce the impacts that outbreaks of climate-sensitive infectious diseases and overlapping social/health emergencies on health systems. Due to climate-sensitive infectious disease outbreaks, healthcare facilities' capacity level can suffer a shock that, depending on their resilience capacity, can lead to different outcomes from collapse to transformation (WHO 2020). That is why adaptation to climate-sensitive infectious diseases should aim at climate resilience development (Cissé et al. 2022). Consequently, a research priority should be to identify the pathways from health systems that, after an epidemic outbreak, collapse, recover worse than before, recover to the previous state, recover better than before, or transform.

The paths for adapting to climate-sensitive infectious diseases involve conflicts of interest and rights, trade-offs, frustrations of expectations, and different feasibility constraints. In this context, transitional justice offers a framework for adjudicating conflict and identifying the just transitional paths. Identifying the just path for transitioning towards climate resilience development can avoid maladaptation, such as unconsidered trade-offs or shifting vulnerability or exposure in the present or future (New et al. 2022). In a transitional scenario, the lack of research priorities increases the risk of maladaptation and an unjust transition.

At the feasibility level, research on the enablers of successful adaptation needs to be part of the regional adaptation agenda, as they are highly contextual. Two kinds of enablers should be investigated: general enablers and justice enablers; though they are interconnected in ways that need to be explored more deeply.

Regarding the general enablers, how corruption and elite capture limit climate-health adaptation should be part of this research agenda. Moreover, research is needed about the place of infectious disease justice in regional, national, and local climate policies. Health still occupies a minor place in climate policy and planning, while policies related to infectious diseases are still not linked with climate impacts. Mapping and assessing the ethical regulations for research, development, and implementation of climate-health policy for climate-sensitive infectious diseases should be a priority. The normative consideration grounding these regulations should consider the emergency nature of climate change. The role of ethics training for non-health professionals—social organisations, native peoples, and grassroots organisations—should also be included in this research planning, as these actors have an important role during climate-health extreme events.[1]

The importance of knowledge systems as a general enabler of climate health research and policy should also be recognised in building this research agenda. In SA,

[1] This observation was taken from the Oxford-Johns Hopkins Global Infectious Disease Ethics Collaborative project "Women's Participation in Climate-Sensitive Infectious Diseases Policy in Paraguay".

research practices and policies tend to hyper-compartmentalise knowledge generation, and inter-sectoral and inter-disciplinary collaboration is still low. This imposes important challenges to the progress towards climate resilient development. Unfortunately, institutional structures, cultural barriers, trust issues, and political cycles affect and undermine effective collaboration. If countries, communities, and people want further development and progress in this area, then developing a transdisciplinary research approach should be considered a primary objective (Serrao-Neumann et al 2021).

A second group of enablers refers to justice conditions that facilitate successful adaptation. Although climate research lacks a comprehensive and consistent approach to integrate justice considerations (Zimm et al. 2024), justice enablers of adaptation have been identified in the area of climate health policy (Rekers et al. 2025). Here, we identify some justice enablers of climate-health adaptation that are particularly relevant for SA.

As for recognitional justice, the climate-sensitive infectious disease justice agenda should include studies on how to incorporate poverty and precarious livelihoods into climate-health policy. The ways of life and resilience mechanisms of populations living under these conditions are frequently overlooked in the development and implementation of infectious disease strategies due to their lack of representation, hermeneutical injustices, etc. Consequently, climate-health national strategies face limitations, mainly at the level of non-pharmacological and social assistance measures, leading these populations to a situation of greater disadvantage.

The question about who should participate in the definition of research priorities invokes the importance of procedural justice enablers. More research is needed on the impact of participatory health adaptation planning and the needs of different sectors to participate in an effective way. Co-production of knowledge is another important topic in the research agenda and requires recognising the fact that different kinds of epistemic and procedural injustices affect native peoples, women, and other vulnerable groups. Frequently, members of vulnerable groups are invited to participate in the development of National Adaptation Plans, but although their views are recorded, it is not clear how they influenced the results of the process. This practice has been described as *epistemic washing* (Rekers et al. 2023). How to overcome this injustice should be part of the SA research agenda for climate-sensitive infectious disease justice.

Even though distributive justice considerations are the least referred to in national adaptation planning (Rekers et al. 2025), there are many distributive aspects of adapting to climate-infectious disease that should be included in the research agenda. The distribution of the burden of reducing climate-health risks among different sectors is an important research question that has not yet been fully addressed. Identifying the relevant factors for weighing the infectious disease burden is needed to respond to many practical challenges. In addition, more research is needed about the benefits of adaptation and mitigation for SA population health (Hartinger et al. 2023). Measuring these benefits is important to maximise them and for addressing questions about their just allocation.

Finally, to avoid maladaptation, the transition towards resilience to climate-sensitive infectious diseases should incorporate intergenerational justice considerations. This dimension of justice plays a relevant role in avoiding an unjust shift of climate-health risks between present and future generations. Unfortunately, the temporal dimension of adaptation options is understudied in the literature (New et al. 2022).

## *South America's Contributions to Climate-Sensitive Infectious Disease Justice*

The third question of this conversation aims to identify how SA can contribute to the global justice challenges posed by climate-sensitive infectious diseases. One aspect to consider is that currently, the Global South is the main beneficiary of research on climate-sensitive infectious diseases. However, under different future climate scenarios, this research may benefit future generations of countries in the Global North, some of which are starting to experience increasingly climatic suitable conditions for VBDs. For example, climate conditions in Europe, especially in Northern Europe, are becoming suitable for dengue and chikungunya (Laverdeur et al. 2024). This fact has justice implications for the intergenerational and international allocation of climate-health burdens and for the evaluation of necessities and benefits at stake in the ethical assessment of clinical trials, protocols, etc.

SA's experience in addressing overlapping emergencies by resorting to human rights and transitional justice approaches can offer important normative frameworks for climate-sensitive infectious disease justice. These approaches are equipped to address the interests, expectations, and rights conflicts that emerged in the transition toward systems resilient to climate-sensitive infectious diseases. From its rich tradition and history of championing human rights, the region can leverage collective memory and employ the heuristic of human rights to address these pressing issues. Identifying how these tools serve to build resilience to climate-sensitive infectious diseases should be part of the research agenda. This priority is more salient if we consider that the human rights perspective has the potential to enable climate-health adaptation. This recognition underscores that climate change is not only an environmental issue but also a matter of human rights (IACHR & REDESCA 2021).

The practical benefits of the human right to health framework should be explored to foster climate litigation in the area of infectious disease. Climate litigation is gaining attention, while climate-health litigation is focused on the right to a healthy environment (https://climatecasechart.com/non-us-climate-change-litigation/). As a strategy, however, climate litigation in the area of infectious disease can raise justice and ethical concerns regarding power asymmetries, since access to legal advice and representation is distributed unequally (Kinder 2023). It can also lead to unjust or inefficient allocation of benefits and burdens, because one case resolution can lead to an unjustified shift of climate-health vulnerabilities. So, more research on the benefits and risks of climate litigation in this area has the potential to cope with the climate-health adaptation goal and reduce the risk of maladaptation.

SA governments have the obligation to adapt to climate change to protect the human right to health of their residents (Rekers et al. 2023). This obligation includes the duty to fund research and set research priorities. The human rights heuristic is a tool for identifying principles of climate justice that should inform states of their obligations to protect the human right to health. A prominent example of this is the process of the Advisory Opinion on Climate Emergency and Human Rights at the Inter-American Court of Human Rights (IACtHR 2017). This process aimed to clarify the scope of states' obligations regarding the protection of the human right to health in the climate emergency context, in line with international human rights regional system.

Finally, we should leverage SA rich tradition in social sciences and humanities research grounded in an anti-colonial perspective. Colonisation, neo-colonial extractivism, and climate change shape climate risks and climate health policy. The colonising history of each country, for example, correlates with the level of vulnerability to climate change (Táíwò 2022). In this context, reparations for colonisation can play an important role in the climate transition (Táíwò 2022). Identifying the forms that reparations for colonisation can adopt in the area of adaptation to climate-sensitive infectious diseases is a pending task.

## Research Limitations

Although this study focuses on SA, we also drew on studies with a Latin American scope. The diversity of participants in this type of philosophical conversation can be improved in future studies. More pluralistic and transdisciplinary conversations need to be developed for co-setting research priorities on climate-sensitive infectious disease justice, including representatives of vulnerable groups and future generations.

## Conclusions

Climate-sensitive infectious diseases pose a special challenge for climate-health policy in SA. Since climate-sensitive infectious disease justice is a problem-responsive value, identifying the background conditions that make it necessary and possible should be the first step for identifying research priorities. Secondly, the just transition framework offers a map for identifying research priorities in the area of climate-sensitive infectious disease justice. Finally, the human rights and decolonisation frameworks' heuristics can contribute to the global challenge of climate-sensitive infectious diseases.

## References


Bárcena-Ibarra A, Samaniego J, Galindo LM, Ferrer J, Alatorre JE, Stockins P, Reyes O, Sánchez L, Mostacedo J (2018) Economics of climate change in Latin America and the Caribbean: a graphic view. United Nations publication Rev 1

Benites E, Gisloti LJ, de Oliveira RF (2021) Brazil: boost COVID-19 vaccine uptake in Indigenous people. Nature 591(7850):369

Benza G, Kessler G (2022)Benza G, Kessler G (2022) The impact of the pandemic on Latin America: social setbacks and rising inequalities. In: Vommaro P (ed) Persistence and emergencies of inequalities in Latin America: a multidimensional approach. Springer International Publishing, pp 33–49

Borges do Nascimento IJ, Pizarro AB, Almeida JM, Azzopardi-Muscat N, Gonçalves MA, Björklund M, Novillo-Ortiz D (2022) Infodemics and health misinformation: a systematic review of reviews. Bull World Health Organ 100(9):544–561

Buedo P, Daly T (2024) A contextual understanding of the high prevalence of depression in Latin America. Lancet Reg Health Am 32:100717

Castellanos E, Lemos MF, Astigarraga L et al (2022) Central and South America. In: Pörtner HO et al (eds.) Climate change 2022: impacts, adaptation and vulnerability. Cambridge University Press

Cissé G, McLeman R, Adams H, Aldunce P, Bowen K, Campbell-Lendrum D, Clayton S, Ebi KL, Hess J, Huang C, Liu Q, McGregor G, Semenza J, Tirado MC (2022) Health, wellbeing, and the changing structure of communities. In: Climate change 2022: impacts, adaptation and vulnerability. Contribution of working group ii to the sixth assessment report of the intergovernmental panel on climate change. Cambridge University Press, pp 1041–1170

CNN Brasil (2023) Ibama emite primeira licença para atividades petrolíferas na Margem Equatorial

Chapman E, Illanes E, Reveiz L, Saenz C (2022) Mapeo de protocolos de investigación, publicaciones y colaboraciones sobre la COVID-19 en América Latina y el Caribe. Rev Panam Salud Publica 46:42

Daly T, Buedo P (2024) Applying ethics to mental health and climate change. Acad Psychiatry 48(1):104–105

de Andrade LO, Pellegrini Filho A, Solar O, Rígoli F, de Salazar LM, Serrate PC, Ribeiro KG, Koller TS, Cruz FN, Atun R (2015) Social determinants of health, universal health coverage, and sustainable development: case studies from Latin American countries. Lancet 385(9975):1343–1351

de Araujo M, Meyer L (2022) Climate change and pandemics: feasibility constraints on mitigation and adaptation. In: Climate change, responsibility and liability. Nomos Verlagsgesellschaft, pp 41–74

de Souza DK, Picado A, Biéler S, Nogaro S, Mathu Ndung'u J (2020) Diagnosis of neglected tropical diseases during and after the COVID-19 pandemic. PLoS Neglected Tropical Diseases, n.8 v.14

Dias M, McCoy T (2024). Brazil's staggering dengue fever crisis is a warning to the world. The Washington Post. https://www.washingtonpost.com/world/2024/03/10/dengue-fever-brazil-outbreak-epidemic/

Dos Santos Costa AC, Ahmad S, Essar MY (2021) Vaccination: Brazil fails Indigenous people again with two-tier scheme. Nature 593(7860)

ECLAC (2021) Panorama social de América Latina. Santiago de Chile

Errazuriz A, Avello-Vega D, Ramirez-Mahaluf JP, Torres R, Crossley NA, Undurraga EA, Jones PB (2023) Prevalence of depressive disorder in the adult population of Latin America: a systematic review and meta-analysis. Lancet Reg Health Am 26:100587

Fernández A, Franchini M, Aragão L, Carte L, Pacheco D, Ochoa Ochoa LM, Preeya SM, Martins do Amaral Cunha AP, Yglesias-González M (2023) Facing climate change across Latin America and the Caribbean. Nat Clim Chang 13(10):1014–1019

Friel S, Collin J, Daube M, Depoux A, Freudenberg N, Gilmore AB, Johns P, Laar A, Marten R, McKee M, Mialon M (2023) Commercial determinants of health: future directions. Lancet 401(10383):1229–1240

Furlan L, Caramelli B (2021) The regrettable story of the "COVID kit" and the "early treatment of COVID-19" in Brazil. The Lancet Regional Health –Americas: v.4.

Global Carbon Project (2021) Are consumption-based $CO_2$ per capita emissions above or below the global average?

Global Climate Change Litigation Database: https://climatecasechart.com/non-us-climate-change-litigation/

Global Witness (2023) Standing firm: the land and environmental defenders on the frontlines of the climate

Halpern B, Ranzani O (2022) Lessons from the COVID-19 pandemic in Latin America: vulnerability leading to more vulnerability. Am J Public Health 112(S6):S579–S580

Hartinger SM, Yglesias-González M, Blanco-Villafuerte L, Palmeiro-Silva YK, Lescano AG, Stewart-Ibarra A, Romanello M (2023) The 2022 South America report of the Lancet countdown on health and climate change: trust the science. Now that we know, we must act. Lancet Regional Health–Americas 20

Harris NL, Gibbs DA, Baccini A, Birdsey RA, De Bruin S, Farina M, Tyukavina A (2021) Global maps of twenty-first century forest carbon fluxes. Nat Clim Chang 11(3):234–240

Hickel J (2020) Quantifying national responsibility for climate breakdown: an equality-based attribution approach for carbon dioxide emissions in excess of the planetary boundary. Lancet Planet Health 4(9):e399–e404. https://doi.org/10.1016/S2542-5196(20)30196-0

Hickmann T, Widerberg O, Lederer M, Pattberg P (2021) The United Nations framework convention on climate change secretariat as an orchestrator in global climate policymaking. Int Rev Adm Sci 87(1):21–38

Husaini D, Manzur K, Medrano J (2024) Indoor and outdoor air pollutants as emerging public health threat in Latin America and the Caribbean: a systematic review. Arab Gulf J Sci Res 42(1):134–145

Hume D (1978) A treatise of human nature. Selby-Bigge LA (ed), 2nd edn. Oxford University Press, Oxford

Hume D (1998) An enquiry concerning the principles of morals, TL Beauchamp (ed). Oxford University Press, Oxford

IACtHR (2017) Advisory Opinion OC-23/17

IACHR & REDESCA (2021) Resolution No. 3/21 climate emergency: scope of inter-American human rights obligations

IOM (2022) Venezuelan refugee and migrant crisis

Kinder RT (2023) Taking climate change to court: how a human rights framework can help plaintiffs overcome power asymmetries in climate litigation. Letter from the executive editors 129

Krisher L, Krisher J, Ambuludi M, Arichabala A, Beltrán-Ayala E, Navarrete P, Ordoñez T, Polhemus ME, Quintana F, Rochford R, Silva M (2016) Successful malaria elimination in the Ecuador-Peru border region: epidemiology and lessons learned. Malar J 15:1–15

Laverdeur J, Desmecht D, Hayette MP, Darcis G. (2024) Dengue and chikungunya: future threats for Northern Europe? Front Epidemiol

Ledford H (2024) Long COVID is a double curse in low-income nations-here's why. Nature 625(7993):20–22

López MS, Müller GV, Sione WF (2018) Analysis of the spatial distribution of scientific publications regarding vector-borne diseases related to climate variability in South America. Spat Spat-Temporal Epidemiol 26:35–93

Markle H, Black G, Khan W, Kelley M, Meritt M, Jamrozik E (2023) Ethics, equity, climate change, and infectious disease research: a scoping review. Wellcome Open Res 8:581

McMichael AJ (2015) Extreme weather events and infectious disease outbreaks. Virulence 6(6):543–547

Mialon M (2020) An overview of the commercial determinants of health. Global Health 16(1):74

Ministerio de Salud de Chile (2021) Complementa Resolución Exenta N° 1138 de 2020, del Ministerio de Salud que aprueba Lineamientos Técnico Operativos Vacunación SARS-CoV-2
Mosegui GBG, Villar FA, de Mello VCM (2022) Burden of disease attributed to acute respiratory infections in South America. J Infect Dev Ctries 16(10):1614–1622
Nagy GJ, Filho WL, Azeiteiro UM, Heimfarth J, Verocai JE, Li C (2018) An assessment of the relationships between extreme weather events, vulnerability, and the impacts on human wellbeing in Latin America. Int J Environ Res Public Health 15:1802
Nature Editorial (2023) Cutting health and science support should not be an option in Argentina's election, vol 623
New M, Reckien D, Viner D, Adler C, Cheong SM, Conde C, Constable A, Coughlan de Perez E, Lammel A, Mechler R, Orlove B, Solecki W (2022) Decision-making options for managing risk. In: Climate change 2022: impacts, adaptation and vulnerability. Cambridge University Press, pp 2539–2654
Parra Saiani P, Ivaldi E, Ciacci A, Di Stefano L (2021) Broken trust. Confidence gaps and distrust in Latin America. Soc Indic Res 1–3
Patel S, Klok E, Steele P, Camara I (2022) After the Paris Agreement, the debt deluge: why lending for climate drives debt distress. IIED, London
Paz-Soldán VA, Valcarcel A, Canal-Solis K, Miranda-Chacon Z, Palmeiro-Silva YK, Hartinger SM, Suárez-Linares AG, Falla-Valdez V, Intimayta-Escalante C, Lehoucq M, Pretell A (2023) A critical analysis of national plans for climate adaptation for health in South America. Lancet Regional Health–Americas 26
Pearson H (2024) The rise of eco-anxiety: scientists wake up to the mental-health toll of climate change. Nature 628(8007):256–258
Pereira A (2021) Los retos de vacunar en los barrios más pobres de Argentina. SWI. Suiza 22
Petrobrás (2023) Petrobras—Fatos e Dados—OTC 2023: Estudamos projeto inédito de hub de captura e armazenamento geológico de $CO_2$ no Brasil
Quichi E (2016) Ecuador–Peru cooperation for climate-informed dengue surveillance: creating an interdisciplinary multinational team. In: Climate services for health: improving public health decision-making in a new climate—Case Studies. World Meteorological Organization and World Health Organization, pp 12–13
Rekers R, Gerbaldo MV, Rekers L, Yabar C, Heinemann J (2023) La protección del derecho humano a la salud en la emergencia climática. La obligación de los estados de adaptarse al cambio climático para garantizar el derecho humano a la salud de sus residentes. Religacion Press
Rekers R, Luna F (2023) Pandemic justice for and from Latin America. In: Luna F, Rekers R, Jamrozik E, Gur-Arie R (eds) Glob Pandemic Justice, Ethic@—An Int J Moral Philos 22:1
Rekers R, Gerbaldo MV, Yabar C, Garat C, Rekers L (2025) Justice enablers of climate-health adaptation in South America. J Clin Change Health 23:100459
Rocklöv J, Dubrow R (2020) Climate change: an enduring challenge for vector-borne disease prevention and control. Nat Immunol 21(5):479–483
Rodrigues M (2021) Scientists reel as Brazilian government backtracks on research funds. Nature https://doi.org/10.1038/d41586-021-02886-9
Romanello M, Di Napoli C, Drummond P, Green C, Kennard H, Lampard P, Scamman D, Arnell N, Ayeb-Karlsson S, Ford LB, Belesova K (2022) The 2022 report of the Lancet countdown on health and climate change: health at the mercy of fossil fuels. The Lancet. 400(10363):1619–1654
Sakhamuri SM, Jankie S, Pereira LMP (2022). Calling on Latin America and the Caribbean countries to recognise the disability from long COVID. Lancet Regional Health–Americas 15
Savulescu J, Gyngell C, Kahane G (2021) Collective reflective equilibrium in practice (CREP) and controversial novel technologies. Bioethics 35(7):652–663. https://doi.org/10.1111/bioe.v35.710.1111/bioe.12869
Semenza JC, Rocklöv J, Ebi KL (2022) Climate change and cascading risks from infectious disease. Infect Dis Ther 11:1371–1390

Serrao-Neumann SS, Moreira FD, Fontana MD, Torres RR, Lapola DM, Nunes LH, Marengo JA, Di Giulio GM (2021) Advancing transdisciplinary adaptation research practice. Nat Clim Chang 3

Silva Rodríguez de San Miguel JA, Martínez Díaz E, Monroy Becerril DM (2021) The relationship between climate change and internal migration in the Americas. Manag Environ Qual: Int J 32(4):822–839

Stabile MCC, Garcia AS, Salomão CSC, Bush G, Guimarães AL, Moutinho P (2022) Slowing deforestation in the Brazilian amazon: avoiding legal deforestation by compensating farmers and ranchers. Front For Glob Change

Suk JE, Van Cangh T, Beauté J, Bartels C, Tsolova S, Pharris A, Ciotti M, Semenza JC (2014) The interconnected and cross-border nature of risks posed by infectious diseases. Glob Health Action 7:1

Táíwò O (2022) Reconsidering reparations. Oxford University Press

Taylor L (2021) We are being ignored': Brazil's researchers blame anti-science government for devastating COVID surge. Nature 593(7857):15–16

UNICEF (2021) 114 million children still out of the classroom in Latin America and the Caribbean: the world's largest number of children without face-to-face schooling

van Maanen N, Lissner T, Harmsen M, Piontek F, Andrijevic M, van Vuuren DP (2023) Representation of adaptation in quantitative climate assessments. Nat Clim Chang 13(4):309–311

Wolff J, de Shalit A (2023) City of equals. Oxford University Press

WHO (2020) Guidance for climate resilient and environmentally sustainable health care facilities

WHO (2021) Health in national adaptation plans: review

Zimm C, Mintz-Woo K, Brutschin E, Hanger-Kopp S, Hoffmann R, Kikstra JS, Schinko T et al (2024) Justice considerations in climate research. Nat Clim Chang 1–9

# Chapter 14
# Climate Change Impacts on Indigenous Diets, Health Perceptions, and Traditional Medicine: Perspectives from Older Adults in the Indian Himalayan Region

**Senjooti Roy and Kate de Medeiros**

**Abstract** Rural populations in the Indian Himalayan Region (IHR) have customarily depended on traditional diets and herbal medicine for sustenance and health. Older adults have been the primary custodians of traditional agricultural practices and indigenous medical knowledge. In the last few decades, the IHR has faced tremendous ecological pressure from various sources, with climate change being an integral force in re-shaping the agricultural and healthcare practices of residents. To explore the health experiences of older adults within the context of climate change, we undertook a qualitative study in three villages along the Indo-China border to study conceptions of health, health behaviours, healthcare utilisation, and healthy ageing. We found that climate change has impacted the availability, quality, quantity, and efficacy of traditional foods and herbs. This has engendered dependence on non-traditional diets that are believed to have introduced new diseases, which must now be treated with unfamiliar modern medicines. This chain of impacts is largely viewed unfavourably by older adults who have deep faith in traditional practices but are compelled to seek alternative dietary and medical options. This chapter highlights how climate change can acutely impact the health of vulnerable populations that are directly dependent on natural resources for food and medicine.

**Keywords** Himalayas · Older adults · Traditional diet · Traditional medicine · Climate change

S. Roy (✉)
Louis and Gabi Weisfeld School of Social Work, Bar-Ilan University, Ramat Gan, Israel
e-mail: senjoor@biu.ac.il

K. de Medeiros
Department of Sociology and Anthropology, Concordia University, Montreal, Canada



W. Leal Filho and F. Wolf (eds.), *Climate Change and Health: Perspectives from Developing Countries*, Climate Change Management,
https://doi.org/10.1007/978-3-031-93177-2_14

## Introduction

In this chapter we explore the concept of "food as medicine" to see how, around the world, food and medicine have traditionally been viewed as two sides of the same coin, with food offering both nutritional and medicinal benefits. We then look at how the current focus on holistic health is re-introducing this concept to encourage people to eat healthfully so as to prevent and heal from lifestyle-based diseases. We then discuss the impacts of climate change on sustainable agriculture, traditional diets, and herbal medicine, all of which contribute to good health but are being threatened by climate events. Finally, we focus on the Indian Himalayan Region (IHR) and, through the perspectives of older adults, view their relationships with their natural resources, especially local food and healthcare systems. We present this as a study of contrasts between "the olden days" and "these days…" to showcase the impacts of climate change on older adults' health and healthcare preferences.

This study includes and highlights the voices of a vulnerable population inhabiting an ecologically fragile region and bearing heavy burdens of environmental change without having contributed in any significant way to the on-going climate crisis. The experiences of people from low- and middle-income countries (LMIC) are generally underrepresented in global research output. The voices of older adults are additionally suppressed due to ageist assumptions about older adults having lived their lives, with little potential for future contributions. This study, therefore, includes the voices of older adults on the important global issue of climate change, it highlights the unique challenges faced by people in LMICs amidst irreversible environmental change, and it strengthens interdisciplinary literature in the fields of ageing and environmental studies.

## Background and Context

### *Food as Medicine*

The ancient *Ayurvedic* (a branch of traditional medicine in India) proverb, "If diet is wrong, medicine is of no use. If diet is correct, medicine is of no need", captures the essence of this chapter. Around the world, food and medicine have traditionally been viewed as interrelated components of health and well-being, with food believed to possess both preventive and curative healthcare properties. Locally available/cultivated herbs, spices, grains, pulses, fruits, vegetables, oils, dairy, fish, and meat have provided sustenance and health to people for millennia. Research exploring traditional diets have found locally sourced and organically grown nutritious food to be associated with myriad positive health outcomes. For example, the Mediterranean diet, rich in fruits, vegetables, nuts, pulses, whole grains, fish, and olive oil is associated with a lower risk of diabetes, heart disease, metabolic syndrome, neurodegenerative diseases, and certain types of cancer (Lăcătusu et al. 2019). The traditional

Japanese diet, rich in fish, rice, vegetables, and fermented foods like soy, is associated with lower rates of obesity, certain types of cancer, and improved gut and heart health (Tsugane 2021). In cold regions, like the Arctic, the Inuit diet of fish, marine mammals, birds, and fur-bearing animals, rich in protein, antioxidants, and omega-3 fatty acids, has traditionally protected people from type 2 diabetes, heart disease, and risk of depression (McGrath-Hanna et al. 2003).

In recent times, however, changes to traditional diets have led to adverse health outcomes. For example, a study found that among Mediterranean populations, those who had recorded high consumption of ultra-processed foods over an 18-year period showed a 40% higher risk of all-cause mortality compared to those who ate unprocessed or minimally processed foods (Torres-Collado et al. 2024). Similarly, increased consumption of red meat in Japan has led to an increase in diet-related, non-communicable diseases that are becoming the leading cause of mortality in the country (Nomura et al. 2023). Changes in the traditional Inuit diet—with the incorporation of high levels of saturated fats and carbohydrates and low levels of omega-3 fatty acids—have led to increased obesity, cardiovascular diseases, diabetes, dental diseases, and mental health issues like depression, anxiety, seasonal affective disorder, and suicidal tendencies (McGrath-Hanna et al. 2003).

Studies from around the world have shown that diets that include high amounts of ultra-processed foods have higher levels of saturated fat, total fat, sodium, added sugar, and carbohydrates. They also have lower levels of fibre, protein, vitamins, and micronutrients. Consumption of such food products are linked to cardiovascular diseases, obesity, hypertension, higher BMI, gastrointestinal disorders, metabolic syndrome, cancer, and several other health problems (Koiwai et al. 2019). Currently, because of such findings, food in the form of "medically prescribed diets", "therapeutic diets", "produce prescription programs" and "food is medicine" initiatives is increasingly being recommended by healthcare professionals to prevent, delay, manage, or reverse chronic diseases linked to modern dietary practices. Some such diets include the DASH diet for hypertension, the MIND diet to prevent/delay cognitive decline, the cardiac diet for heart health, and the anti-inflammatory diet for chronic, systemic inflammation, among many others.

Dietary modifications alone, however, may not always be adequate for the management of health needs. Other factors such as physical activity and medicine are also integral to holistic health management. Although modern medicine (Western biomedicine) leads the global healthcare industry in terms of visibility, infrastructure, scientific research, technological advancement, financing, and policy, various branches of traditional medicine continue to support over 85% of the global population in some form, with the latter often being the first port of call, especially for chronic illnesses (World Health Organisation 2024a). Traditional medical systems, whether used alongside or in place of modern medicine, are often grouped together under the umbrella term "complementary and alternative medicine" (CAM) to distinguish them from modern medicine in medically pluralist settings (World Health Organisation 2024b).

Traditional medicine comprises many healthcare traditions from around the world and their use of a variety of diagnostic methods, remedies, treatments, and therapies.

It also includes the use of herbal medicine. Medicinal herbs may be consumed as part of daily nutrition and dietary practices or in the form of pills, concentrated liquid extracts (tinctures), and powders. They may be gathered from the wild or be cultivated in farms and domestic settings. Many medicinal plant species are endemic to specific regions in the world. The IHR, one of 36 "biodiversity hotspots" around the world and the focus area for this study, is one such region with over 1,700 known medicinal plant species (Uttarakhand Forest Department 2024). Many of these are categorised as "high value" for their potency and high quality. Unfortunately, several species are also now categorised as "endangered" or "threatened" due to very high demand and consequent overharvesting, coupled with lower regeneration of herbs in their natural habitat.

## *Climate Change Impacts on Agriculture, Traditional Diets, and Medicinal Plants*

The efficacy of traditional diets and medicinal plants rests on the maintenance of robust ecosystems, local accessibility, and the use of sustainable production methods. Climate change, however, has disrupted traditional agricultural practices and the natural regeneration of plants in the wild, which in turn have compromised the food, health, life, and livelihood security of millions of people, especially in LMICs. Climate change has both direct and indirect impacts on fragile ecosystems. Warmer temperatures, pest attacks, and unpredictable rainfall patterns weaken environmental health, reduce crop yield and nutrition, endanger livestock survival, and lead to the migration and/or extinction of (medicinal) plant species, resulting in a decreased potency and quality of their medicinal properties. Extreme weather events like floods, droughts, and wildfires can destroy entire ecosystems and displace people from their local habitats, forcing them to adopt different dietary, health, and livelihood practices.

## *Indian Himalayan Region*

The Himalayan region is amongst the most biodiversity-rich, yet vulnerable, regions of the world. Also known as the "third pole" or the "water tower of Asia", it directly or indirectly supports over a billion people across several countries. Therefore, the consequences of climate change in this area, in the form of increased glacier melt, floods, droughts, irregular rainfall, glacial lake outburst floods, cloudbursts, and other natural disasters can endanger the lives of large numbers of people. This is already apparent in the increased frequency and intensity of recent natural disasters. However, even when climate events do not constitute a disaster or a "sudden onset event", climate change can still lead to "slow onset events" and processes such as land and forest degradation, increasing temperatures, desertification, glacial retreat, and loss of biodiversity, among other consequences. The people of the IHR are susceptible to impacts of both slow and sudden onset events; this is increasingly being reflected in their evolving dietary patterns and healthcare choices.

Traditional diets in the IHR are characterised by their local availability/cultivation potential, organic farming methods, lack of processing, limited use of technology, and farm-to-table consumption patterns. Diets are influenced by geographic and environmental conditions, local ecosystems, and sociocultural norms. Food systems in the IHR may include one or more of the following: harvesting, foraging, gathering, fishing, and hunting, depending on the ethnicity of a populace (Ahirwar and Khan 2023). Regardless of how food is acquired, traditional consumption patterns prioritise the conservation of local environment and biodiversity; sustainability is inherent in people's lifestyles due to high dependence on natural resources.

The use of herbal medicine in the IHR is as varied as the diets of the people. Like traditional foods, the continued use of herbal medicine is also dependent on altitude, local ecosystems, availability of herbs in the wild, potential for domestic cultivation, environmental conditions, and sociocultural norms. Herbal medicine is commonly used and practiced in individual households; they sometimes also overlap with ingredients that are used for daily cooking (e.g., seeds, herbs, spices). Herbal medical knowledge is transmitted orally and experientially from one generation to another, with family and/or community elders usually being the custodians of such knowledge. For uncommon health issues, however, people visit professional herbalists with specialised knowledge.

The sustainability of both traditional diets and herbal medicine in the IHR is impacted by climate change, and alterations in the ecological balance of the region acutely impacts the lives of people. Climate change threatens the symbiotic relationship between the people and their environment, with consequences ranging from modifications in diet, livelihood, and healthcare to climate-induced migration, from changes in disease profiles to forced adjustments in family composition, caregiving, and living arrangements. Moreover, climate change affects different groups of people differently. People with fewer financial and material resources, women, people with disabilities, and older adults are more likely to shoulder disproportionate burdens of climate change in the IHR.

## *The Case of Three Himalayan Villages*

In this chapter, we focus on a study that was carried out in three villages located in Uttarkashi district in the northern Indian state of Uttarakhand, one of thirteen states and union territories in the IHR, sharing borders with Nepal to the east and China to the north. Uttarkashi is the least urbanised district in the state, with over 90% of the population residing in rural areas. The people of Uttarkashi hold their land sacred; it is home to numerous species of endemic and native medicinal flora that populate its 60–80% forest-covered land across varied altitudes and geoclimatic conditions (Government of Uttarakhand 2024). Herbal medicine is widely practiced and favoured by the residents of the area.

The goal of our study was to explore the health experiences of older adults within the context of climate change in the selected villages. We had three broad research questions:

a. How had climate change impacted the availability and cultivation of traditional food/crops in the selected villages?
b. How had climate change impacted the traditional healthcare systems practised by older adults in the selected villages?
c. How had changes in diet and traditional healthcare, if any, shaped older adults' conceptions of health, health behaviours, healthcare utilisation patterns, and experiences of healthy ageing?

## Methods

We used a qualitative-ethnographic approach, as it is greatly beneficial for research requiring intensive fieldwork, including observation of daily life/practices, immersion in community life, informal interactions with residents, key informant interviews, group discussions, and extensive recording of memos and field notes. We interviewed long-term residents over the age of 55 years (age range: 55–87 years), as they were best suited to answer the research questions. We used a purposive sampling method to select "information rich cases" for in-depth insights into our research questions (Patton 1990). We conducted 32 key informant interviews (25 interviews with older residents, 11 men and 14 women; 7 interviews with practitioners of different branches of medicine in the villages). This was in addition to the informal conversations that were being held and recorded in field notes for over eight weeks. The study included representations from a diverse population in terms of age, gender, caste, tribe, religion, educational background, and occupation. Data analysis was undertaken over several months and included translation, transcription, verification, and de-identification of transcripts, followed by multiple readings of transcripts, memos, and field notes, extensive coding, creation of themes, extraction of illustrative quotes, and creation of two schemas for the visual representation of relationships between codes.

## Results

This section comprises three sub-sections: (1) "understanding illness", (2) "defining health", and (3) "In the olden days…" versus "These days…": A study of contrasts. The first two sub-sections present findings from the study as well as provide context for the third sub-section.

## *Understanding "Illness"*

Participants conceptualised health issues as falling into four primary categories: "small illnesses", "big illnesses", "new illnesses", and "illnesses of old age", each of them signifying the extent of people's knowledge and familiarity with cause of illness, commonality of occurrence, method of diagnosis, treatment plan, typical duration of ailment, and availability of herbs needed for treatment.

**"Small illnesses"** were defined by their common occurrence; they were characterised as "*minor*" and "*light*" issues that did not cause much concern to patients and their families. They were generally tied to seasonal changes and the local climate. They included cold, cough, stomachache, headache, fever, upset stomach, and breathing difficulties while walking uphill. "Small illnesses", although called "illnesses", were mostly conceived of as "health inconveniences" that could get cured with or without medicine. They affected people of all age groups and had been part of people's health experiences for generations.

In contrast to "small illnesses", **"big illnesses"** caused fear, concern, and anxiety among participants, primarily due to their lack of familiarity with cause, treatment, and prognosis. "Big illnesses" were believed to affect the internal organs of the body and impact an individual's longevity and quality of life. They included stroke, heart disease, cancer, diabetes, tuberculosis (TB), and kidney stones. Notably, participants mentioned that "big illnesses" were still rare among older adults in the villages and that the few people who had been diagnosed belonged to younger age groups.

The third category mentioned by the participants was that of **"new illnesses"**. These overlapped with "big illnesses" and were considered new because there were no tried-and-tested herbal remedies or indigenous medical knowledge surrounding these illnesses. "New illnesses", in addition to illnesses categorised as "big illnesses", also included dengue, tumours, and hypertension. These were illnesses that participants had learned of only in the last decade or so and believed them to be "*dangerous*" and "*fear-inducing*". "New illnesses" sometimes caused participants to incorrectly associate cause with effect. For example, one participant shared that,

> …we are sheep farmers, we work with wool…the fibre that flies around goes inside through our breath. What that does is damage the lungs…that is why, for those who work with wool, you will find that 2% of people will complain of TB.

This claim was later refuted by a traditional medicine practitioner who said,

> …If fibre goes in, it will lead to asthmatic conditions, not tuberculosis…TB is a bacterial disease. Until a bacterium affects, it won't happen.

Another participant opined that, "*TB happens…it does not need much treatment*", demonstrating how lack of familiarity with "big" and "new" illnesses could potentially lead to ambiguity and misunderstanding, resulting in delayed (or lack of) diagnosis and treatment.

The final category of illnesses was called **"illnesses of old age"**. These were not illnesses per se but encompassed various deteriorations in health status that the participants associated with the process of ageing. These did not cause alarm or worry as they were both expected and accepted as one grew older. "Illnesses of old age" primarily included knee pain/arthritis, asthma/breathing difficulties while walking uphill, and hypertension. These were widely prevalent among the older population. Other common "illnesses of old age" included general weakness in the body, decreasing body mass, deteriorating eyesight and hearing, aches and pains, weakened nerves, reduced appetite, toothache, discomfort with loud sounds, softening of voice, forgetfulness, and an overall slowing down of the mind. Cognitive and psychological illnesses were very rare in the villages; none were reported in the study. In addition to accepting "illnesses of old age" as natural accompaniments to the ageing process, participants were also accepting of death as naturally following old age and related illnesses.

## *Defining Health*

Participants conceived of good health as a combination of several facets of an individual's life. These included quality of diet, daily lifestyle habits/practices, environment, availability of social support, and interaction with sociocultural norms and traditions. When these factors complement one another, an individual is likely to be in good health, they can manage the occurrence of "small illnesses" efficiently and safely, and they are largely protected from "big illnesses". If these factors do not align, an individual is likely to experience ill health and consequent "*sadness*". The participants especially acknowledged the role of their environment as a key determinant of good health, particularly psychological health. They believed that the residents' ability to "*roam freely*", to "*do things freely*", to live in a place that is "*open*", where "*air and water are clean*", and where people "*can see the nature, the mountains, and the snow*", it is hard to have any kind of mental illness because "*the heart will stay happy.*"

Another key aspect of determining good health lay in the participants' ability to work. If they were able to live independently, eat well, and undertake the various physically strenuous tasks that accompanied life in the mountains, they considered themselves to be healthy. Some common tasks performed by the participants included farming; cooking; cleaning homes and barns; washing clothes, bed linen, and animals' blankets; walking long distances, often uphill, for various types of work or to meet with others; caring for animals; and harvesting, collecting, and carrying grass, firewood, and fodder from the forest. Participants also engaged in knitting, weaving, and preserving food for the winter season. Additionally, they undertook repairs to their homes and barns depending on their physical ability.

## *"In the olden days…" versus "These days…": A study of contrasts*

### *"In the olden days…"*

It is in light of the background above that this sub-section will be discussed. This sub-section comprises two parts, "in the olden days…" and "these days…" to show how participants perceived their past with respect to their present. Specifically, this section highlights the impacts of changing environmental and climatic conditions on older adults' traditional diets, their health status, and availability of herbal healthcare.

A common refrain among the participants was that their lifestyles had undergone irreversible changes in the past two to three decades. As shown in Fig. 14.1, in the past, participants consumed "pure food", lived in "pure environments", were physically strong and able (as they had been younger), and possessed traditional medical knowledge. The third and the fourth factors—physical strength and traditional medical knowledge—allowed them to harvest medicinal herbs from the high reaches of the Himalayan mountains and use them in various ways to cure "small illnesses", which were the only type prevalent in the region. In rare instances, if somebody was afflicted by an illness that could not be cured at home or by expert herbalists, the participants had to set off on long journeys, covering large distances, a good part of it on foot, to visit doctors in bigger towns and cities.

"Pure food" for the participants comprised local, natural, organically farmed beans, potatoes, apples, and a few varieties of vegetables. Rice, a staple, did not grow at their altitude and was brought from lower altitudes in exchange for potatoes, kidney beans, and high-altitude medicinal herbs. Edible oils and some spices were also exchanged/purchased from relatives, friends, or the marketplace. Seeds and roots were dried and processed at home. Several varieties of wild vegetables and tubers were harvested from the jungles. Typical meals comprised a combination of the following: unleavened barley or millet bread, rice, foraged greens, vegetables, potatoes, kidney beans, and wild or farmed fruits. Residents who had domestic

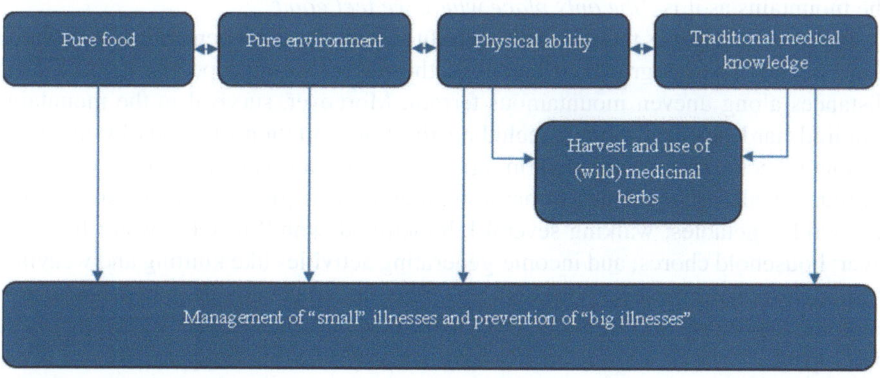


**Fig. 14.1** "In the olden days…"

animals also consumed and sold cow/goat/sheep milk, yogurt, butter, ghee (clarified butter), and mutton (goat and sheep meat). The limited, yet fresh and healthy diet, kept people physically and mentally fit. As one participant shared, "*...as long as my breathing trouble gets cured, I'm fine. I eat green vegetables and that cures it...I eat lentils and beans*". For older adults in the villages, healthy food serves as medicine.

Participants, however, pointed out that eating well was not enough on its own. To enjoy good health, one must also refrain from consuming unhealthy foods, drinks, and addictive substances. Participants believed that it was easier for them to stay healthy and active in the past due to unavailability/limited availability of addictive substances, fertilisers for farming, and processed/packaged foods and beverages. According to one participant, "*we [would] use cow dung...we would bring sheep/goat droppings and add that in our farms...we would grow...our original crops. That is what we ate. We did not add anything to our potatoes or apples...*" Several participants believed that they were still reaping the benefits of their past diets and health behaviours as they continued to age in relatively good health.

The second factor to which the participants attributed their good health and active lifestyles was their "pure environment". Located at an altitude of approximately 8000 feet, the participants received fresh drinking water from glacier melt. According to the participants, the glacial runoffs came in contact with medicinal herbs while flowing downstream, thereby enriching the mineral content and medicinal properties of the water. In addition to pure water, participants enjoyed clean, fresh, pollution-free mountain air; the lack of roads and vehicles allowed them this privilege. The participants also valued their natural surroundings of mountains and dense forests comprising hundreds of species of native flora, including medicinal plants, that purified the air. Moreover, the cool to cold temperatures throughout the year kept mosquitoes and vector-borne illnesses at bay. According to the participants, "*there is no pollution here. The air is pure, and the water is clean*", "*the climate is good...people work all day*", *and* "*the mountain air keeps the mind sane*". Pure environment was believed to have kept the participants healthy and physically active well into old age. One problem, however, was common among the older population—that of shortness of breath in the rarefied air. Even then, participants preferred to live in the mountains as it is "*the only place where we feel good*".

Physical ability was the third factor credited with the older generation's sustained good health. The geographical layout of the region required people to walk long distances along uneven mountainous terrain. Moreover, survival in the mountains required hard physical labour, including the transportation of goods by mules or on one's back; cattle rearing; farming; medicinal plant gathering from very high altitudes; walking to jungles to harvest/gather/collect (fire)wood, fodder, manure, and wild vegetables; walking several kilometres downhill to fetch water from the river; household chores; and income-generating activities like knitting and weaving. Although some of these activities are no longer undertaken by the participants, they continue to perform the remaining tasks to the best of their abilities. According to one participant, "*...since we stay mostly busy, we stay healthy. Too much leisure is not good. It becomes a difficulty.*" Participants believed that sustained physical activity can offer several health benefits, the most important being the ability to live and work independently well into old age, along with the prevention of "big illnesses".

The fourth and final component of good health in the past was that of traditional (medical) knowledge that had been passed down from older to younger generations. Consequently, almost all of the participants possessed skills and understanding about harvesting methods and the correct usage of medicinal herbs. One participant shared, "*my mother told me that I should eat this [herb]...she must have been told by her mother or mother-in-law that eating this [herb] may lead to this [cure]... somebody must have passed on this traditional knowledge, so we are also doing the same. It can only benefit*". This intergenerational transfer of knowledge contributed to the collective body of traditional medical knowledge that ensured the health and survival of the population. Traditional medical knowledge, however, formed only one part of the large body of indigenous knowledge that helped people to harness natural resources for energy needs, food production, clothmaking, construction, cattle rearing, and other activities that ensured the survival of the people in harsh terrains.

Traditional medical knowledge, when combined with physical ability, allowed the traditional medical system to flourish. This is because the most potent herbs are found in the highest reaches of the mountains. Accessing these areas required one to trek for a day, sometimes two, and spend a night in the area before descending the following day. The treks were often dangerous and required immense endurance, strength, agility, and stamina. Physical fitness alone, however, was not enough; people needed to possess sufficient medical knowledge in order to identify herbs and harvest them in the correct manner. Therefore, both physical fitness and medical knowledge were needed to ensure timely cures for "small illnesses".

***"These days…"***

As participants recounted their lifestyles of the past, they contrasted them with their present. According to them, much had changed in a short span of 20–30 years, mostly for the worse. Although several factors were responsible for altering the lifestyles of the people in the region, climate change had played a direct and significant role as it impacted the fundamental pillars of their health security: food and herbal medicine. Climate change in the region manifested as warmer temperatures, erratic and unseasonal rainfall, reduced snowfall, introduction of mosquitoes and pests, and climate-induced natural disasters. Figure 14.2 illustrates the chain of impacts that have altered the dietary composition, health status, and healthcare options available to the participants.

The impact of climate change on food cultivation has been immense. In general, the villages have limited cultivable land due to the steep and rocky geographical terrain. Moreover, natural disasters in the last decade have transformed portions of cultivable land into arable land due to the large quantities of silt and boulders that were washed down from higher altitudes during floods. Local climatic conditions support the production of only a limited variety of crops, primarily potatoes, kidney beans, apples, and some herbs and green vegetables. These continue to be supplemented with herbs, spices, salt, oil, rice, and wheat from marketplaces at lower altitudes. This dietary pattern has served the residents of the villages for several generations.

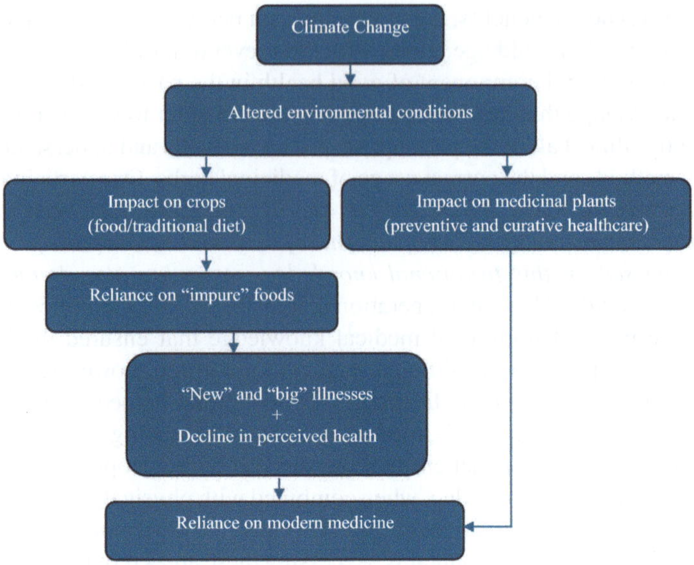


**Fig. 14.2** "These days…"

Climate change is now impacting the cultivation of these limited varieties of crops and forcing people to incorporate new foods, often packaged and processed, into their diet. The participants, however, preferred homegrown food as it was more nutritious, "*strengthening*", and "*pure*" due to organic farming methods. Homegrown food was also cheaper, fresher, and unadulterated.

Apple cultivation has long been the primary source of income for most households in the area due to the suitability of the climate and the soil. Climate change, however, has severely impacted this activity. Earlier, the villages received up to 6 feet of snow that froze and melted gradually, keeping the soil underneath cool and moist for several months. This was very favourable for the growth of sweet, juicy apples; it also protected the apple trees from pests. Now, however, there is not enough snowfall; it starts later than usual, and melts before it has had the opportunity to freeze. As a result, apple growers are forced to use chemical fertilizers and pesticides to protect the apples and grow them to full size in order to fetch better prices in the markets. According to one participant,

> Earlier the whole house would be buried…under the snow…we had to break it and enter the house…when it snows, the crops grow well. If there is no snow, there will be no crop. There is less softness…when it snows, it stays frozen on the ground. If it stays frozen it will melt gradually [into the soil]…that helps the crops to grow well.

Other food crops, mostly grown for personal consumption, are also being affected by climate change, especially by erratic rainfall patterns. Too much or too little rainfall, coupled with early or delayed onset of the monsoon season can destroy an entire season's harvest. As one participant shared, "*It should rain…the three months of monsoons…this time it didn't happen…this time the rain came [late]. That is why all our crops got destroyed…too much water filled up, the crops could not flower*". Crops are also being attacked by new pests that are migrating upward due to warmer temperatures. Untimely and inadequate rainfall is also impacting the regeneration of medicinal plants in their natural habitats, resulting in lower growth and supply of herbs. Additionally, changes in rainfall patterns are impacting the regeneration of grazing pastures for cattle, making them weak and thin, leading to reduced milk and meat yield. According to one participant, "*there isn't enough grass on the ground…we rear sheep…if there is less grass then our animals will be weak. It is a loss for us…*". In the villages, food, health, and income security are closely tied to natural resources, and any change in the climate affects people in multiple ways.

The impact of climate change on agriculture has directly increased dependence upon food purchased from markets. All participants agreed that there was a significant difference in the variety and quality of foods that they had consumed in the past compared to the foods that are available at present. They had also noticed that the incidence and prevalence of "big" and "new" illnesses had increased in tandem with increased consumption of market-bought foods. Additionally, the participants expressed disapproval of modern farming methods that have increasingly strayed from natural organic cultivation practices to high dependence on chemical fertilisers for maximum yield in minimum time. The participants attributed all "new" and "big" illnesses to farming practices that rely on "*chemical medicines*" to force the natural growth process of a food crop to adapt to an unnatural, artificially engineered growth cycle. According to one participant,

> Everything is because of food. It is all processed food these days….they put poison. That is why people fall sick…humans today are the same as the humans of bygone days. So, it is a problem with the food. They put fertilizer in everything. Take vegetables, for example…in the evening the plants are small…they add fertilizer, and the next morning the plant grows…that will obviously make us sick.

Participants also expressed suspicions regarding the adulteration of powdered spices and ingredients used in (ultra)processed snacks and beverages that are increasing gaining popularity in the mountains, especially among younger generations.

> Earlier, we were very happy with these herbs…now food and drinks are poisonous. As far I am concerned, even the medicines are poisonous…the illnesses are poisonous—all kinds of illnesses are being discovered…it happens from the food that we consume. Earlier we used to eat finger millet and local mountain grains. We were doing fine. Now we're consuming food from the plains—wheat, rice, processed food…no one knows what is added to these things…there has to be something poisonous, that why all this [big and new illnesses] is happening…we never had such things before…now people eat all kinds of food, drink all kinds of water [beverages], which is why they suffer from all kinds of illnesses.

Additionally, the participants shared concerns about food inflation and its impacts on the food products that they were able to purchase. For example, good quality cooking oil was often prohibitively expensive for some households that were forced to compromise on quality and buy cheaper alternatives which, according to them, led to "new illnesses".

Dependence on "*impure*", "*poisonous*" foods and the onset of "new" and "big" illnesses was driving people, old and young, to seek cures within the modern healthcare system. For most participants, the modern healthcare setup was unfamiliar, depersonalised, expensive, and replete with administrative formalities, medical jargon, tests, intimidating healthcare staff, long waits, and "*scary*" diagnoses. For people who were accustomed to addressing health issues within their homes and communities, with the help of their families and social network, and using trusted methods and treatments that had worked for centuries, using the modern healthcare system was not the preferred choice. Most participants initially tried to cure health issues with herbal remedies, failing which they reluctantly visited a modern medicine practitioner, often at the coaxing of younger members of the family who had more faith in modern medicine than traditional healthcare.

Modern medicine was not wholly trusted by the participants who believed that "*poisonous things*" were added to them. According to one participant,

> Now these [modern] medicines, they are not medicines. They mix poisonous things in them. That is how diseases get cured so easily. Earlier, medicinal herbs did not have poison. They would cure illnesses from the very root. These [modern] medicines will provide relief for some time and then [people] will suffer from some new illness.

The fact that a medicine used to cure one illness could cause a new illness (side effect) was difficult for the participants to understand. Consequently, many of them had developed a mistrust of modern medicine, with some even believing that modern medicine had gained undeserved recognition because it could not have progressed without the existence of medicinal herbs. According to one participant, "*all these medicines that are made by the government, they are all made with medicinal herbs. They mix things like powder and make medicines from them*".

Overall, participants believed that "*impure*" foods and increased leisure had given rise to illnesses that only modern medicine could cure. They were of the opinion that "*people these days do not survive without English medicine*" and that "*it has become fashionable to take tablets*" as soon as someone feels even slightly unwell. Participants also felt that modern medicine, unlike herbal medicine, may help some people but not others, thereby providing no guarantee of (healthy) longevity.

## Discussion

The aim of this chapter was to share the perspectives of older adults in the IHR about their struggles with climate change impacts on their diet, health, and healthcare systems. Older adults are especially vulnerable to climate impacts like heatwaves, natural disasters, and weather disruptions due to greater physiological susceptibility, disability, pre-existing health conditions, social vulnerability, and resource poverty (Help Age International 2015). Climate change also affects older adults' mental health by increasing the vulnerabilities of those with pre-existing conditions and triggering new conditions like trauma, shock, and survivor's guilt after natural disasters

(Ayalon et al. 2021). Additionally, older adults are likely to experience "solastalgia" or distress caused by environmental degradation in one's home environment, causing depression or physical illness (Albrecht 2006). While these are some direct impacts of climate change, this study highlights some indirect risks and challenges, like altered diets and healthcare choices, experienced by older adults who inhabit a region that is vulnerable to both slow and sudden onset climate events.

This study also demonstrates how rapidly climate change can alter the lives of people who interact closely with their natural environments. According to the participants, all the changes that they have witnessed have come about in the last 20–30 years. In that time, participants have had to adapt to several new realities while giving up lifestyle practices that had sustained generations of their ancestors. Now, within the span of one generation, their traditional knowledge and skills are becoming obsolete in the modern world. Their descendants, most of whom have migrated to cities for education and employment, have neither interest in nor use for older adults' traditional knowledge. This growing social and intellectual separation of generations in a deeply traditional society and the resulting decline in the roles of older adults as protectors, guides, and custodians of society are reminiscent of Cowgill's modernisation theory (1974) that suggests that as a society modernises, the status of its older population declines, especially in relation to health technology, economics, urbanisation, and education. According to Cowgill (1974), modernisation and urbanisation separates families, and younger members rarely regret the loss of centuries worth of accumulated wisdom, indigenous knowledge, and life skills. However, as the world increasingly looks to indigenous solutions for climate action, it may be hoped that intergenerational transmission of traditional knowledge will be revived soon.

## Limitations and Future Directions

This study provides insight into the lived experiences of a small population of older adults. The IHR, however, is an extremely diverse region and home to over 52 million people (Government of India 2019). This study, therefore, is limited in scope. Similar studies, may be undertaken with populations living in different Himalayan States, at varying altitudes, engaged in diverse traditional occupations, exhibiting varied patterns of migration and living arrangements, bound by locale-specific cultural traditions, and residing in very remote areas to identify similarities and differences of experiences. Future research that explores variations in the life experiences of older adults can only enhance the rather limited knowledge that exists about the hardships faced by this population in the IHR. In some cases, studies like this one could also inspire interventions to foster intergenerational learning whereby older generations can transfer traditional knowledge and skills to aid climate action, and younger generations can help older adults with the use of technology, revival of organic farming methods using modern techniques, and access to information.

## References

Ahirwar K, Khan J (2023) Medicinal properties of traditional foods and associated indigenous knowledge system: a case study of the Himalayan region, India. In: Ghosh S, Kumari Panda A, Jung C, Singh Bisht S (eds) Emerging solutions in sustainable food and nutrition security. Springer, Cham. https://doi.org/10.1007/978-3-031-40908-0_14

Albrecht G (2006) Solastalgia. Altern J 32(4/5):34–36

Ayalon L, Keating N, Pillemer K, Rabheru K (2021) Climate change and mental health of older persons: a human rights imperative. Am J Geriatr Psychiatry 29(10):1038–1040

Cowgill DO (1974) The aging of populations and societies. Ann Am Acad Pol Soc Sci 415(1):1–18

Government of India (2019) Status of ecosystem health in the Indian Himalayan region. https://dst.gov.in/sites/default/files/DST_Status%20of%20health%20of%20ecosystem%20in%20IHR%202.pdf

Government of Uttarakhand (2024) Uttarkashi. https://uttarkashi.nic.in/

Help Age International (2015) Climate change in an ageing world. https://reliefweb.int/sites/reliefweb.int/files/resources/COP21_HelpAge_PositionPaper_Final_0.pdf

Koiwai K, Takemi Y, Hayashi F, Ogata H, Matsumoto S, Ozawa K, Machado PP, Monteiro CA (2019) Consumption of ultra-processed foods decreases the quality of the overall diet of middle-aged Japanese adults. Public Health Nutr 22(16):2999–3008

Lăcătuşu CM, Grigorescu ED, Floria M, Onofriescu A, Mihai BM (2019) The Mediterranean diet: from an environment-driven food culture to an emerging medical prescription. Int J Environ Res Public Health 16(6):942

McGrath-Hanna NK, Greene DM, Tavernier RJ, Bult-Ito A (2003) Diet and mental health in the Arctic: Is diet an important risk factor for mental health in circumpolar peoples?—A review. Int J Circumpolar Health 62(3):228–241

Nomura M, Yamaguchi M, Inada Y, Nishi N (2023) Current dietary intake of the Japanese population in reference to the planetary health diet-preliminary assessment. Front Nutr 10:1116105

Patton MQ (1990) Qualitative evaluation and research methods. Sage Publications, Thousand Oaks, CA

Torres-Collado L, Rychter A, González-Palacios S, Compañ-Gabucio LM, Oncina-Cánovas A, de la Hera MG, Vioque J (2024) A high consumption of ultra-processed foods is associated with higher total mortality in an adult Mediterranean population. Clin Nutr 43(3):739–746

Tsugane S (2021) Why has Japan become the world's most long-lived country: Insights from a food and nutrition perspective. Eur J Clin Nutr 75(6):921–928

World Health Organisation (2024a) WHO global traditional medicine centre. https://www.who.int/initiatives/who-global-traditional-medicine-centre/about-us

World Health Organisation (2024b) Traditional, complementary and integrative medicine. https://www.who.int/health-topics/traditional-complementary-and-integrative-medicine#tab=tab_1

Uttarakhand Forest Department (2024) Medicinal and aromatic plants. https://forest.uk.gov.in/medicinal-and-aromatic-plants#:~:text=According%20to%20Samant%20et%20al,species%20are%20medicinal

# Chapter 15
# Developing an Urban Overheating Vulnerability Index in Addis Ababa, Ethiopia

**Seyoum Melese Eshetie**

**Abstract** Urban overheating poses significant health risks to populations residing in urban areas. The rise in temperatures in urban environments can adversely affect human health in various ways. The heat stress caused by urban overheating manifests in an increase of heart rate, blood pressure, and the risk of cardiovascular events such as heart attacks and strokes. The impact of urban overheating extends beyond physical health to mental well-being. Addis Ababa, the capital city of Ethiopia, is facing a significant heat-health challenge due to the increasing impacts of heat exposure on its population. This study aims to develop an urban overheating vulnerability index for human health, to locate hotspots of vulnerable areas to pinpoint locations with heightened susceptibility to heat-related health risks, and to examine the spatially varying impact of urban overheating on public health. The study identified that the spatial epidemiology of urban overheating impacts the urban population unfairly and unevenly. Secondly, geographically weighted regression was used to explore the spatially varying impact of urban overheating on population health. Accordingly, the sub cities of Addis Kalema, Arada, and Lideta exhibit the highest vulnerability to urban overheating, with a significant population at risk. Gulelie, Nifas Silk, and Akaki Kality follow closely as the second most vulnerable areas in terms of heat hazards. Bole and Yeka are identified as the sub cities with the third highest vulnerable population in this context. The findings call for targeted public health interventions focused on areas with heightened vulnerability to urban overheating, including preparedness plans and community-based heat mitigation strategies. Public awareness campaigns are also required to educate residents about the health risks associated with urban overheating and the importance of adopting heat mitigation measures.

**Keywords** Climate change · Human health · Urban overheating · Vulnerability index

S. M. Eshetie (✉)
Space Science and Geospatial Institute of Ethiopia, Addis Ababa, Ethiopia
e-mail: eshetieseyoum@gmail.com



W. Leal Filho and F. Wolf (eds.), *Climate Change and Health: Perspectives from Developing Countries*, Climate Change Management,
https://doi.org/10.1007/978-3-031-93177-2_15

## Introduction

Urban overheating poses significant health risks to populations residing in urban areas. The rise in temperatures in urban environments can adversely affect human health in various ways. The heat stress caused by urban overheating manifests in an increase in heart rate, blood pressure, and the risk of cardiovascular events such as heart attacks and strokes (Nazarian et al. 2022; Verdonck et al. 2019). The impact of urban overheating extends beyond physical health to mental wellbeing. Living in excessively hot urban environments can contribute to stress, anxiety, and sleep disturbances (Feng et al. 2023). Vulnerable populations, including the elderly and those with pre-existing cardiovascular conditions, are particularly at risk (Dushi and Berila 2022). The escalating urban overheating in Addis Ababa poses a significant threat to the health, wellbeing, and productivity of its residents. The city's rapid expansion, characterised by dense built-up areas, lack of green spaces, and inefficient urban planning, has contributed to the creation of an urban heat island effect (Report 2021). This phenomenon disproportionately affects vulnerable populations, such as the elderly, children, and low-income communities. The motivation for developing an urban overheating vulnerability index in Addis Ababa stems from the urgent need to address the city's vulnerability to urban heat islands and associated impacts on human health and wellbeing. The city's rapid urbanisation, coupled with the lack of comprehensive research on climate change vulnerability, underscores the necessity for quantifying, mapping, and prioritising adaptation measures to mitigate the effects of urban overheating. Furthermore, the integration of vulnerability assessment models and the identification of climate change hotspots in urban centres are crucial for preparing and implementing targeted adaptation strategies. Despite the increasing acknowledgment of urban overheating as a significant concern, there is a lack of thorough investigations evaluating vulnerability levels at a detailed level in Addis Ababa. This study seeks to fill this knowledge gap by creating an urban overheating vulnerability index (UOVI) to pinpoint hotspots and guide specific interventions (Eshetie 2024).

The objectives of this research are as follows: (a) to develop a map of an urban overheating vulnerability index for human health; (b) to identify hotspots and vulnerable areas to pinpoint locations with heightened susceptibility to heat-related health risks; and (c) to examine the spatially varying impact of urban overheating on public health. Developing an urban overheating vulnerability index for Addis Ababa is crucial for allowing targeted interventions and resource allocation.

## Research Method

### *Data Source*

Developing a robust Urban Overheating Vulnerability Index (UOVI) requires a comprehensive dataset encompassing various physical, environmental, and socio-economic factors. The following data sources can be considered:

1. Satellite remote sensing images obtained from the NASA Earth Data portal (https://lpdaac.usgs.gov/) were analysed to extract information about urban overheating, normalized difference vegetation index (NDVI), and normalized difference built-up index (NDBI).

(a) Extracting urban overheating from Landsat data (Li et al. 2013).

i. Calculation of TOA (Top of Atmospheric) spectral radiance.

$$\mathrm{TOA(L)} = \mathrm{ML} * \mathrm{Q_{cal}} + \mathrm{AL}$$

ML = Band-specific multiplicative rescaling factor from the metadata;
$Q_{cal}$ = corresponds to band 10.
AL = Band-specific additive rescaling factor from the metadata.

ii. TOA to Brightness Temperature conversion (Jeevalakshmi et al. 2017).

$$\mathrm{BT} = (\mathrm{K_2}/(\ln(\mathrm{K_1}/\mathrm{L}) + 1)) - 273.15$$

$K_1$ = Band-specific thermal conversion constant from the metadata; $K_2$ = Band-specific thermal conversion constant from the metadata.

iii. NDVI Calculation (Guha et al. 2018).

$$\mathrm{NDVI} = (\mathrm{Band\ 5} - \mathrm{Band\ 4})/(\mathrm{Band\ 5} + \mathrm{Band\ 4})$$

Note that the calculation of the NDVI is important because, subsequently, the proportion of vegetation ($P_v$), which is highly related to the NDVI, and emissivity (ε), which is related to the $P_v$, must be calculated.

iv. Calculate the proportion of vegetation $P_v$.

$$\mathrm{P_v} = [(\mathrm{NDVI} - \mathrm{NDVI_{min}}) / (\mathrm{NDVI_{max}} - \mathrm{NDVI_{min}})]^2$$

v. Emissivity ε.

$$\varepsilon = 0.004 * \mathrm{P_v} + 0.986$$

vi. Calculate urban overheating using the land surface temperature (see Fig. 15.1).

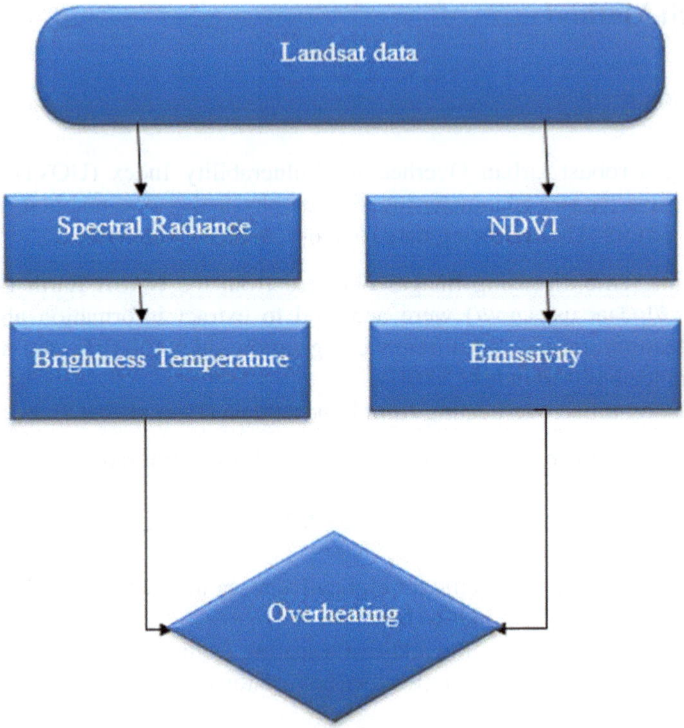


**Fig. 15.1** Flowchart of remote sensing data processing

$$\text{LST} = (\text{BT}/(1 + (0.00115 * \text{BT}/1.4388) * \text{Ln}(\varepsilon)))$$

2. Normalized Difference Vegetation Index (NDVI) was calculated from assessed vegetation cover and its impact on urban cooling.

$$\text{NDWI} = \frac{(\text{NIR}-\text{SWIR})}{(\text{NIR} \ + \text{SWIR})}$$

3. Normalized Difference Built-up Index (NDBI) was extracted to identify built-up areas and their contribution to heat storage.

$$\text{NDBI} = \frac{(\text{NIR}-\text{SWIR})}{(\text{NIR} + \text{SWIR})}$$

4. Population density has been calculated to identify areas with high population concentration and potential heat vulnerability. Demography data was acquired from World Pop (https://www.worldpop.org/).

## *Development of Urban Overheating Vulnerability Index*

### Spatial Weighted Overlay Approach

The spatial weighted overlay method, utilising the adaptive layer, exposure layer, and sensitivity layer, is employed to produce an urban overheating vulnerability index. This approach integrates various spatial characteristics, environmental exposure, and human sensitivity factors to assess the vulnerability of different areas to urban overheating. By combining the adaptive capacity layer (represented by the normalized difference vegetation index, NDVI), the exposure layer (urban overheating data), and the sensitivity layer (population density data and normalized difference built-up index, NDBI), this method aims to create a comprehensive index that accounts for both environmental and socio-economic factors.

### Geographically Weighted Regression (GWR)

Geographically Weighted Regression was used to explore the spatially varying impact of urban overheating on population health. GWR can help identify spatial patterns and variations in the relationship between urban overheating and public health (Czarnota et al. 2015). GWR takes into account the spatial proximity of areas and the influence of neighbouring zones with similar exposure to urban overheating. GWR can provide a more nuanced understanding of how urban overheating affects population density in different areas (Dushi and Berila 2022).

In GWR models, the spatial variability between the response variable (urban overheating) and the explanatory variable is taken into consideration. The GWR is computed as:

$$\gamma = \beta_{(0)}(u_i, v_i) + \sum_k [\beta_k(u_i, v_i)X_{ik} + ]\varepsilon_i$$

where $u_i$ and $v_i$ represent the point coordinates of ith in space. Thus, the GWR equation distinguishes that spatial variations in relationships might exist and provides a way in which they can be measured (Zhao et al. 2018).

## Limitation of the Research

One of the most significant challenges in developing an urban overheating vulnerability index for Addis Ababa is the availability and quality of data. While remote sensing data has advanced significantly, there are still limitations in terms of spatial and temporal resolution, particularly for historical data. Moreover, socioeconomic data, which is crucial for assessing vulnerability, often lacks the necessary detail and

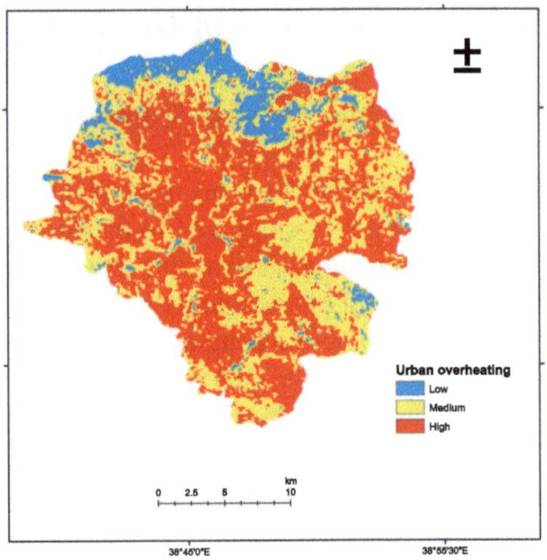


**Fig. 15.2** Urban overheating exposure map (map created by the author)

granularity. For instance, data on income levels, population density, and building characteristics at a micro-level have been found to be limited or unavailable.

## Results

### *Urban Overheating Vulnerability Index*

The composite urban overheating vulnerability index revealed distinct spatial patterns of vulnerability within Addis Ababa. Areas with high population density and extensive built-up surfaces exhibited heightened vulnerability to urban overheating, indicating the influence of urbanisation and land use on the exposure and sensitivity of the population to heat-related risks (Fig. 15.2). Furthermore, the analysis highlighted the role of vegetation cover, as indicated by NDVI, in mitigating urban overheating vulnerability, particularly in areas with high population density.

### *Impacts on Human Health*

The assessment of the composite urban overheating vulnerability index elucidated the heightened vulnerability of specific areas within Addis Ababa to heat-related health outcomes. The findings underscored the significant association between high urban overheating vulnerability and an increased incidence of heat-related illnesses among residents in these areas. Additionally, the analysis emphasised the potential

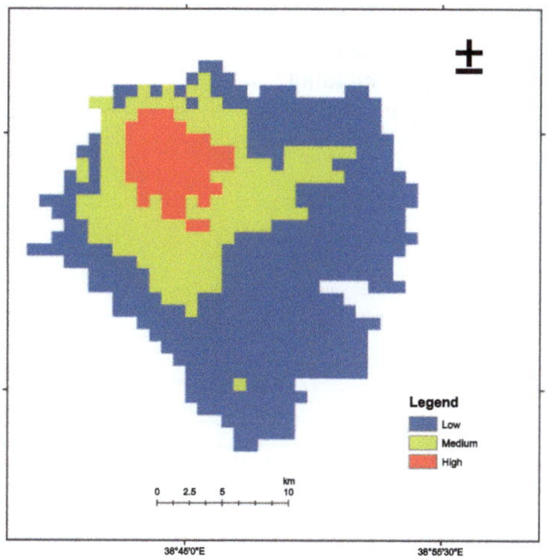


**Fig. 15.3** Level of population sensitivity to urban overheating (map created by the author)

of vegetation cover, as indicated by NDVI, to act as a protective factor against heat-related health risks, thereby highlighting the importance of green infrastructure in reducing vulnerability to urban overheating.

The results of this study have significant implications for urban planning and public health strategies in Addis Ababa. The spatial analysis of the composite urban overheating vulnerability index provides crucial evidence for the development of targeted interventions to mitigate the adverse effects of urbanisation on public health. The identification of areas with heightened vulnerability to urban overheating and associated health risks underscores the need for incorporating these findings into urban development policies and public health initiatives.

By integrating exposure, sensitivity, and adaptive capacity layers through a weighted overlay approach to create a composite urban overheating vulnerability index, this research has contributed to the understanding of the complex relationships between urban overheating and human health outcomes (see Figs. 15.3, 15.4, 15.5, 15.6).

## *Geographically Weighted Regression*

The research findings indicate that the impact of overheating on public health varies significantly across different locations. This is evident from the geographically weighted regression local r-squared values, which range from 0.34 to 0.78. These findings suggest that the relationship between overheating and public health outcomes is not uniform and is influenced by local contextual factors. According to the research finding, the sub cities of Addis Kalema, Arada, and Lideta exhibit the highest vulnerability to urban overheating, with a significant population at risk. Gulelie, Nifas Silk, and Akaki Kality follow closely as the second most vulnerable

Fig. 15.4 Level of sensitivity to urban overheating by building density (map created by the author)

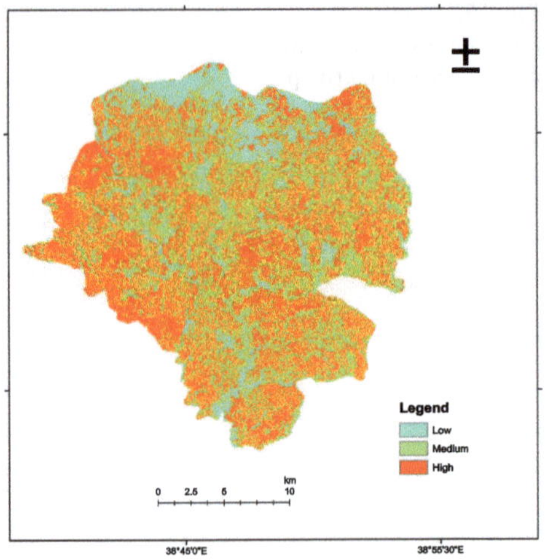


Fig. 15.5 Level of adaptive capacity by vegetation coverage (map created by the author)

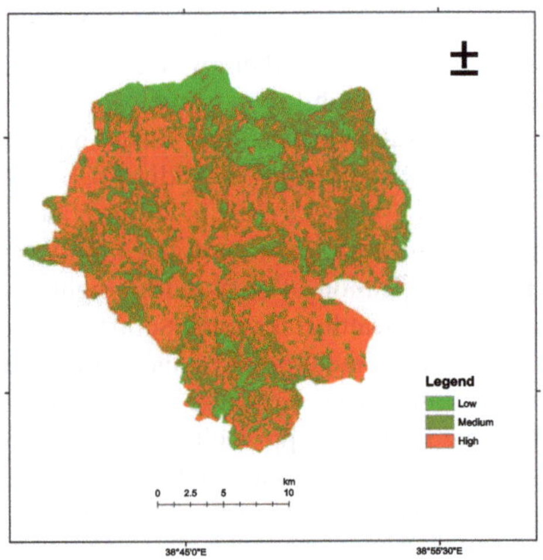


areas in terms of heat hazards. Bole and Yeka are identified as the sub cities with the third highest vulnerable population in this context.

The variation in the local r-squared values implies that certain areas experience a stronger association between overheating and public health issues. In these locations, overheating poses a more significant threat to public health, leading to a higher proportion of adverse health outcomes. Additionally, socio-economic factors and demographic characteristics of a location can also contribute to the variation in the impact of overheating on public health. Vulnerable populations, such as the elderly,

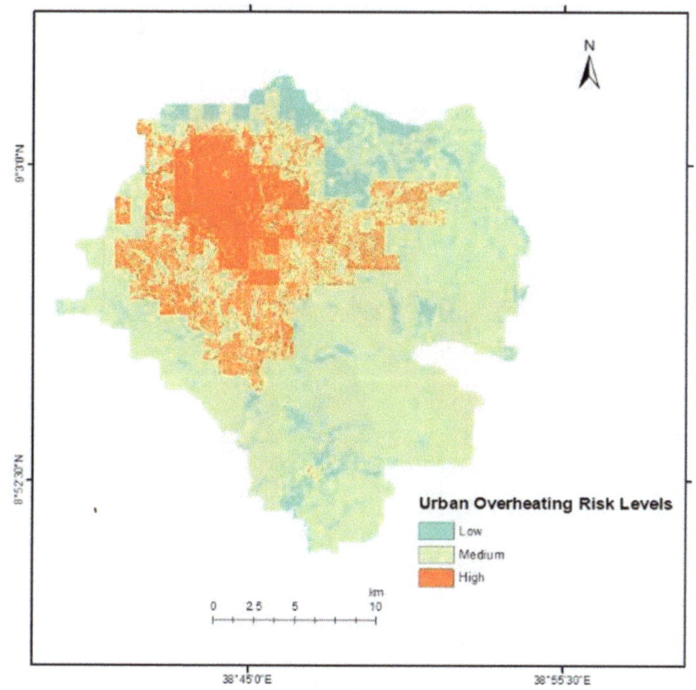


**Fig. 15.6** Urban overheating risk level (map created by the author)

young children, and individuals with pre-existing health conditions, are more susceptible to the adverse effects of overheating. Therefore, areas with a higher concentration of these vulnerable groups may experience a stronger association between overheating and public health issues.

It is crucial to consider the variation in the impact of overheating on public health when developing strategies and policies to address this issue. A one-size-fits-all approach may not be effective in mitigating the health risks associated with overheating. Instead, targeted interventions tailored to the specific needs and characteristics of each location should be implemented. The research findings highlight the varying impact of overheating on public health across different locations. The range of geographically weighted regression local r-squared values indicates that the relationship between overheating and public health outcomes is influenced by local factors. Understanding and addressing this variation is essential for developing effective strategies to protect public health in the face of increasing temperatures. GWR has been employed to model the relationship between urban overheating and population density at a local scale. This technique allows for the estimation of spatially varying regression coefficients, enabling a more nuanced understanding of the relationship within different areas of Addis Ababa (Fig. 15.7).

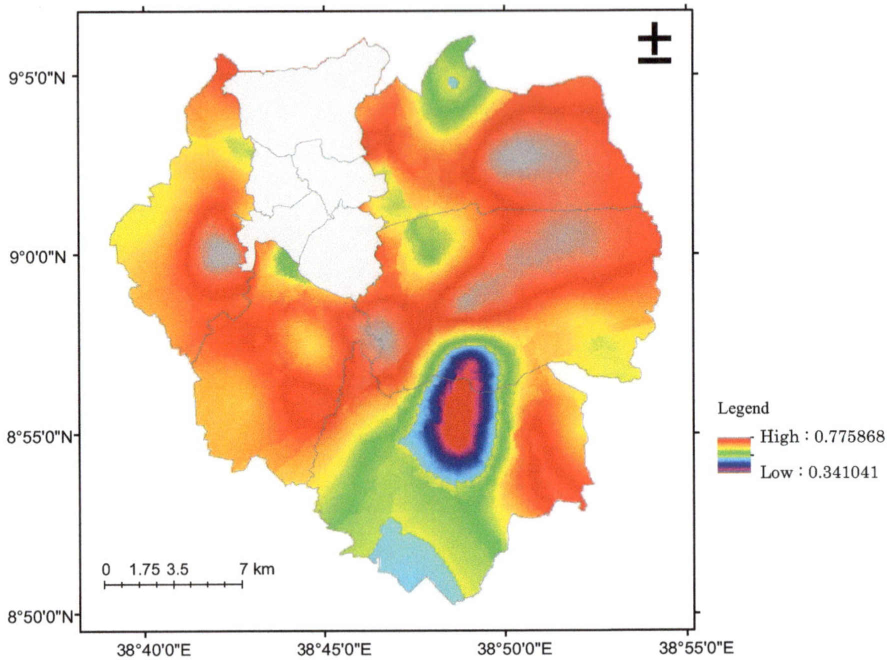


**Fig. 15.7** Urban overheating vulnerability levels (map created by the author)

## Conclusion and Recommendation

The comprehensive urban overheating vulnerability index and its impacts on human health was created through a weighted overlay approach and revealed critical insights into the spatial dynamics and implications of urban overheating for public health. The integration of the exposure layer, sensitivity layer, and adaptive capacity layer data has provided an understanding of the complex relationships in the vulnerability of the urban population to heat-related health risks.

The spatial mapping of urban overheating, coupled with the assessment of sensitivity to urban overheating and the evaluation of adaptive capacity, has highlighted the heightened vulnerability of specific areas within Addis Ababa to the impacts of urban overheating on human health. The identification of hotspots and vulnerable areas through the composite vulnerability index underscores the urgency of targeted interventions and evidence-based policies to mitigate the adverse effects of urban overheating on the wellbeing of the city's residents. Based on the identified hotspots and vulnerable areas, it is crucial to integrate urban overheating considerations into urban planning strategies, emphasising the need for sustainable urban design, green infrastructure, and the preservation of green spaces to mitigate the impacts of urban overheating on human health.

The findings call for targeted public health interventions focused on areas with heightened vulnerability to urban overheating, including preparedness plans, community-based heat mitigation strategies, and health education programmes to raise awareness about heat-related health risks. Encouraging the expansion and preservation of green spaces and vegetation cover within urban areas can serve as a critical adaptive measure to reduce vulnerability to urban overheating, contributing to improved public health outcomes and enhanced urban livability.

Public awareness campaigns are required to educate residents about the health risks associated with urban overheating and the importance of adopting heat mitigation measures. Collaborations with local community organisations and schools are also needed to promote sustainable practices, such as water conservation and tree planting, to mitigate the urban heat island effect.

Future research could focus on conducting longitudinal studies to assess the evolving nature of urban overheating and its consequences over time. Advancing data-driven modelling techniques, such as machine learning and spatial analysis, to predict urban overheating vulnerability at a fine scale would improve the precision of vulnerability assessments. This could involve leveraging high-resolution satellite imagery and advanced modelling algorithms.

## References

Czarnota J, Wheeler DC, Gennings C (2015) Evaluating geographically weighted regression models for environmental chemical risk analysis. Cancer Inform 14:117–127. https://doi.org/10.4137/CIN.S17296

Dushi M, Berila A (2022) Determining the influence of population density on the land surface temperature based on remote sensing data and GIS techniques: application To Prizren, Kosovo. Sci Rev Eng Environ Sci 31(1):47–62. https://doi.org/10.22630/srees.2324

Eshetie SM (2024) Exploring urban land surface temperature using spatial modelling techniques: a case study of Addis Ababa city, Ethiopia. Sci Rep. https://doi.org/10.1038/s41598-024-55121-6

Feng J, Gao K, Khan H, Ulpiani G, Vasilakopoulou K, Young Yun G, Santamouris M (2023) Overheating of cities: magnitude, characteristics, impact, mitigation and adaptation, and future challenges. Annu Rev Environ Resour 48:651–679. https://doi.org/10.1146/annurev-environ-112321-093021

Guha S, Govil H, Dey A, Gill N (2018) Analytical study of land surface temperature with NDVI and NDBI using Landsat 8 OLI and TIRS data in Florence and Naples city, Italy. Eur J Remote Sens 51(1):667–678. https://doi.org/10.1080/22797254.2018.1474494

Jeevalakshmi D, Narayana Reddy S, Manikiam B (2017) Land surface temperature retrieval from LANDSAT data using emissivity estimation. Int J Appl Eng Res 12(20):9679–9687

Li Z-L, Tang B-H, Wu H, Ren H, Yan G, Wan Z, Trigo IF, Sobrino JA (2013) Satellite-derived land surface temperature: current status and perspectives. Remote Sens Environ 131:14–37. https://doi.org/10.1016/j.rse.2012.12.008

Nazarian N, Krayenhoff ES, Bechtel B, Hondula DM, Paolini R, Vanos J, Cheung T, Chow WTL, de Dear R, Jay O, Lee JKW, Martilli A, Middel A, Norford LK, Sadeghi M, Schiavon S, Santamouris M (2022) Integrated assessment of urban overheating impacts on human life. Earth's Future 10(8). https://doi.org/10.1029/2022EF002682

Report T (2021) NASA DEVELOP National program San Diego urban development

Verdonck M, Demuzere M, Hooyberghs H, Priem F, Van Coillie F (2019) Heat risk assessment for the Brussels capital region under different urban planning and greenhouse gas emission scenarios. J Environ Manag 249(July 2018):109210. https://doi.org/10.1016/j.jenvman.2019.06.111

Zhao C, Jensen J, Weng Q, Weaver R (2018) A geographically weighted regression analysis of the underlying factors related to the surface urban heat Island phenomenon. Remote Sens 10(9). https://doi.org/10.3390/rs10091428

# Chapter 16
# Local Strategies to Improve Urban Resilience and Adaptation to Climate Change to Reduce the Health Vulnerabilities in a Metropolitan City of India

**Shailendra K. Mandal, Ruchita Rashmi, and Tarun Pant**

**Abstract** Climate change poses significant threats to human health, particularly in densely populated metropolitan areas like Patna, India. This study provides a comprehensive analysis of health vulnerabilities associated with climate change in Patna and identifies effective local strategies to enhance the city's resilience. An interdisciplinary approach, integrating public health, environmental science, and urban planning, is employed. Using Geographic Information Systems (GIS), the study incorporates Land Use and Land Cover (LULC) analysis, Land Surface Temperature (LST) analysis, rainfall pattern analysis, and isochrone mapping. LULC analysis reveals changes in land use over the decades while LST analysis identifies temperature variations across the region. Rainfall pattern analysis explores shifts in precipitation, offering insights into flood risks and water management challenges. Isochrone mapping assesses access to critical services during extreme weather events. Recommendations include increasing green spaces, improving stormwater and sanitation management, and enhancing healthcare accessibility. The study highlights the city's vulnerability to extreme weather, vector-borne diseases, and disproportionate effects on vulnerable populations such as the elderly, children, and low-income areas. By integrating GIS-based analyses into urban planning, this research presents a framework for climate resilience, offering actionable recommendations to inform policymaking, urban planning, and public health interventions, ultimately improving Patna's capacity to adapt to climate change and safeguard its residents' health.

S. K. Mandal (✉) · R. Rashmi · T. Pant
Department of Architecture and Planning, National Institute of Technology, Patna, India
e-mail: shailendra@fulbrightmail.org

R. Rashmi
e-mail: ruchitar.pg22.ar@nitp.ac.in

T. Pant
e-mail: tarun.ar18@nitp.ac.in

 
W. Leal Filho and F. Wolf (eds.), *Climate Change and Health: Perspectives from Developing Countries*, Climate Change Management,
https://doi.org/10.1007/978-3-031-93177-2_16

**Keywords** Climate change · Urban planning · Climate resilience · Health vulnerabilities · Geographic information system (GIS)

## Introduction

Climate change, a pressing global challenge of our time, poses significant threats to human health and wellbeing (McMichael and Kovats 2000). Its far-reaching effects are being felt across the globe, characterised by escalating temperatures, a surge in the frequency and intensity of extreme weather events, altered precipitation patterns, and significant shifts in ecosystems (Berrang-Ford et al. 2021; Debnath et al. 2024; Ebi 2024; Marsland and Staples 2024; McMichael and Kovats 2000; Patz and Olson 2006; Tong and Ebi 2019). These changes not only disrupt the intricate balance of natural systems but also pose substantial risks to human health, exacerbating existing health disparities and presenting unprecedented challenges to healthcare systems worldwide.

**Climate Change and Health**: Climate change, characterised by long-term shifts in global weather and rising temperatures, presents a major global challenge with significant impacts on human health (IPCC 2022). This complex issue has gained increasing scientific attention due to its varied and interconnected health effects (Shivanna 2022). Direct impacts include heat-related illnesses and an increased frequency of extreme weather events, which contribute to injuries, displacement, and mental health issues like anxiety and PTSD (Cruz et al. 2020).

Indirect health impacts are troubling, affecting communities through less obvious pathways such as changes in disease transmission, food and water security, and air quality. Rising temperatures and altered precipitation promote vector-borne diseases, while disruptions in agriculture can cause food insecurity and malnutrition (Rocklöv and Dubrow 2020; Mirzabaev et al. 2023).

**Vulnerable Populations and Health Disparities**: While climate change impacts everyone, certain populations suffer more, worsening social, economic, and political inequalities. Low- and middle-income countries, especially in tropical areas, face severe effects due to limited resources and infrastructure, with marginalised groups experiencing increased health risks and deepened disparities (People Who Are Vulnerable to Climate Change n.d.).

Studies highlight the socio-economic factors driving these differential vulnerabilities, such as poverty, limited access to healthcare, and inadequate housing. Urban slum dwellers, for example, are more prone to heat-related illnesses due to overcrowded living conditions, lack of green spaces, and poor access to clean water and sanitation. Similarly, indigenous communities in remote or environmentally sensitive regions experience heightened risks of food insecurity, waterborne diseases, and mental health challenges due to disruptions in their traditional livelihoods and cultural practices (Friesen et al. 2020).

**Adaptation and Resilience Strategies**: In response to the health impacts of climate change, governments and organisations worldwide have developed various adaptation and resilience strategies. These include improving healthcare infrastructure, enhancing disease surveillance systems, establishing early warning mechanisms for extreme weather, and promoting sustainable agriculture and water management (US EPA 2024).

Resilience-building focuses on bolstering the ability of individuals, communities, and systems to manage and recover from climate-related health effects. Efforts include strengthening social networks, investing in education, and integrating climate considerations into healthcare policies (Mosadeghrad et al. 2023).

**Climate change and health in Indian context**: In India, the intersection of climate change and health is notably complex due to the country's vast population and climatic diversity. From the snow-capped Himalayas to the coastal plains of the Indian Ocean, India's varied landscapes host over a billion people, many of whom are highly vulnerable to climate-related health impacts (Bush et al. 2011; Dhiman et al. 2010; Dholakia and Garg 2018; Dimitrova et al. 2021; Karmakar and Pradhan 2020; Majra and Gur 2009; Moors et al. 2013; Nomani and Parveen 2020; Picciariello et al. 2021). These disruptions lead to a range of health risks, including:

a. **Increased Infectious Diseases**: Climate change influences the distribution and transmission of vector-borne diseases such as malaria, dengue, and chikungunya. Warmer temperatures, changes in rainfall patterns, and altered habitats create favourable conditions for disease vectors, leading to increased incidence and geographic spread of these diseases.
b. **Heat-Related Illnesses**: Rising temperatures and heatwaves pose significant health risks, particularly for vulnerable populations such as the elderly, children, and those with pre-existing health conditions. Heat-related illnesses, including heat exhaustion and heatstroke, can lead to increased morbidity and mortality.

   Nearly 100 people were killed in northern India in 2023 due to a severe heatwave, with temperatures reaching up to 45 °C. Hundreds were hospitalised, especially in Uttar Pradesh and Bihar, as heat-related illnesses surged (Sharma 2023).
c. **Respiratory Problems**: Air pollution, worsened by climate change factors like dust storms, wildfires, and industrial emissions, leads to respiratory issues such as asthma, bronchitis, and lung cancer. Prolonged exposure to PM2.5 and PM10, due to their tiny size and toxic chemical composition, exacerbates chronic respiratory and immune problems, particularly in those with existing conditions (Manisalidis et al. 2020). In Patna, severe air pollution persists despite two years of the Clean Air Action Plan, with high PM2.5 levels and challenges in managing vehicular emissions and construction dust. The worsening air quality, especially in winter, underscores the need for more effective measures (Hindustan Times 2021).
d. **Mental Health Disorders**: The psychological impacts of climate change, including anxiety, depression, and post-traumatic stress disorder (PTSD), are often overlooked but can have profound effects on mental wellbeing.

e. **Nutritional Deficiencies**: Changes in agricultural yields and food production patterns due to climate change can impact food security and nutrition, leading to deficiencies in essential nutrients.

## *Research Questions*

a. How do various climatic factors co-relate with LULC changes, and what is their impact on health vulnerabilities in a metropolitan city of India?
b. How do urban rainfall patterns aggravate health risks in vulnerable populations, and how can local adaptation strategies mitigate these effects?

## *Methodology*

The methodology employed a remote sensing-based approach to analyse the spatial and temporal dynamics of urban environments and their effects on health vulnerabilities in the context of climate change. Satellite imagery-high-resolution data from Landsat was used to produce Land Use and Land Cover (LULC) maps. Supervised classification techniques were applied to track LULC changes over time. Land Surface Temperature (LST) data, extracted from thermal infrared bands, was used to evaluate temperature variations across different land cover types. Satellite-derived precipitation data was analysed in conjunction with LULC and LST to assess their combined impact on urban health risks. By integrating remote sensing data with Geographic Information System (GIS) tools, spatial overlays and analyses were conducted to identify high-risk areas. These insights were then correlated with health data to evaluate the effects of climate-related factors on vulnerable populations, informing targeted strategies to enhance urban resilience and adaptation.

### Significance of Study

As climate change increasingly impacts cities, especially in developing nations like India, the need for localised strategies becomes critical. Urban centres are particularly vulnerable due to high population density, inadequate infrastructure, and socio-economic disparities. This study emphasises the importance of developing context-specific adaptation strategies that enhance urban resilience, particularly in safeguarding public health. By focusing on local conditions, the research highlights the necessity of tailored approaches that can effectively mitigate health risks and improve the overall wellbeing of city residents. The study's significance lies in its potential to inform policymakers and urban planners, enabling them to implement more effective, sustainable measures to combat the adverse effects of climate change in metropolitan environments.

## Case Study: Metropolitan City Patna

Patna (25.4948° N, 85.3550° E), the capital city of Bihar, India, is situated on the southern bank of the river Ganga and is home to over 3.62 million people. The city has a tropical climate, with hot summers and mild winters. Patna Metropolitan Area (Fig. 16.1) encompasses five municipal councils and one municipal corporation within the Patna district. As per the 2011 Census of India (as the official Census of India for 2021 was not yet available), the population of the Patna Metropolitan Region stood at approximately 3.62 million. Spanning an area of 1156 km$^2$, the study area includes 13 blocks: Patna Rural, Phulwari, Punpun, Sampatchak, Bihta, Daniawan, Dhanarua, Danapur, Khusrupur, Maner, Masaurhi, Naubatpur, and Fatwah. In recent years, Patna has been experiencing the effects of climate change, which is leading to a number of health problems for its residents (Neeti et al. 2023). These include heat-related illnesses, respiratory problems, and waterborne diseases. In addition, climate change is also affecting the mental health of Patna's residents (Nezami 2023a, b).

### *Land Use Land Cover Change Analysis*

Land Use and Land Cover (LULC) changes are linked to global climate change and urban and agricultural development (Srivastava et al. 2012). Understanding the

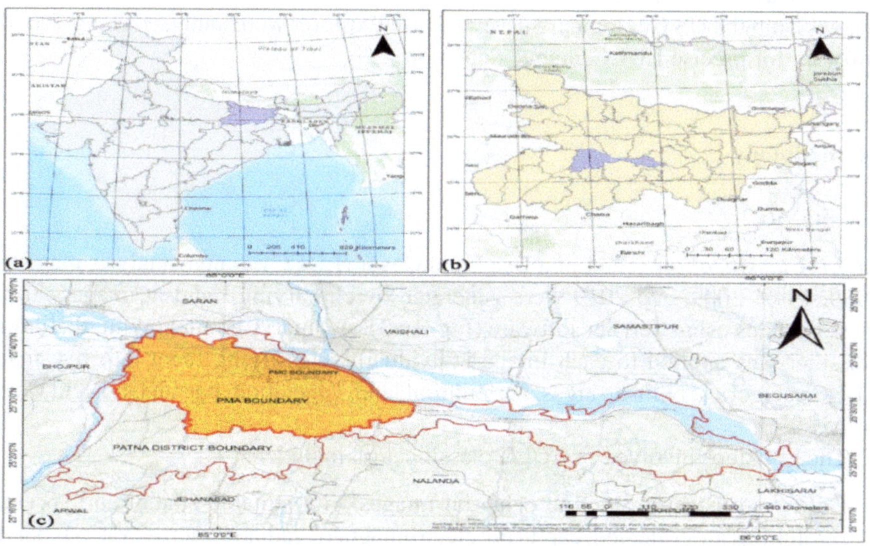


**Fig. 16.1** Map of Patna metropolitan area (PMA), **a** Outline Map of India showing location of Bihar **b** Outline Map of Bihar Districts showing Patna District **c** Administrative boundaries of Patna District (3202 sq. km), Patna Metropolitan Area (1156 sq. km), Patna Municipal Area (109 sq. km). *Source* Author

complex relationship between LULC changes and regional climate is key for environmental adaptation and management. These changes affect land-surface-atmosphere interactions, altering weather patterns and climate conditions. Remote sensing of LULC through satellite multi-spectral imagery has become a major focus in research, providing accurate maps that are crucial for monitoring climate change, biodiversity, and environmental factors. Remote sensing and Geographical Information Systems (GIS) are recognised as effective tools for managing land resources and natural landscapes (Mishra and Rai 2016).

## Materials and Methods

Satellite remote sensing data is a critical tool for change detection, offering frequent temporal coverage and a broad range of spatial and spectral resolutions (Hussain et al. 2013). This study utilised multi-decadal Landsat data to classify land use and land cover (LULC) within the study area. The temporal aspect of this data is essential for identifying regions experiencing significant changes. Geoinformatics is used to correlate spatial contexts with LULC changes, considering key factors like population growth, climate variations, and terrain characteristics, which drive these transformations (Roy and Roy 2010). Landsat imagery, known for its high spectral, spatial, and temporal resolutions, is widely employed for environmental monitoring and urban planning (Sadidy et al. 2012). Stratified random sampling was applied to select 30–35 representative training pixels per class, following established guidelines (Srivastava et al. 2012). Landsat-5 TM (1989, 1999, 2009) and Landsat-8 OLI (2019) images from the US Geological Survey's Earth Explorer were analysed using TerrSet software for precise LULC change detection.

## Methodology

See Fig. 16.2.

***LULC Classification***: The Land Use Land Cover (LULC) change analysis maps for 1989, 1999, 2009, and 2019 were generated by classifying Landsat imagery from these decades using TerrSet software (Fig. 16.3). A supervised maximum likelihood classifier was applied to each image individually, classifying them into five main LULC categories: built-up areas, agricultural land, vegetation, barren land, and water bodies.

The key steps involved in LULC classification include:

- **Data Acquisition**: Satellite or aerial images with suitable spatial and spectral resolutions are collected, capturing electromagnetic radiation across various bands (visible, infrared, thermal) based on the study's timeframe, season, and land cover characteristics.
- **Pre-processing**: The images are corrected for atmospheric, sensor, and geometric distortions and spatially referenced to the UTM projection system (zone 45 north)

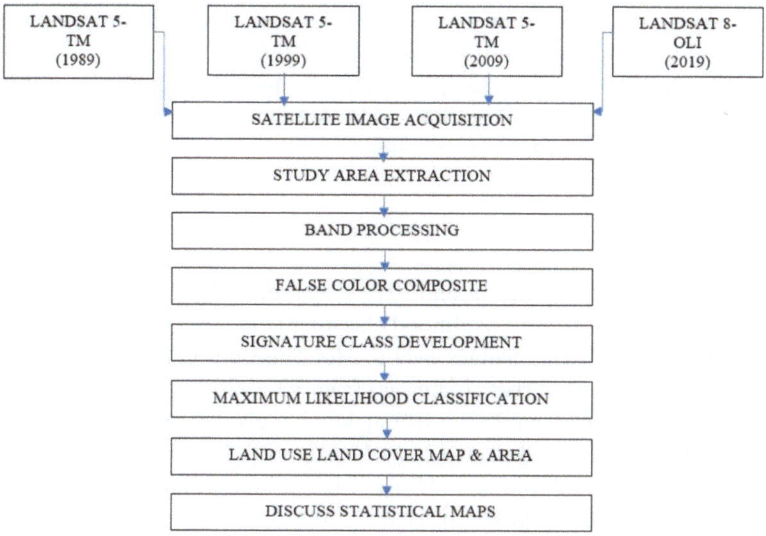


**Fig. 16.2** Flowchart of methodology—LULC analysis. *Source* Author

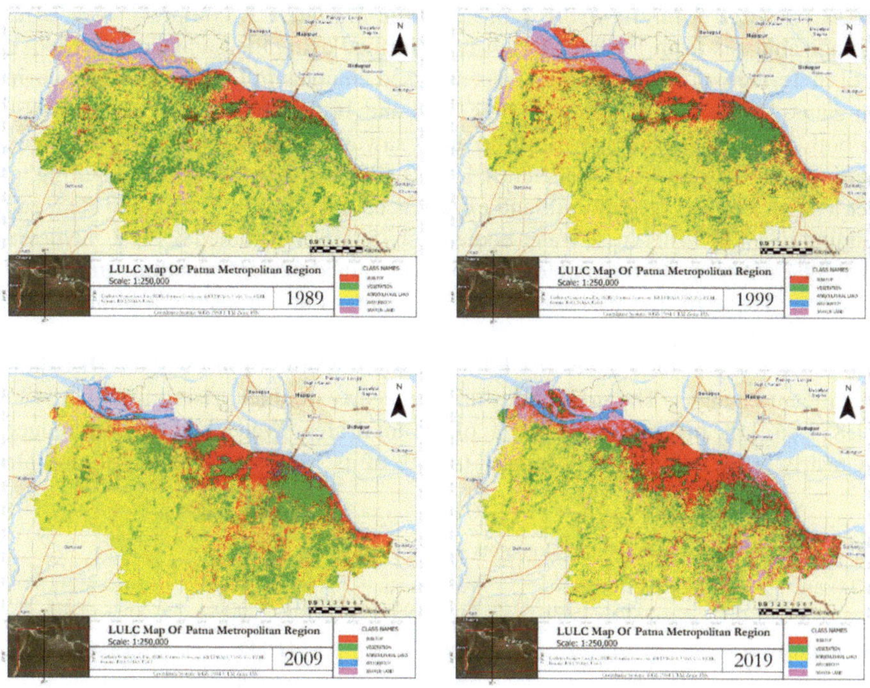


**Fig. 16.3** LULC maps generated for the years 1989, 1999, 2009 and 2019. *Source* Author

using WGS 1984 datum with a 30-m pixel size. False color composites are generated for selecting training samples for LULC classes.

- **Training Inputs**: Training samples for each land cover class are gathered for supervised classification. Algorithms like Maximum Likelihood classify pixels to ensure accurate LULC categorisation.

## Result

LULC maps for 1989, 1999, 2009, and 2019 were created by classifying Landsat TM/OLI images with a maximum likelihood classifier (Fig. 16.3). The analysis of these maps shows the following changes over the four decades (1989–2019) (Table 16.1 and Fig. 16.4): **Built-up Areas**: Increased from 93 to 200 sq. km, nearly doubling. **Barrenland**: Decreased from 133 to 109 sq. km, largely due to conversion to built-up areas. **Vegetation**: Declined from 225 to 177 sq. km, due to settlement expansion. **Agricultural Land**: Reduced from 690 to 647 sq. km, primarily due to conversion for development. **Water Bodies**: Increased from 13 to 20 sq. km, possibly due to altered precipitation patterns, temperature changes, extreme weather, and land use changes.

***LULC Change Analysis using Land Change Modeler***: A comprehensive LULC change analysis from 1989 to 2019 was performed using the Land Change Modeler tool in Terrset software. This analysis assessed changes in LULC classes, focusing on gains, losses, and persistence (Figs. 16.4 and 16.5). Most classes showed losses in both area and percentage. Built-up areas saw a 48.08% loss but a 76.17% gain, netting a 28.09% increase. Vegetation lost 72.17% but gained 64.58%, resulting in a net loss of 7.59%. Agricultural land experienced a 31.44% loss and a 26.89% gain, leading to a 4.55% net loss. Water bodies lost 75.04% but gained 83.79%, achieving an 8.39% net gain. Barrenland lost 69.35% but gained 62.75%, resulting in a net loss of 6.6%. These results indicate significant changes and transitions among LULC classes during this period.

A land use and land cover (LULC) analysis of Patna from 1989 to 2014 shows significant changes. The urban built-up area nearly doubled, rising from 31.343 to 61.859 sq. km, making up 61.521% of the total area. Water bodies were reduced by nearly 50%, fallow land decreased to 20% of its 1989 size, and agricultural land

**Table 16.1** Class areas in square kilometres for the years 1989–2019

| Classes | Area in sq. km—1989 | Area in sq. km—1999 | Area in sq. km—2009 | Area in sq. km—2019 |
|---|---|---|---|---|
| 1. Builtup | 91.329 | 134.533 | 142.142 | 199.002 |
| 2. Vegetation | 225.442 | 146.516 | 194.366 | 177.094 |
| 3. Agricultural land | 689.672 | 756.628 | 732.114 | 646.779 |
| 4. Waterbody | 13.135 | 18.581 | 26.874 | 20.228 |
| 5. Barrenland | 132.699 | 96.020 | 56.782 | 109.174 |

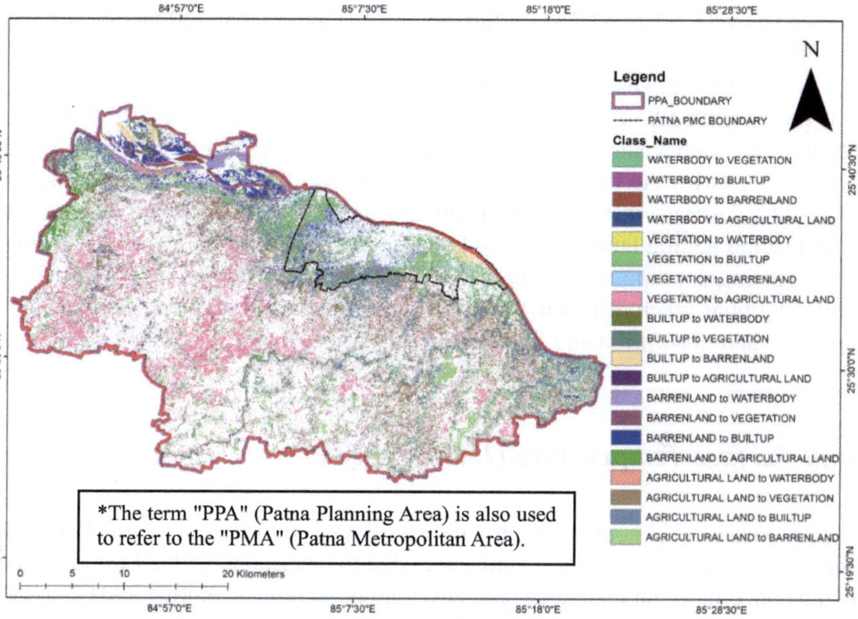


Fig. 16.4 Gains, losses and persistence analysis for each class and combined LULC change (1989–2019). *Source* Author

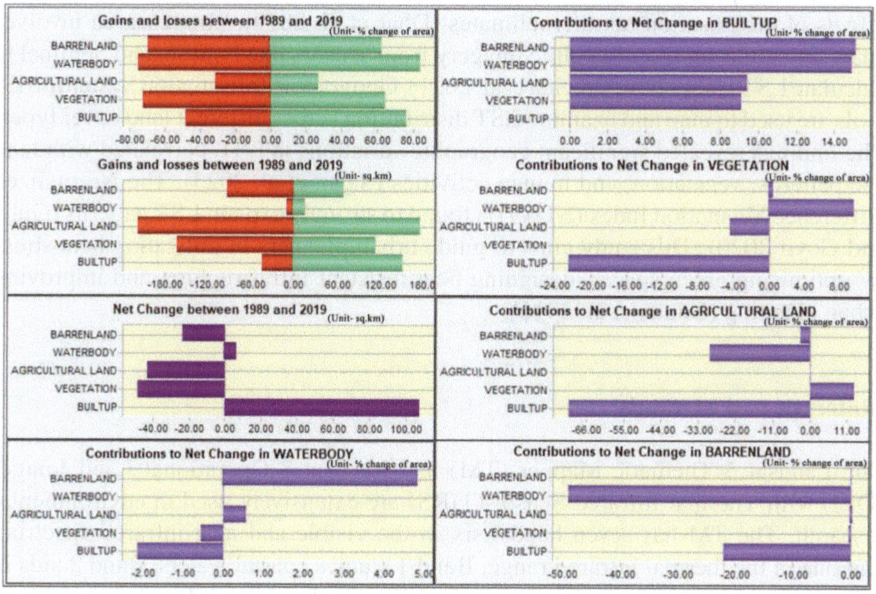


**Fig. 16.5** Change analysis—gains and losses by category and classes that contributed to net loss/gain for each class (1989–2019). *Source* Author

dropped from 53.708 to 32.274 sq. km, while vegetation increased by over 1.6 times (Raf 2014).

A study of Patna District from 1988 to 2013 revealed a similar pattern: built-up areas grew from 31.85 to 185.36 sq. km, agricultural land decreased from 621.11 to 555.06 sq. km, and water bodies fell from 11.12 to 5.44% (Mishra and Rai 2016). Our LULC analysis for the PMA from 1989 to 2019 also shows an increase in built-up areas and a decrease in vegetation, from 225 to 177 sq. km. This urban expansion has converted natural landscapes into built environments, reducing natural drainage and increasing flooding, as seen in the 2019 Patna flood (Rashiq and Prakash 2023). Enhancing blue-green infrastructure, such as wetlands and permeable surfaces, can improve water absorption and flood resilience.

## *Land Surface Temperature (LST) Analysis*

In recent years, rapid population growth and urban development in Patna have led to significant alterations in natural landscapes due to unsustainable land-use decisions, raising concerns about ecological and environmental threats (Kumar and Chaudhary 2021). This study focuses on analysing Land Surface Temperature (LST) variations in the Patna Metropolitan Area to better understand urban thermal properties and their environmental impacts. LST data is crucial for climate research, offering insights into local temperature variations, supporting weather forecasting, and revealing the effects of urbanisation on microclimates (Dhar et al. 2019). The research involves analysing multi-temporal satellite imagery from sensors like Landsat and Sentinel to calculate LST values (De Almeida et al. 2021). Geographic Information System (GIS) tools are used to map and examine LST distribution across different land cover types. The findings revealed significant geographic variations in LST, correlated with land use patterns, vegetation, and human activities (Sajan et al. 2023). The Normalized Difference Vegetation Index (NDVI) is found to strongly explain LST patterns (Guha and Govil 2020). This study aims to guide urban planners in mitigating heat stress by optimising green spaces, designing heat-resistant infrastructure, and improving urban planning (Coutts et al. 2016).

### Materials and Methods

The Landsat 5 Thematic Mapper (TM) and Landsat 8 Operational Land Imager (OLI) with Thermal Infrared Sensor (TIRS) are extensively used in environmental research. The TM has seven bands: six in the visible and near-infrared spectrum and one in the thermal infrared range. Band 1 studies coastal waters, Band 2 aids in crop and vegetation monitoring, Bands 3 and 4 calculate NDVI, Bands 5 and 7 help identify clouds, ice, and geological formations, and Band 6 is used for Land Surface Temperature (LST) extraction.

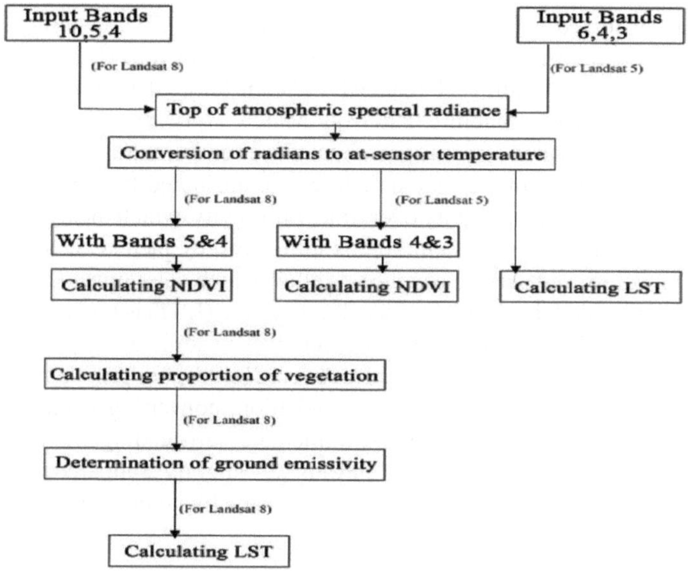


**Fig. 16.6** Flowchart for methodology—LST analysis. *Source* Author

The imagery captured by Landsat 8 OLI and TIRS comprises nine spectral bands, with a spatial resolution of 30 m for Bands 1–7 and 9. The introduction of Band 1 (ultra-blue) caters to coastal and aerosol studies, while Band 9 aids in cirrus cloud detection. Band 8 (panchromatic) offers a finer resolution of 15 m. Thermal bands 10 and 11, crucial for precise surface temperature retrieval, are initially collected at 100 m but are resampled to 30 m during Level-1 product generation. Satellite data from Landsat-5 and Landsat-8 spanning the summer months of March, April, and May for the years 1989, 1999, 2009, and 2019 were acquired from the US Geological Survey's official website (Survey U. U. G. (n.d.). *EarthExplorer*. None. https://earthexplorer.usgs.gov/). For the retrieval of LST, the process outlined by Avdan and Jovanovska (2016) was implemented using ArcGIS raster processing techniques.

## Methodology

See Fig. 16.6.

## Result

The Land Surface Temperature (LST) of the Patna Metropolitan Area was derived by processing data from Landsat 8 (Band 10) and Landsat 5 (Band 6) using the raster calculator tool in ArcGIS (Fig. 16.6). The calculated LST values represent the temperature of mixed land cover types within each pixel. The resulting LST maps reveal the spatial temperature distribution across the study area (Fig. 16.7). Analysis indicates that the maximum LST reached 49 °C in May 2019, while the minimum was 21 °C in May 1999. Satellite data from March, April, and May in 1989, 1999, 2009, and 2019 provided key insights into these temperature variations, detailed in Table 16.2. Changes in thermal signatures are attributed to evolving land cover characteristics and their emissivity over time. Emissivity, which measures a surface's efficiency in emitting thermal radiation, is influenced by factors such as vegetation type, soil moisture, and surface roughness, all of which can impact LST readings.

The study revealed that Land Surface Temperature (LST) is higher in agricultural lands compared to built-up areas due to differences in surface composition and land use practices. Agricultural lands, typically covered by crops or grasslands, have lower albedo, meaning they absorb more solar radiation, converting it into heat. While transpiration from plants can cool the immediate surroundings, it also increases moisture content in the air, which may elevate LST slightly compared to urban areas. Conversely, built-up areas with impervious surfaces like concrete and asphalt have

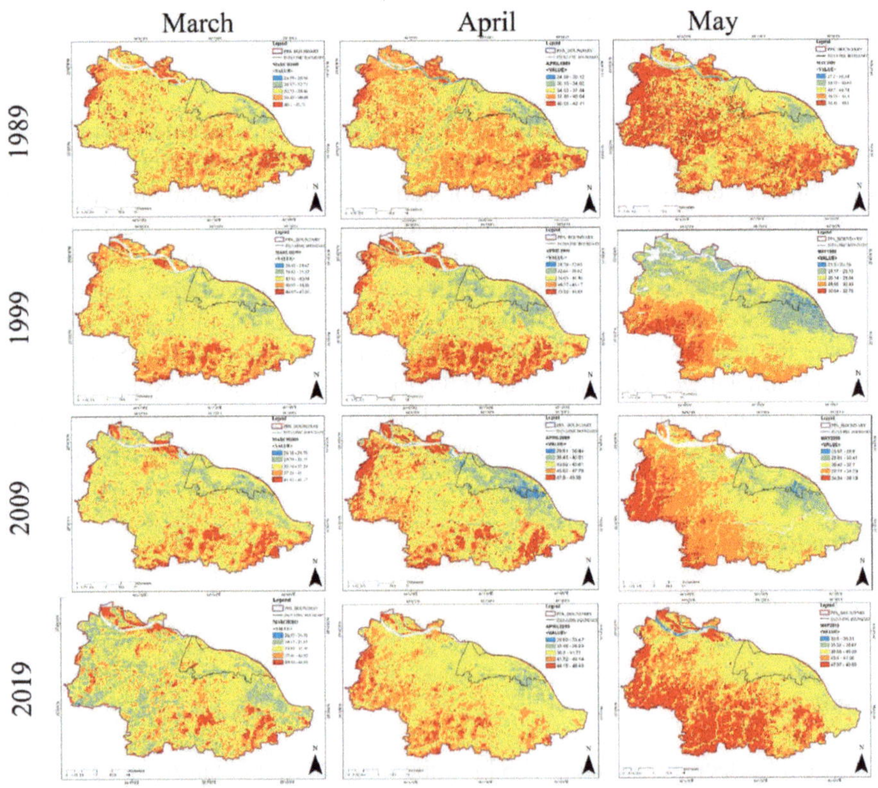


**Fig. 16.7** Land surface temperature (LST) maps of months March, April and May for the years 1989, 1999, 2009 and 2019. *Source* Author

**Table 16.2** Minimum and maximum land surface temperatures obtained in the months of March, April and May for the years 1989, 1999, 2009 and 2019

| Year | March | | April | | May | |
|---|---|---|---|---|---|---|
| | Min in °C | Max in °C | Min in °C | Max in °C | Min in °C | Max in °C |
| 1989 | 24.5 | 42.7 | 24.5 | 42.7 | 27.2 | 49.1 |
| 1999 | 26.4 | 47.0 | 28.7 | 44.8 | 21.3 | 32.7 |
| 2009 | 26.3 | 45.17 | 29.8 | 49.3 | 25.5 | 38.1 |
| 2019 | 26.9 | 44.6 | 29.9 | 46.4 | 30.6 | 49.9 |

higher albedo, reflecting more sunlight and absorbing less solar energy (Tahooni et al. 2023). The urban heat island effect, caused by human activities, can locally increase temperatures but may not significantly impact LST on a broader scale. The study also used the Normalized Difference Vegetation Index (NDVI) (Fig. 16.8) to measure vegetation health over the years, showing variations from 1989 to 2019, with NDVI values declining, reflecting changes in land cover (Pettorelli 2013) (Table 16.3).

In the study "Effects of Urbanization on the Progressive Rise in Land Surface Temperature of Patna Municipal Corporation in the New Millennium (2000–2018)" by Raf (2014), the research focused on evaluating the impact of urbanisation on Land Surface Temperature (LST) in Patna Municipal Corporation after 2000. The analysis of satellite data revealed a significant correlation between the expansion

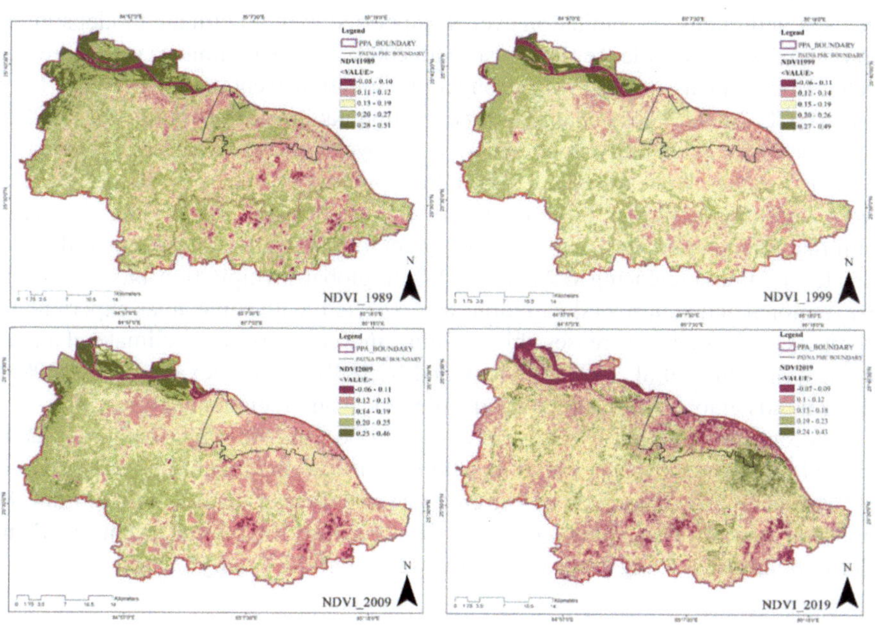


**Fig. 16.8** Normalized difference vegetation index (NDVI) maps generated for the years 1989, 1999, 2009 and 2019. *Source* Author

**Table 16.3** Minimum and maximum NDVI values obtained for the years 1989, 1999, 2009 and 2019

| 1989 | | 1999 | | 2009 | | 2019 | |
|---|---|---|---|---|---|---|---|
| Min NDVI | Max NDVI | Min NDVI | Max NDVI | Min NDVI | Max NDVI | Min NDVI | Max NDVI |
| − 0.05 | 0.51 | − 0.06 | 0.49 | − 0.06 | 0.46 | − 0.07 | 0.43 |

of built-up areas and the rise in LST. The maximum recorded LST increased from 35.11 °C in 2000 to 39.05 °C by 2013, reflecting the impact of urban sprawl. Our study also tracked LST trends from 1989 to 2019 using Landsat 5 and Landsat 8 data, showing a consistent upward trend. Additionally, the research explored the Normalized Difference Vegetation Index (NDVI) over the same period, finding a decline in vegetation (Fig. 16.8), which correlated with the rising LST. These findings underscore the critical importance of preserving vegetation in urban areas to mitigate the adverse effects of urbanisation, such as increased temperatures, and to promote sustainable growth in Patna.

## *Rainfall*

The increase in greenhouse gases and global warming is expected to lead to more frequent and intense precipitation events, supported by both climate model simulations and empirical evidence (Rajeevan et al. 2008). Warmer climates, with increased water vapour, are associated with stronger precipitation events and heightened flood risks (IPCC 2007). These changing precipitation patterns have significant implications for water availability, food security, and water-related disasters triggered by extreme weather (McCarthi et al. 2001). Precipitation is crucial to the hydrological cycle, and any alterations in its distribution could directly affect water resources, with rising temperatures likely altering evaporation rates and impacting soil moisture levels, thus increasing flood risks. While global precipitation is projected to rise, regional and continental variations are expected, with some areas experiencing increases and others decreases, highlighting the complexity of climate change's impact (Groisman et al. 2001; Haylock and Nicholls 2000; Alexander et al. 2006). These trends emphasise the need for detailed local and regional analyses.

A recent investigation by Goswami et al. (2006), utilising high-resolution daily gridded rainfall data (Rajeevan et al. 2006), affirmed significant upward trends in the frequency and magnitude of extreme rain events over central India during the monsoon season. This study examined the variability of rainfall events in Patna Metropolitan Area using 50 years (1970–2020) of gridded daily rainfall data.

### Materials and Methods

The CRU Global Climate Dataset, accessible through the IPCC DDC (Data Distribution Centre of the Intergovernmental Panel on Climate Change), comprises a multidimensional 0.5° latitude by 0.5° longitude resolution average monthly climate data set for global land regions, excluding Antarctica. The CRU TS4.07 dataset presents monthly gridded fields derived from observed data calculated from daily or sub-daily records gathered by National Meteorological Services and other external sources. Interpolation tools generate a continuous or predictive surface based on sampled point values. Kriging has become a widely adopted interpolation technique for estimating the spatial distribution of climate factors, including rainfall.

### Result

Patna's increasing rainfall trend over the past five decades is driven by climate change and regional meteorological shifts (Fig. 16.9). Global warming has raised atmospheric moisture levels, resulting in more intense and frequent rainfall. Additionally, changes in sea surface temperatures in the Indian Ocean and Bay of Bengal have altered monsoon patterns, causing heavier and more erratic rainfall in the region. Urbanisation, including reduced green spaces and poor drainage, has worsened the impact, leading to frequent flooding and waterlogging, highlighting the broader effects of climate change.

Climate change is profoundly affecting human health in Patna, with rising temperatures and erratic weather worsening existing health issues. Warmer conditions have increased the incidence of vector-borne diseases like malaria and dengue by creating favourable environments for mosquito breeding. Extreme weather events, such as floods and heatwaves, have become more frequent, heightening the risk of waterborne diseases and heat-related illnesses (Nezami 2023a, b). Urbanisation from 2013 to 2019 led to decreased vegetation, increased runoff, and severe urban flooding in 2019 due to heavy monsoon rains and poor drainage (Rashiq and Prakash 2023). Flooding contaminates water supplies with sewage and waste, exacerbating waterborne diseases, while stagnant water increases the risk of malaria and dengue. Additionally, floods can cause respiratory infections from mold, physical injuries, and mental health issues from displacement and stress, with poor sanitation and disrupted healthcare further compounding public health challenges in densely populated areas.

## *Isochrone Map*

An isochrone map (Fig. 16.10) represents areas accessible within a specific time frame from a particular location, useful in urban planning, transportation analysis, and emergency response. It visualises travel time or distance, helping to analyse accessibility and service coverage. The creation of isochrone maps typically

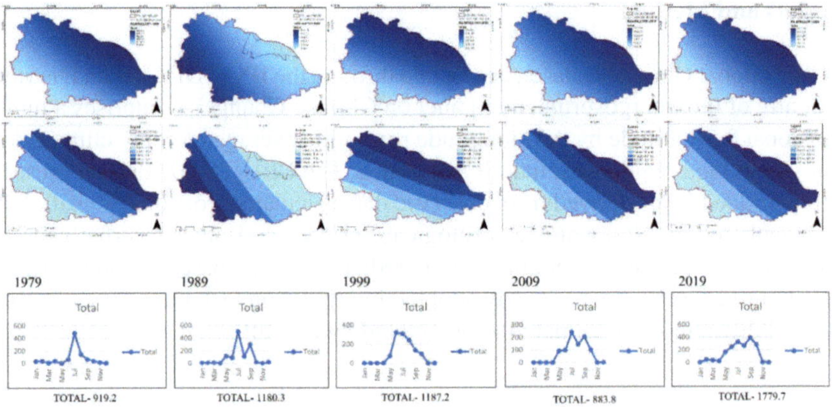


(a) Yearly Rainfall analysis, PMA for the years- 1979,1989,1999, 2009 and 2019

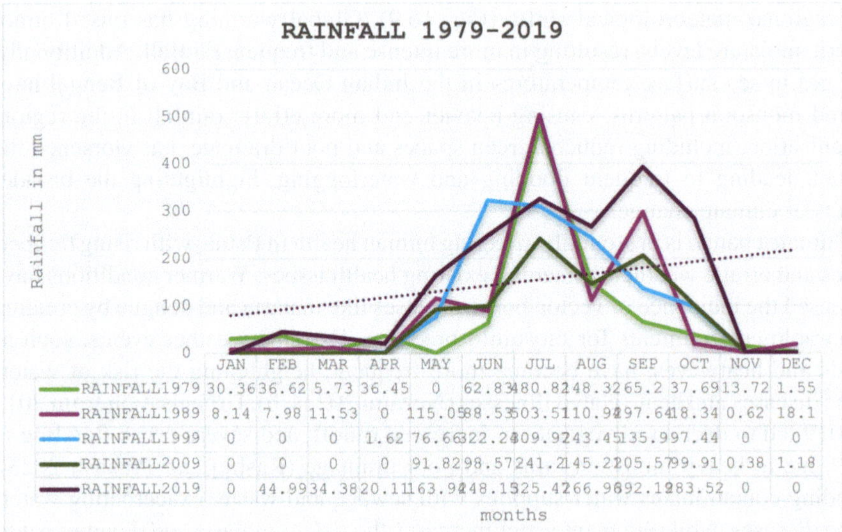


(b) Monthly Rainfall analysis, PMA for the years- 1979,1989,1999, 2009 and 2019 (Source: Author)

**Fig. 16.9** **a** Yearly rainfall analysis, PMA for the years—1979, 1989, 1999, 2009 and 2019. **b** Monthly rainfall analysis, PMA for the years—1979, 1989, 1999, 2009 and 2019. *Source* Author

involves network analysis tools that account for factors like road networks, traffic conditions, and transport modes (Weber and Kwan 2002). These maps are particularly effective in identifying underserved areas, optimising resource allocation, and enhancing decision-making in spatial planning (Miller 1999). Isochrone mapping in

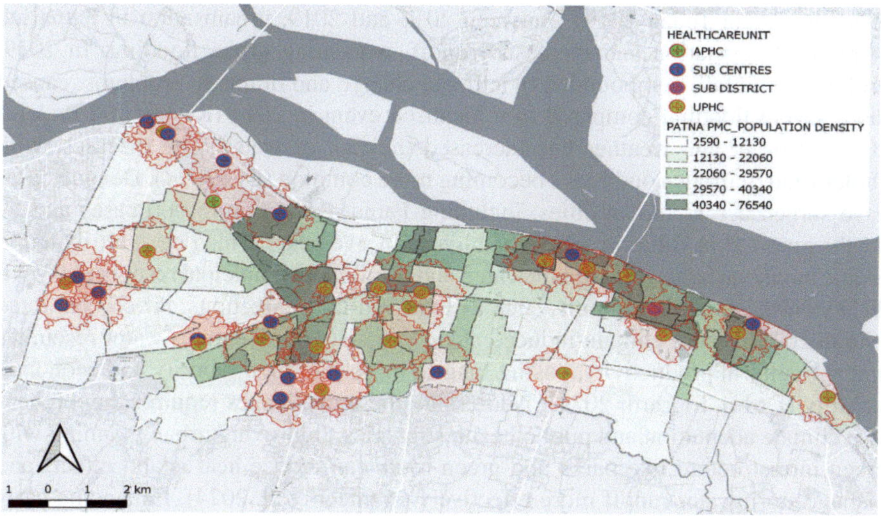


**Fig. 16.10** Isochrone map—15 min' walk from healthcare infrastructure. *Source* Author

GIS continues to evolve with advancements in spatial analysis techniques and data availability.

Isochrone maps depicting a 15-min walk from healthcare infrastructure are critical in evaluating the accessibility of medical services, particularly in urban areas. These maps help identify populations with sufficient access to healthcare facilities and those in underserved regions, facilitating targeted interventions to improve equity in health services (Neutens et al. 2010). By visualising walkable areas, planners can assess the impact of new healthcare infrastructure and optimise locations for new facilities, ensuring that vulnerable populations are within a reasonable walking distance (McGrail and Humphreys 2014). This approach is especially important for the elderly and those without access to private transport, as it directly influences health outcomes and quality of life.

The isochrone map (Fig. 16.10) displays the major healthcare facilities in the Patna Municipal area and a 15-min walking buffer distance.

## Discussion

Climate change is increasingly impacting Patna, the capital of Bihar, through rising temperatures and erratic weather patterns. Recent reports highlight a significant rise in extreme heatwaves, with summer temperatures often surpassing 40 °C (104 °F), which is affecting public health and increasing energy consumption (The Hindu 2023). The city has also encountered unpredictable rainfall, leading to severe flooding during the monsoon season. This has intensified infrastructure issues and displaced residents (India Today 2019). Between 2013 and 2019, urbanisation in Patna led to reduced vegetation and increased runoff, worsening urban flooding. In 2019, prolonged rainfall, compounded by climate change and drainage problems, caused more severe flooding compared to a localised event in 2013 (Rashiq and Prakash 2023). This erratic weather has increased the risk of waterborne diseases, with cholera and diarrhoea outbreaks becoming more common (Das 2024). Dengue cases have surged across Indian cities, including Patna, with over 15,000 cases and 59 deaths reported in Bihar in 2023, the highest in five years. This rise is attributed to erratic monsoon patterns, stagnant water, and insufficient municipal responses (Wakil 2023). New dengue symptoms, such as liver and lung infections, have been noted (Singh 2023). These climate-induced changes are straining Patna's water resources and agricultural productivity, posing long-term risks to food security and economic stability (Down To Earth 2024). Addressing these challenges requires comprehensive climate adaptation and public health strategies. Improving urban planning with green infrastructure like parks and green roofs can reduce heat island effects and manage stormwater runoff more effectively (Azmeer et al. 2024). Enhancing early warning systems for extreme weather and investing in climate-resilient healthcare facilities will better prepare Patna for heatwaves and disease outbreaks (Lowe et al. 2011). Strengthening community health programmes and ensuring access to clean water and sanitation are also crucial (Lugten and Hariharan 2022). Implementing these strategies, alongside proactive policies and community engagement, will help Patna become more resilient to the impacts of climate change.

## Conclusion

The study highlights the critical role of local strategies in enhancing urban resilience and adaptation to climate change, particularly in the context of health vulnerabilities in a metropolitan city of India. By analysing land use and land cover (LULC) changes, land surface temperature (LST) variations, and rainfall patterns, the study provides a comprehensive understanding of the intricate linkages between urbanisation and climate-induced health risks limited to LULC, LST and rainfall. The findings underscore the necessity of integrating LULC and LST data into urban planning and urban design (blue-green infrastructure) to mitigate the adverse effects of climate change at city level. Increased urbanisation and altered LULC patterns have exacerbated the urban heat island effect, as reflected in rising LST, which in turn elevates heat-related health risks. Additionally, erratic rainfall patterns, influenced by changing climate conditions, have further aggravated public health vulnerabilities through the proliferation of vector-borne diseases and water-related health issues. The research advocates for targeted interventions such as green infrastructure, improved drainage and sanitation systems, and community-based early warning systems to address these challenges. Ultimately, the study calls for a holistic and interdisciplinary approach, combining GIS-based analyses with public health initiatives, to build a more resilient urban environment that safeguards the health and wellbeing of its inhabitants.

This study has certain limitations, primarily due to its reliance on remote sensing data without any physical or ground verification. The absence of on-site validation may introduce inaccuracies in interpreting the results, as satellite data can sometimes misrepresent local conditions. Furthermore, only three parameters—land use and land cover (LULC), land surface temperature (LST), and rainfall—have been utilised for the analysis. While these are critical factors, they do not provide a comprehensive understanding of the complex environmental dynamics. For a more extensive study in the future, additional parameters such as air pollution and soil moisture could be incorporated. These would help provide a deeper and more refined analysis of environmental impacts and urban resilience. Incorporating ground-truthing methods alongside remote sensing could further enhance the reliability of the results.

## References

Alexander LV, Zhang X, Peterson TC, Caesar J, Gleason B, Klein Tank AM, Haylock M, Collins D, Trewin B, Rahimzadeh F, Tagipour A, Rupa Kumar K, Revadekar J, Griffiths G, Vincent L, Stephenson DB, Burn J, Aguilar E, Brunet M, Taylor M, New M, Zhai P, Rusticucci M, Vazquez-Aguirre JL (2006) Global observed changes in daily climate extremes of temperature and precipitation. J Geophy Res Atmos 111(D5). https://doi.org/10.1029/2005JD006290

Avdan U, Jovanovska G (2016) Algorithm for automated mapping of land surface temperature using LANDSAT 8 satellite data. J Sens 2016(1):1480307

Azmeer A, Tahir F, Al-Ghamdi SG (2024) Progress on green infrastructure for urban cooling: evaluating techniques, design strategies, and benefits. Urban Climate 56:102077. https://doi.org/10.1016/j.uclim.2024.102077

Berrang-Ford L, Sietsma AJ, Callaghan M, Minx JC, Scheelbeek PFD, Haddaway NR, Haines A, Dangour AD (2021) Systematic mapping of global research on climate and health: a machine learning review. Lancet Planet Health 5(8):e514–e525. https://doi.org/10.1016/S2542-5196(21)00179-0

Bush KF, Luber G, Kotha SR, Dhaliwal RS, Kapil V, Pascual M, Brown DG, Frumkin H, Dhiman RC, Hess J, Wilson ML, Balakrishnan K, Eisenberg J, Kaur T, Rood R, Batterman S, Joseph A, Gronlund CJ, Agrawal A, Hu H (2011) Impacts of climate change on public health in india: future research directions. Environ Health Perspect 119(6):765–770. https://doi.org/10.1289/ehp.1003000

Cruz J, White PCL, Bell A, Coventry PA (2020) Effect of extreme weather events on mental health: a narrative synthesis and meta-analysis for the UK. Int J Environ Res Public Health 17(22):8581. https://doi.org/10.3390/ijerph17228581

Coutts AM, Harris RJ, Phan T, Livesley SJ, Williams NS, Tapper NJ (2016) Thermal infrared remote sensing of urban heat: hotspots, vegetation, and an assessment of techniques for use in urban planning. Remote Sens Environ 186:637–651. https://doi.org/10.1016/j.rse.2016.09.007

Das T (2024) Diarrhoea to cholera: monsoon diseases to stay safe from. Hindustan Times. https://www.hindustantimes.com/lifestyle/health/diarrhoea-to-cholera-monsoon-diseases-to-stay-safe-from-101722416676307.html

De Almeida CR, Teodoro AC, Gonçalves A (2021) Study of the urban heat Island (UHI) using remote sensing data/techniques: a systematic review. Environments 8(10):105. https://doi.org/10.3390/environments8100105

Debnath KB, Jenkins D, Patidar S, Peacock AD, Bridgens B (2024) Climate change, extreme heat, and south asian megacities: impact of heat stress on inhabitants and their productivity. ASME J Eng Sustain Build Cities 4(041006). https://doi.org/10.1115/1.4064021

Dhar RB, Chakraborty S, Chattopadhyay R, Sikdar PK (2019) Impact of land-use/land-cover change on land surface temperature using satellite data: a case study of Rajarhat block, North 24-Parganas District, West Bengal. J Indian Soc Remote Sens 47(2):331–348. https://doi.org/10.1007/s12524-019-00939-1

Dhiman RC, Pahwa S, Dhillon GPS, Dash AP (2010) Climate change and threat of vector-borne diseases in India: are we prepared? Parasitol Res 106(4):763–773. https://doi.org/10.1007/s00436-010-1767-4

Dholakia HH, Garg A (2018) Climate change, air pollution and human health in Delhi, India. In: Akhtar R, Palagiano C (eds) Climate change and air pollution: the impact on human health in developed and developing countries. Springer, Berlin, pp 273–288. https://doi.org/10.1007/978-3-319-61346-8_17

Dimitrova A, Marois G, Kiesewetter G, KC S, Rafaj P, Tonne C (2021) Health impacts of fine particles under climate change mitigation, air quality control, and demographic change in India. Environ Res Lett 16(5):054025. https://doi.org/10.1088/1748-9326/abe5d5

Down To Earth | Latest news, opinion, analysis on environment & science issues | India, South Asia. (2024, December 15). Down to Earth. https://www.downtoearth.org.in/magazine/archives/2024-climate-change

Ebi KL (2024) Climate changes health. J Infect Dis, jiad570. https://doi.org/10.1093/infdis/jiad570

Friesen J, Friesen V, Dietrich I, Pelz PF (2020) Slums, space, and state of health—a link between settlement morphology and health data. Int J Environ Res Public Health 17(6):2022. https://doi.org/10.3390/ijerph17062022

Goswami BN, Venugopal V, Sengupta D, Madhusoodanan MS, Xavier PK (2006) Increasing trend of extreme rain events over India in a warming environment. Science 314(5804):1442–1445. https://doi.org/10.1126/science.1132027

Groisman PY, Knight RW, Easterling DR, Karl TR, Hegerl GC, Razuvaev VN (2005) Trends in intense precipitation in the climate record. J Climate 18(9):1326–1350. https://doi.org/10.1175/JCLI3339.1

Guha S, Govil H (2020) Land surface temperature and normalized difference vegetation index relationship: a seasonal study on a tropical city. SN Appl Sci 2(10). https://doi.org/10.1007/s42452-020-03458-8

Haylock M, Nicholls N (2000) Trends in extreme rainfall indices for an updated high quality data set for Australia, 1910–1998. Int J Climatol J Royal Meteorolog Soc 20(13):1533–1541. https://www.vsamp.com/resume/publications/Haylock_Nicholls.pdf

Hindustan Times (2021) Patna grapples with 'severe' air pollution after 2 years of clean action plan. Hindustan Times. https://www.hindustantimes.com/india-news/patna-grapples-with-severe-air-pollution-after-2-years-of-clean-action-plan-101610359045951.html

Hussain M, Chen D, Cheng A, Wei H, Stanley D (2013) Change detection from remotely sensed images: from pixel-based to object-based approaches. Int J Photogrammetry Remote Sens 80:91–106. https://doi.org/10.1016/j.isprsjprs.2013.03.006

India Today (2019) Worst monsoon in 25 years kills 148 across India, floods ravage Patna|10 points. India Today. https://www.indiatoday.in/india/story/worst-monsoon-in-25-years-kills-over-100-across-india-floods-rains-in-patna-10-points-1604967-2019-09-30

IPCC (2007) Climate change 2007: Synthesis report. In: Pachauri RK Reisinger A (eds) Contribution of working groups I, II and III to the fourth assessment report of the intergovernmental panel on climate change. IPCC, Geneva, Switzerland, 104 pp

IPCC (2022) Climate change: a threat to human wellbeing and health of the planet. Taking action now can secure our future. IPCC. https://www.ipcc.ch/2022/02/28/pr-wgii-ar6/

Karmakar M, Pradhan MM (2020) Climate change and public health: a study of vector-borne diseases in Odisha. India. Natural Hazards 102(2):659–671. https://doi.org/10.1007/s11069-019-03594-4

Kumar C, Chaudhary R (2021) Environmental sustainability practices in hospitals of Bihar. Curr Res Environ Sustain 3:100106. https://doi.org/10.1016/j.crsust.2021.100106

Lugten E, Hariharan N (2022) Strengthening health systems for climate adaptation and health security: key considerations for policy and programming. Health Secur 20(5):435–439. https://doi.org/10.1089/hs.2022.0050

Lowe D, Ebi KL, Forsberg B (2011) Heatwave early warning systems and adaptation advice to reduce human health consequences of heatwaves. Int J Environ Res Public Health 8(12):4623–4648. https://doi.org/10.3390/ijerph8124623

Majra JP, Gur A (2009) Climate change and health: why should India be concerned? Indian J Occup Environ Med 13(1):11. https://doi.org/10.4103/0019-5278.50717

Manisalidis I, Stavropoulou E, Stavropoulos A, Bezirtzoglou E (2020) Environmental and health impacts of air pollution: a review. Front Public Health 8. https://doi.org/10.3389/fpubh.2020.00014

Marsland R, Staples J (2024) Time for a focus on climate change and health. Med Anthropol 43(1):1–4. https://doi.org/10.1080/01459740.2023.2293125

McCarthi JJ, Canziani OF, Leary NA, Dokken DJ, White KS (2001) Climate change 2001: impacts, adaptation, and vulnerability. Cambridge University Press eBooks, p 1042. http://ci.nii.ac.jp/ncid/BA53287338

McGrail MR, Humphreys JS (2009) Measuring spatial accessibility to primary care in rural areas: improving the effectiveness of the two-step floating catchment area method. Appl Geogr 29(4):533–541. https://doi.org/10.1016/j.apgeog.2008.12.003

McGrail MR, Humphreys JS (2014) Measuring spatial accessibility to primary health care services: utilising dynamic catchment sizes. Appl Geogr 54:182–188. https://doi.org/10.1016/j.apgeog.2014.08.005

McMichael AJ, Kovats RS (2000) Climate change and climate variability: adaptations to reduce adverse health impacts. Environ Monit Assess 61(1):49–64. https://doi.org/10.1023/A:1006357800521

Mishra VN, Rai PK (2016) A remote sensing aided multi-layer perceptron-Markov chain analysis for land use and land cover change prediction in Patna district (Bihar), India. Arab J Geosci 9(4). https://doi.org/10.1007/s12517-015-2138-3

Mirzabaev A, Kerr RB, Hasegawa T, Pradhan P, Wreford A, Von Der Pahlen MCT, Gurney-Smith H (2023) Severe climate change risks to food security and nutrition. Clim Risk Manag 39:100473. https://doi.org/10.1016/j.crm.2022.100473

Miller HJ (1999) Measuring space-time accessibility benefits within transportation networks: basic theory and computational procedures. Geogr Anal 31(2):187–212. https://doi.org/10.1111/j.1538-4632.1999.tb00976.x

Mosadeghrad AM, Isfahani P, Eslambolchi L, Zahmatkesh M, Afshari M (2023) Strategies to strengthen a climate-resilient health system: a scoping review. Globalization Health 19(1). https://doi.org/10.1186/s12992-023-00965-2

Moors E, Singh T, Siderius C, Balakrishnan S, Mishra A (2013) Climate change and waterborne diarrhoea in northern India: impacts and adaptation strategies. Sci Total Environ 468–469:S139–S151. https://doi.org/10.1016/j.scitotenv.2013.07.021

Neeti K, Faisal MM, Singh R (2023) Air pollution, its health effects on residents of Patna: a case study. https://doi.org/10.3390/ecas2023-16346

Neutens T, Schwanen T, Witlox F, De Maeyer P (2010) Equity of urban service delivery: a comparison of different accessibility measures. Environ Plann Econ Space 42(7):1613–1635. https://doi.org/10.1068/a4230

Nezami S (2023a) 152 new dengue cases in city. The Times of India. https://timesofindia.indiatimes.com/city/patna/152-new-dengue-cases-in-city/articleshow/104638446.cms

Nezami S (2023b) Number of dengue cases in Bihar reduces with fall in temperature. The Times of India. https://timesofindia.indiatimes.com/city/patna/dengue-cases-in-state-reduction-with-fall-in-temperature/articleshow/105608640.cms

Nomani MdZM, Parveen R (2020) Prevention of chronic diseases in climate change scenario in India. Environ Justice 13(4):97–100. https://doi.org/10.1089/env.2019.0032

Patz JA, Olson SH (2006) Climate change and health: global to local influences on disease risk. Ann Trop Med Parasitol 100(5–6):535–549. https://doi.org/10.1179/136485906X97426
People who are vulnerable to climate change (n.d.) National Institute of Environmental Health Sciences. https://www.niehs.nih.gov/research/programs/climatechange/health_impacts/vulnerable_people
Pettorelli N (2013) The normalized difference vegetation index. https://doi.org/10.1093/acprof:osobl/9780199693160.001.0001
Picciariello A, Colenbrander S, Roy R (2021) The costs of climate change in India: a review of the climate-related risks facing India, and their economic and social costs. https://odi.org/en/publications/the-costs-of-climate-change-in-india-a-review-of-the-climate-related-risks-facing-india-and-their-economic-and-social-costs/
Srivastava PK, Han D, Rico-Ramirez MA, Bray M, Islam T (2012) Selection of classification techniques for land use/land cover change investigation. Adv Space Res 50(9):1250–1265. https://doi.org/10.1016/j.asr.2012.06.032
Public health adaptation strategies for climate change|US EPA (2024). US EPA. https://www.epa.gov/arc-x/public-health-adaptation-strategies-climate-change
Raf MA (2014) An assessment of land use land cover change pattern in Patna Municipal corporation over a period of 25 Years (1989–2014) using remote sensing and GIS techniques. Int J Innov Res Sci Eng Technol 03(10):16782–16791. https://doi.org/10.15680/ijirset.2014.0310053
Rajeevan et al (2006) High resolution daily gridded rainfall data for the Indian region analysis of break and active monsoon spells. Curr Sci 91:296–306. References—Scientific Research Publishing (n.d.). https://www.scirp.org/reference/referencespapers?referenceid=2373981
Rajeevan M, Bhate J, Jaswal AK (2008) Analysis of variability and trends of extreme rainfall events over India using 104 years of gridded daily rainfall data. Geophys Res Lett 35(18). https://doi.org/10.1029/2008gl035143
Rashiq A, Prakash O (2023) Urban floods: a case study of Patna floods 2019—natural or anthropogenic? Springer eBooks, pp 57–78. https://doi.org/10.1007/978-3-031-21618-3_4
Rocklöv J, Dubrow R (2020) Climate change: an enduring challenge for vector-borne disease prevention and control. Nat Immunol 21:479–483. https://doi.org/10.1038/s41590-020-0648-y
Roy P, Roy A (2010) Land use and land cover change: a remote sensing & GIS perspective. J Indian Inst Sci 90:489–502
Sadidy J, Firouzabadi PZ, Entezari A (2012) The use of radarsat and landsat image fusion algorithms and different supervised classification methods to improve landuse map accuracy–case study: sari plain–Iran
Sajan NB, Kanga NS, Singh NSK, Mishra NVN, Durin NB (2023) Spatial variations of LST and NDVI in Muzaffarpur district, Bihar using Google earth engine (GEE) during 1990–2020. J Agrometeorology 25(2):262–267. https://doi.org/10.54386/jam.v25i2.2155
Singh S (2023) Bihar records highest dengue cases and deaths in 5 years: what's behind the surge. The Indian Express. https://indianexpress.com/article/explained/bihar-highest-dengue-cases-deaths-9008868/
Sharma S (2023) Nearly 100 die as scorching heatwave with 45C temperatures sweeps northern India. The Independent. https://www.independent.co.uk/climate-change/news/india-heatwave-deaths-weather-45-temperature-b2360083.html
Shivanna KR (2022) Climate change and its impact on biodiversity and human welfare. Proc Indian Natl Sci Acad 88:160–171. https://doi.org/10.1007/s43538-022-00073-6
Tahooni A, Kakroodi A, Kiavarz M (2023) Monitoring of land surface albedo and its impact on land surface temperature (LST) using time series of remote sensing data. Eco Inform 75:102118. https://doi.org/10.1016/j.ecoinf.2023.102118
The Hindu (2023) Heatwave in U.P. and Bihar kill nearly 100. The Hindu. https://www.thehindu.com/news/national/other-states/heatwave-up-bihar-kill-100-india-heatwave/article66982736.ece
Tong S, Ebi K (2019) Preventing and mitigating health risks of climate change. Environ Res 174:9–13. https://doi.org/10.1016/j.envres.2019.04.012

Wakil ZM (2023) Surge in dengue cases across multiple Indian cities—Bhopal, Hoshiarpur and Patna witness sharp spike. Financialexpress. https://www.financialexpress.com/healthcare/news-healthcare/surge-in-cases-across-multiple-indian-cities-bhopal-hoshiarpur-and-patna-witness-sharp-spike-bkg/3314154/

Weber J, Kwan MP (2002) Bringing time back in: a study on the influence of travel time variations and facility opening hours on individual accessibility. Prof Geogr 54(2):226–240. https://doi.org/10.1111/0033-0124.00328

Website. Survey U. U. G. (n.d.) EarthExplorer. None. https://earthexplorer.usgs.gov/

# Chapter 17
# Urban Agriculture for Climate Change Adaptation, Food Security and Public Health in Addis Ababa, Ethiopia

**Zenebe Mekonnen**

**Abstract** In response to climate change and the COVID-19 pandemic, urban agriculture has become an important part of food and nutrition security in many cities globally. During the COVID-19 pandemic, food prices have increased due to blockages that have reduced the supply of food items in Addis Ababa. Urban agriculture not only offers the benefits of food security and economic gain but also improves ecosystem services—water quality improvement, microclimate amelioration, biodiversity enhancement, and waste and flood management—that positively impact public health. A detailed literature review and rapid field observations were used. This study portrays the diversified types of farms and species accustomed to urban farming in and around Addis Ababa, the role of urban agriculture during and post-COVID-19 pandemic times, and the opportunities and challenges that prevail. Despite some difficulties, sustainable urban agriculture allows for the attainment of the economic, social, and ecological sustainability dimensions. In doing so, during the COVID-19 pandemic, the inhabitants of Addis Ababa were encouraged to cultivate every open area in the home compounds, school, office compounds, and pre-urban farmlands to grow different types of vegetables, fruits, and cereals to adapt to the challenge. Urban agriculture needs to be encouraged to continue in the future in a more advanced way.

**Keywords** Biodiversity · Climate change · Food price · Food security · Public health · Urban agriculture

## Introduction

The population of Addis Ababa has increased from nearly 2.73 million in 2007 (CSA 2007) to 5.7 million in 2024 as reported in **Macrotrends**. Climate change and pandemics like COVID-19 have caused social disruptions. In response to these, in

Z. Mekonnen (✉)
Ethiopian Forestry Development, Addis Ababa, Ethiopia
e-mail: zenebem@efd.gov.et



W. Leal Filho and F. Wolf (eds.), *Climate Change and Health: Perspectives from Developing Countries*, Climate Change Management,
https://doi.org/10.1007/978-3-031-93177-2_17

many cities around the globe, including Addis Ababa, urban agriculture has become an essential part of food and nutrition security and is growing in importance in post-COVID-19 times (FAO 2021, Hirvonen et al. 2021, Ju et al. 2022, John and Johan 2022, Mengistu et al. 2022). New York and Atlanta in the USA, Victoria, B.C. and Toronto in Canada, London in the UK, Paris in France, Dubai in UAE, Singapore in Singapore, Brussels in Belgium, and Manila in the Philippines are the **top ten cities** in the world paving the way forward for urban agriculture, with long-time advocates and experience as well as innovative activities (Laddha 2021).

Food insecurity in Ethiopia's urban areas has been increasing due to skyrocketing food prices compounded by high population pressure and high housing rents (Nigussie et al. 2021; McCann et al. 2023). This was particularly exacerbated during the COVID-19 pandemic due to blockages that have reduced the supply of food items in Addis Ababa. In the global as well as Ethiopian context, the COVID-19 pandemic boosted urban agriculture as one of the means of maintaining household food security (Artmann et al. 2021; Turnšek et al. 2022) and alerts a call for future pandemics. **Urban agriculture**, which commonly encompasses horticulture, animal husbandry, and aquaculture in urban and peri-urban settings, is one way to improve nutrition and income in cities. Urban agriculture is one of the instruments that contribute to food security, and the government of Ethiopia has encouraged this through the urban productive safety net programme (Abate 2018; Tadesse et al. 2018; Pauleit et al. 2019). Indeed, urban agriculture not only has the benefits of food security and economic gain but also improves ecosystem services such as water quality improvement, microclimate amelioration, biodiversity enhancement, and waste and flood management in cities, thereby contributing to health benefits (Deksissa et al. 2021; Ebissa and Desta 2022). The inhabitants of Addis Ababa are experiencing climate change-related risks in flooding from unpredictable rainfall, rising heat during the day, and lowering cold during the evenings. Public health problems (e.g., malaria and diarrhoea), poverty, and poor housing have compounded the impacts. As reported by CDKN (2021), urban and pre-urban agriculture are options to support the livelihoods of inhabitants, with more than 16,000 individual farmers engaged in vegetable farming in Addis Ababa.

High consumer prices amalgamated with skyrocketing house rents have challenged the living conditions of the inhabitants of Addis Ababa. Lockdowns during the COVID-19 have exacerbated these conditions severely. In this regard, the objectives of this assessment are to portray the diversified types of farms and species accustomed to urban farming in and around Addis Ababa (what type of urban farms are found in and around Addis Ababa?), the role of urban agriculture during and post-COVID-19 pandemic times (What role has urban agriculture played during and post-COVID in Addis Ababa?), and the opportunities and challenges that prevail in urban farming (What are the opportunities and challenges for urban agriculture in Addis Ababa?). Answering these questions to the corresponding objectives could highlight the significance of this paper in providing directions on what should be done in policy and strategy development if similar pandemics happen in the future, highlighting the role of urban agriculture in urban food security and sustainable livelihood. The paper also assesses the challenges and opportunities in urban agriculture in and around Addis Ababa and discusses how to utilise the opportunities and minimise the challenges for the betterment of sustainable urban agriculture to benefit the city's inhabitants from the economic, environmental and social perspectives.

## Materials and Method

### *Study Area*

The study focuses on Ethiopia's capital city of Addis Ababa (Fig. 17.1). Addis Ababa is the centre of gravity for Ethiopia's political, social, and economic momentum, where its pre-urban areas and areas in the Oromia Region's Sheger City are booming with urban expansion composed of many fast-growing emerging cities. Sheger City surrounds the pre-Addis Ababa areas in the Oromia Region, where there are high economic and cultural exchanges with Addis Ababa. Sheger City encompasses emerging towns such as Sebeta, Sululta, Burayu, Gelan, Alem Gena and Lega Tafo. The pre-Addis Ababa areas are under pressure from urban expansion from both sides of Addis Ababa and Sheger Cities. It is prone to illegal house construction and is a hot spot of ownership between the two administrations. Many poor inhabitants reside in the pre-Addis Ababa and Sheger City due to the inability to afford house rents in Addis Ababa. Addis Ababa is the economic and political centre of the country and home to many international communities. The city's population has increased from around 3 million in 2007 to about 6 million in 2024.

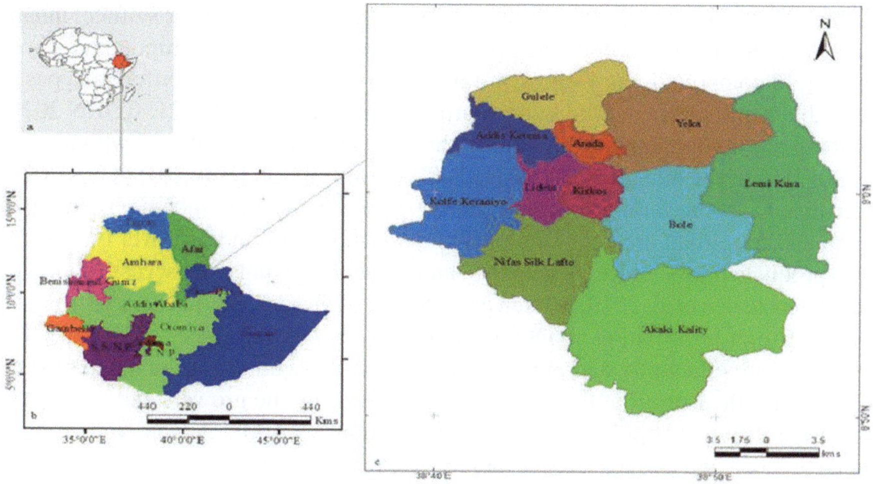


**Fig. 17.1** Map of Addis Ababa and its sub cities relative to map of Ethiopia. *Source* Ayele et al. (2022)

### *Assessment Approaches*

A detailed literature review of published and grey literature was conducted using the following Google search engine keywords: urban agriculture, urban food security, urban food price, urban biodiversity, pre-urban agriculture, urban agriculture, public health and climate, COVID-19 and urban agriculture. The date of publication since 2015 was one of the selection criteria, unless and otherwise the literature was found to be thematically important to the topic in context. The city's administration documents were also included in the selection. In doing so, from more than 300 articles found in the search engine, only 104 were selected for the review, and those that were thematically not relevant to the topic in context were excluded. The author also made rapid field observations in and around Addis Ababa and took photos to supplement the review visually. The photos were taken by consulting producers at the field and office compounds at different government bureaus to get their consent. Thematic analysis was made qualitatively in the form of narration and interpretation.

## Results and Discussions

### *Types and Diversity in Urban Agriculture*

Urban agriculture is not a new approach for the inhabitants of Addis Ababa. However, during the COVID-19 pandemic, it has boomed to its highest level. This was because the supply of commodities from other parts of the country had been lowered due to blockages of transportation compounding the skyrocketing food prices. With the pushing force from the city's administration and the prime minister's office, inhabitants were encouraged to cultivate every open area in their home compounds, school, and office compounds as well as pre-urban farmlands to grow different types of vegetables, fruits, cereals, and so on (Further Africa 2020; CAP 2021; CDKN 2021). In doing so, the inhabitants, especially the poor and the unemployed who account for 30–45% of the inhabitants (UN-HABITAT 2008; World Bank 2015; CSA 2020), were able to secure family food and sell the excess to generate income. Being in an urban and pre-urban area can minimise the transport cost and/or travel time to sell farm products (dairy, crops, vegetables, and fruits) by giving easy access to markets (Vandercasteelen et al. 2019; Minten et al. 2020a, b). In addition, farmers/growers are more aware of information about farm inputs and marketing prices so that they can produce more efficiently and are not liable to brokers who can deceive the growers to maximise their profit at the expense of profit loss by the growers and/or farmers. There are different topologies of urban agriculture in Addis Ababa, with varying characteristics and acreage sizes (Table 17.1). Differentiating between the different types of urban agriculture and knowing their common characteristics within each type is very important in future planning and strategy setting in urban and pre-urban agriculture, as well as in urban greening and urban forestry (Bechauf et al. 2022, Cutler et al. 2022, FAO et al. 2022) (Fig. 17.2).

Different types of herbal plant species are grown by urban communities in Addis Ababa that are commonly used for medicinal purposes and food flavouring ingredients (Fig. 17.3). Studies show that garlic, ginger, and citrus fruits were the most commonly consumed foods during the COVID-19 pandemic in Addis Ababa to reduce the risks of the pandemic (Hirvonen et al. 2020). The use of traditional medicine has a long history among Ethiopians to heal from different illnesses including COVID-19 (Haile 2021, Fekadu et al. 2023, Tuasha et al. 2023). Several traditional healers in Addis Ababa provide alternative medical options, especially for the poor inhabitants (Tuasha et al. 2023). In their study, Woldeamanuel et al.

**Table 17.1** Major topologies of urban agriculture observed in Addis Ababa

| Urban agriculture typology | Description | Acreage |
|---|---|---|
| Community-based gardening | • Practised often on public, vacant or open land in the city<br>• Food products such as vegetables, fruits and herbs are produced for home consumption<br>• Both the poor and the rich could be involved in this type of gardening | Very small |
| Home-based gardening | • Usually practised on small areas in or on the house as well as around the house<br>• Performed with limited economic resources and it is pro-poor focused<br>• Contributes to household food security and nutrition by providing direct access to fresh and nutritious food that can be harvested, prepared, and fed to family members | Zero to very small |
| Commercial crop and livestock production | • This type of UA is mostly practised in the pre-urban areas of Addis Ababa<br>• Farmers are involved in horticultural and cereal crop production to fill the demand gaps for fresh vegetables, fruits, teff and wheat to benefit from the comparative advantage of the nearer markets | Small to medium |
| Urban greening and urban forestry | • It is mainly practised in home and office compounds, street sides, parks, and embassy compounds for ornamental purposes and environmental amelioration, providing physical and mental health benefits<br>• The Green Legacy Initiative has provided a good opportunity to enhance urban greening and urban forestry. This in turn will increase biodiversity, water retention and groundwater recharge, property value and carbon sequestration; moreover, it decreases exposure to heat and air pollution | Small to medium |

**Fig. 17.2** Different types of urban agriculture in and around Addis Ababa (top left and right different lettuce, cabbage and herbal plants in different arrangements in an office compound, middle left different vegetables in the form of urban agroforestry in an office compound, middle right pre-urban teff and wheat cultivation, and bottom horizontal and green wall farming of different lettuce and cabbage). *Source* Author

**Fig. 17.3** Commonly grown herbal medicinal plants in urban farms in Addis Ababa. The annotations of the pictures are: **a** 'Besobela' (Ethiopian Holy Basil (Ocimum spp.), **b** 'Tena Adam' (Ruta chalepenesis), **c** Ethiopia Rosemary (Rosmarinus officinalis), **d** 'Koseret' (Lippia abyssinica), **e** 'Demakese' (Ocimum lamifolium), and **f** Mint (Mentha spicata). *Source* Author

(2022) identified 81 different medicinal plants in the Gulele Botanical Garden that were collected by more than 60 traditional healers to treat different diseases. In general, medicinal plants, as an integral component of urban agriculture, have a great contribution to public health.

## *Climate Change, Food Security and COVID-19*

As climate change can alter natural vegetation and cross-species interaction, virus and host environment and distribution, and gene mutation and can exacerbate other natural hazards, it is assumed that it can predispose for the emergence, transmission, and expansion of COVID-19 pandemics and will continue to do so for other pandemics in the future (Gupta et al. 2021). Among the key agendas under the endorsement of 123 countries at the COP28 UAE Declaration on Climate and Health were shifts to sustainable healthy diets, a transformation of health systems to be climate-resilient, maximising the health gains from mitigation and adaptation actions, preventing worsening health impacts from climate change, and understanding the contribution of healthy populations to climate resilience and successful adaptation across different sectors such as food and agriculture. Addis Ababa is facing growing climate change risks with a high probability risk level with severe consequences from drought, flooding, landslides, heatwaves, extreme hot days and hot nights, rainstorms and rain variability (Feyissa et al. 2018; Zegeye 2018; CAP 2021). Focusing on public health and food security, the Addis Ababa city administration has determined to maintain a 30% allocation of the city's land for urban green spaces to reduce the urban heat island effect and flooding as well as to increase the GDP share of urban agriculture production and productivity from 1% in 2021 to 50% by 2030 (CAP 2021). Studies have shown that climate change has posed a reduction in crop and livestock production, leading to a corresponding impact on nutrition availability and access, which in turn leads to health impacts, especially on children (Hagos et al. 2014, Gebreyesus et al. 2014, Asfew et al. 2023). The COVID-19 pandemic has affected not only the food production supply chain in the agricultural sector but also the medical supply chain in the health sector, which imposes challenges on both food security and public health (Minten et al. 2020a, b; Hirvonen et al. 2021; Tefera 2021; Zhang et al. 2022; Abay et al. 2023; McCann et al. 2023). To this end, the role of urban agriculture in improving food security and public health in Addis Ababa was paramount during the COVID-19 pandemic (Coulton 2020, Further Africa 2020, Anberbir 2022).

Studies show that urban agriculture has a positive impact on climate change adaptation and food security as well as on public health (Fig. 17.4). On one hand, climate change has disrupted crop production leading to malnutrition, and on the other hand, it has several negative effects on mental health such as a rise in depression, anxiety, and suicide attempts (IPCC 2023; Wright et al. 2024). The most significant effects of climate change on human health include, but are not limited to, the greater spread of insect-borne diseases (El-Sayed and Kamel 2020; Biswas 2022; Khan et al. 2023), water and food scarcity leading to malnutrition and greater disease (Arsiso et al. 2017, Myers et al. 2017, Worku 2017, Ouda et al. 2020, Zakar et al. 2020), increasingly poor air quality that can cost lives (Tarekegn and Gulilat 2018; Fikeraddis and Endeshaw (2020; Martelletti and Martelletti 2020; Frontera et al. 2020; Domingo and Rovira 2020; Ali et al. 2021), climate forced displacement and migration putting people at risk (Abel et al. 2019; McCarney and Kent 2020; Serbessa et al. 2023; Vanhala and Calliari 2022), and impacts on mental health (Hayes et al. 2018; Ardalan et al. 2019; Cianconi et al. 2020; Palinkas and Wong 2020; Lawrance et al. 2021; Newbury et al. 2021).

## The Role of Urban Agriculture

Certainly, climate change has posed challenges to urban agriculture, and this in turn impacts human health negatively (Fanzo et al. 2018; Chari and Ngcamu 2022; Chaudhry 2023; Mirzabaev et al. 2023). However, as farmland size becomes smaller even in the rural areas of Ethiopia, urban agriculture creates an opportunity to utilise efficiently small land spaces in cities. Urban agriculture helps to build a resilient urban food system in times of climate change and the emergence of pandemics like COVID-19 (Dubbeling et al. 2019; Dorr et al. 2021; Mensah 2023). It is also an

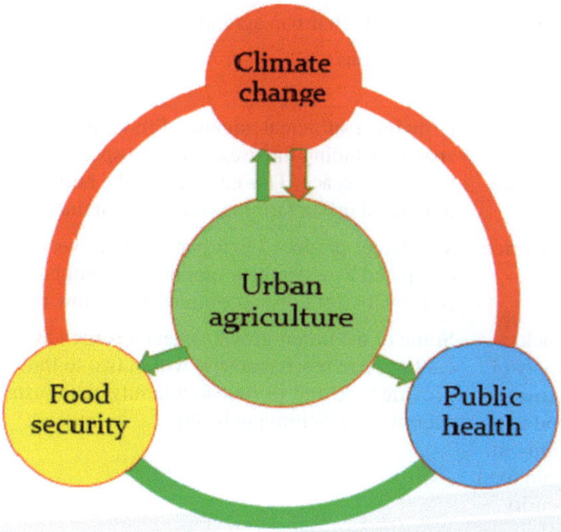


**Fig. 17.4** The climate change-food security-public health nexus and the role of urban agriculture in this nexus (the red lines and arrow show negative impacts on the other and the green line and arrows show positive impacts or correlations) (developed by the author)

option for supplying food to the urban centres within a minimum distance and with minimum transport cost (Table 17.2). Sustainable urban agriculture has a role to play in the three sustainability dimensions (Fig. 17.5). First, it creates opportunities for alternative food access, availability, utilisation, and stability, and by increasing off-farm benefits, it maintains economic sustainability. Second, it helps in pollution control, increases biodiversity, and helps to adapt to climate change, working together to maintain ecological sustainability. Finally, from the social sustainability angle, it enhances community engagement and health benefits.

## *Challenges of Urban Agriculture*

Irrespective of its benefits and significance to attain food security, urban agriculture in and around Addis Ababa is challenged by different factors. These include, but are not limited to: (i) low understanding among the society, as agriculture is a rural activity (Alemu et al. 2024; Delelegn and Mulugeta 2018; Etana et al. 2019); (ii) use of every parcel of small land for legal and/or illegal construction (Gebremichael et al. 2014), which makes horizontal urban agriculture more challenging in the future (Farinati et al. 2022, Ju et al. 2022); (iii) disease, insect, and pest damage to vegetables (e.g. potato, tomato and garlic rust) (Haile et al. 2016; Damte and Tabor 2015; Worku and Sahela 2018); and (iv) high concentrations of heavy metals in rivers used for irrigation that are built up in the vegetables through the process of nutrient uptake and are expected to impose health risks to the consumers (Bahiru 2021; Aschale et al. 2015; Woldetsadik et al. 2017). The nutritional status of people determines the status of their health conditions, which was showcased by the food system disruption

**Table 17.2** Ways in which urban agriculture could establish food security

| Benefits of UA | Descriptions | References |
|---|---|---|
| Efficient space utilisation | Growing a sufficient amount of food in a small available space, including the use of vertical space instead of horizontal space. This encourages the building of integrated urban farming, such as container gardens | Alemu et al. (2024), Ju et al. (2022) |
| Low price | Vegetables produced in urban areas are very cheap as compared to vegetables supplied by farm business entities or farmers from distant sources due to low transport costs | Mezgabu and Tolossa (2015), Further Africa (2020) |
| Quick supply of nutritious food to a family to attain food security | Some of the urban-grown vegetables such as lettuce and cabbage can reach maturity within two to three weeks to provide food and income to a family, thus sustaining food security and improving health | Anberbir (2022), Gebremichael et al. (2014), Further Africa (2020), Kennard and Bamford (2020), Serbessa et al. (2023) |
| Rich diversity | Urban agriculture is more diverse than large farms. They encompass different types of vegetables and fruits. In addition, considerable medicinal plants and spices are grown in urban farming in very congested areas. They are used as ingredients in food preparation to give a sweet scent and tasty flavour and are also used for medicine as a treatment for different diseases including COVID-19, catarrh, headache, and stomach-ache (Fig. 17.5) | Regassa (2016), Mengistu and Alemayehu (2017), Kebebew (2018) |
| Meeting many of the SDGs | As a solution to advance multiple SDGs, urban agriculture is being identified to attain food security, climate and ecosystem resilience, health and wellbeing, job creation, and social equity. In general, sustainable urban agriculture enables the attainment of the three sustainability dimensions in the urban and pre-urban landscapes | Tefft et al. (2017), Ayuk et al. (2021) |

during the COVID-19 pandemic (Bron et al. 2023; McCann et al. 2023). That is, food insecurity, exacerbated by climate change, has a serious impact on human health and the performance of labour (Simane et al. 2016, Schnitter and Berry 2019, Bhopal et al. 2021, Chari and Ngcamu 2022, Baars et al. 2023). Unless strategic solutions are sought proactively, the challenges of climate change can impact urban and pre-urban agriculture as they impact rural agriculture. This in turn affects food security and public health through the climate change-food security-human health nexus (Fig. 17.6).

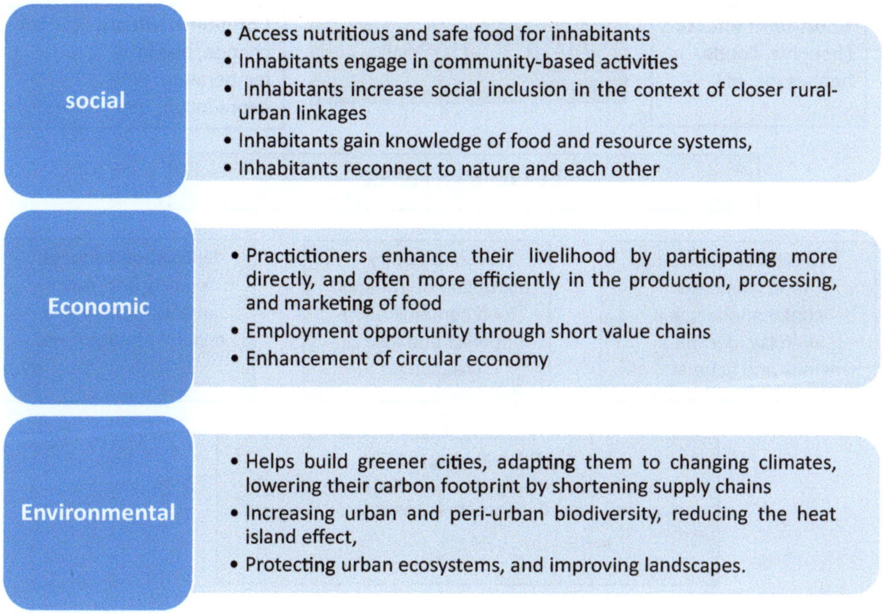


**Fig. 17.5** The role of urban agriculture in the three sustainability dimensions. Adapted from FAO et al. (2022)

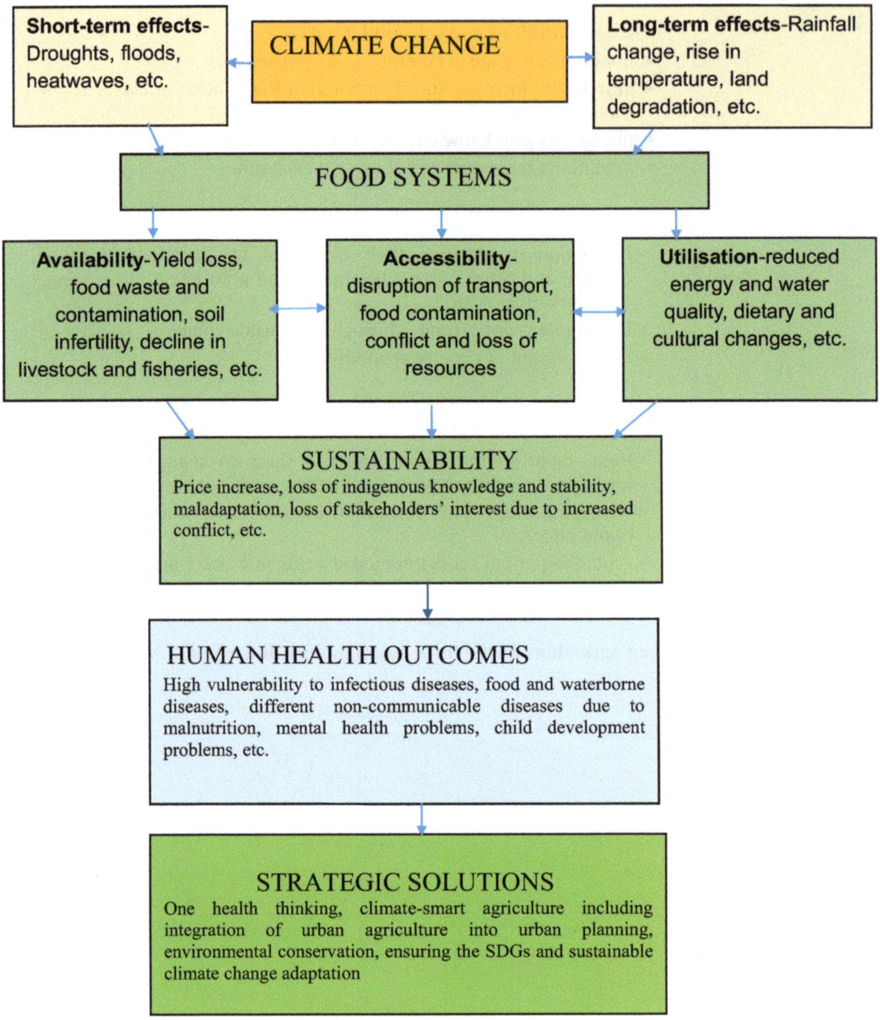


**Fig. 17.6** The likelihood of impacts between food security, climate change and human health. Adapted from Baars et al. (2023)

## Conclusion

The population increase in Addis Ababa has resulted in a parallel increase in the demand for food, housing, and different services. Poverty has also increased concurrently and has been exacerbated by the COVID-19 pandemic and climate change. During and in the post-COVID-19 pandemic, food prices have reached their highest level due to blockages and lockdowns. The COVID-19 pandemic has allowed the boosting of urban agriculture by the inhabitants to minimise the burden of food insecurity. It signals a wake-up call on what should be done in the future if other pandemics occur. Even though urban agriculture in and around Addis Ababa is still at its traditional stage and mostly ground-based, competing with the booming construction, it has enabled the inhabitants of Addis Ababa to obtain their subsistence food consumption during and in the post-COVID-19 pandemic periods, thereby also helping to maintain food security and health benefits. The practice of urban agriculture in and around Addis Ababa has continued aggressively in the post-COVID-19 period as the awareness and perception of the people have been enhanced. It is recommended that urban agriculture should be integrated into urban planning and in the one health programme approach, and that lessons should be taken from the COVID-19 pandemic experience to apply towards the expectations of future pandemics. This study highlights a significant policy agenda for policymakers on how to integrate urban agriculture with urban planning and development. The study is limited to a literature review and very quick field observations to supplement the review visually. To this end, an in-depth study including a detailed field survey is recommended in the future to showcase more on urban agriculture in Addis Ababa.

## References

Abate M (2018) Contribution of urban productive safety net program to households' livelihood improvement and environmental protection in Addis Ababa: case study of Addis Ketema and Arada sub-cities. MA thesis, Addis Ababa University, Addis Ababa, Ethiopia

Abay KA, Berhane G, Hoddinott J, Tafere K (2023) COVID-19 and food security in Ethiopia: do social protection programs protect? Econ Dev Cult Change 71(2):373–402

Abel GJ, Brottrager M, Cuaresma JC, Muttarak R (2019) Climate, conflict and forced migration. Glob Environ Chang 54:239–249

Addis Ababa, Ethiopia Metro Area Population 1950–2022. https://www.macrotrends.net/cities/20921/addis-ababa/population

Alemu T, Teshome D, Mebratu T, Demelash M, Tadele M, Tabor G, Diro S et al (2024) Assessment of urban agriculture in Addis Ababa potentials, practices, challenges, and intervention options. https://doi.org/10.13140/RG.2.2.20371.90400

Ali N, Fariha KA, Islam F, Mishu MA, Mohanto NC, Hosen MJ, Hossain K (2021) Exposure to air pollution and COVID-19 severity: a review of current insights, management, and challenges. Integr Environ Assess Manag 17(6):1114–1122

Anberbir AG (2022) The role of urban agriculture towards livelihood and food security: the case of Addis Ababa, Ethiopia. MSc thesis, SLU, Uppsala, Sweden

Ardalan A, Belay GD, Assen M, Hosseinzadeh-Attar MJ, Khoei EM, Ostadtaghizadeh A (2019) Impact of climate change on community health and resilience in Ethiopia: a review article. Hum Antibodies 27(S1):11–22

Arsiso BK, Tsidu GM, Stoffberg GH, Tadesse T (2017) Climate change and population growth impacts on surface water supply and demand of Addis Ababa, Ethiopia. Clim Risk Manag 18:21–33

Artmann M, Specht K, Vávra J, Rommel M (2021) Introduction to the special issue "a systemic perspective on urban food supply: assessing different types of urban agriculture". Sustainability 13(3798). https://doi.org/10.3390/su13073798

Aschale M, Sileshi Y, Kelly-Quinn M, Hailu D (2015) Assessment of potentially toxic elements in vegetables grown along Akaki river in Addis Ababa and potential health implications. Food Sci Qual Manag 40:42–52

Asfew M, Mitiku F, Gemechu A, Bekele Y, Lemma T (2023) Do climate change and political instability affect crop production in sub-Saharan Africa countries? J Agric Food Res 12:100576

Ayele BY, Megento TL, Habetemariam KY (2022) Assessing green infrastructure spatial plans in Addis Ababa, Ethiopia. Socio-Ecol Pract Res 4(2):85–101

Ayuk E, Ramaswami A, Teixeira I, Akpalu W, Eckart E, Ferreira J, Kirti D, de Souza Leao V (2021) Urban agriculture's potential to advance multiple sustainability goals: an international resource panel think piece. In: A think piece of the international resource panel, United Nations Environment Programme, Nairobi, Kenya

Baars C, Barbir J, Paulino Pires Eustachio JH (2023) How can climate change impact human health via food security? A bibliometric analysis. Environments 10(11):196

Bahiru DB (2021) Assessment of some heavy metals contamination in some vegetables (tomato, cabbage, lettuce and onion) in Ethiopia: a review. Am J Environ Prot 10(2):53–58

Bechauf R, Cutler E, Bassi AM, Casier L, Kapetanakis M, Pallaske G, Simmons B (2022) The value of incorporating nature in urban infrastructure planning. International Institute for Sustainable Development (IISD), Winnipeg, Manitoba, Canada

Bhopal A, Medhin H, Bærøe K, Norheim OF (2021) Climate change and health in Ethiopia: to what extent have the health dimensions of climate change been integrated into the Climate-Resilient Green Economy? World Med Health Policy 13(2):293–312

Biswas BK (2022) Effect of climate change on vector-borne disease. In: Emerging issues in climate smart livestock production. Academic Press, pp 263–316

Bron GM, Siebenga JJ, Fresco LO (2023) In the age of pandemics, connecting food systems and health: a global one health approach. Science and innovations for food systems transformation. Springer, Cham, pp 869–875

CAP (2021) Addis Ababa climate action plan 2021–2025. Towards a climate resilient, inclusive, carbon neutral city by 2050. Addis Ababa environmental protection and green development commission in collaboration with C40 cities climate leadership group. Addis Ababa, Ethiopia

CDKN (2021) Addis Ababa: a community alliance faces up to Covid-19. https://cdkn.org/story/feature-a-community-alliance-faces-up-to-covid-19?loclang=en_gb

Chari F, Ngcamu BS (2022) Climate change and its impact on urban agriculture in Sub-Saharan Africa: a literature review. Environ Socio-Econ Stud 10(3):22–32

Chaudhry D (2023) Climate change and health of the urban poor: the role of environmental justice. J Clim Change Health 100277

Cianconi P, Betrò S, Janiri L (2020) The impact of climate change on mental health: a systematic descriptive review. Front Psych 11:74

Coulton J (2020) Sustainable urban agriculture starts at school in Ethiopia. Slow Food. https://www.slowfood.com/blog-and-news/sustainable-urban-agriculture-starts-at-school-in-ethiopia/

CSA (2007) Ethiopian population census. Addis Ababa, Ethiopia

CSA (2020) Key findings on the 2020 urban employment unemployment survey (a comparative analysis with 2014–2016 and 2018 survey results). Addis Ababa, Ethiopia

Cutler E, Gouett M, Guzzetti M (2022) Sustainable asset valuation (SAVi) of tree planting in Addis Ababa, Ethiopia. International Institute for Sustainable Development. https://nbi.iisd.org/report/savi-tree-planting-addis-ababa-ethiopia/

Damte T, Tabor G (2015) Small-scale vegetable producers' perception of pests and pesticide uses in East Shewa zone, Ethiopia. Int J Pest Manag. https://doi.org/10.1080/09670874.2015.1036822

Deksissa T, Trobman H, Zendehdel K, Azam H (2021) Integrating urban agriculture and stormwater management in a circular economy to enhance ecosystem services: connecting the dots. Sustainability 13(8293). https://doi.org/10.3390/su13158293

Delelegn M, Mulugeta M (2018) The status of urban agriculture in and around Addis Ababa, Ethiopia. J Sustain Dev Africa 20(2):128–147

Domingo JL, Rovira J (2020) Effects of air pollutants on the transmission and severity of respiratory viral infections. Environ Res 187:109650

Dorr E, Goldstein B, Horvath A, Aubry C, Gabrielle B (2021) Environmental impacts and resource use of urban agriculture: a systematic review and meta-analysis. Environ Res Lett 16(9):093002

Dubbeling M, van Veenhuizen R, Halliday J (2019) Urban agriculture as a climate change and disaster risk reduction strategy. Field actions science reports. J Field Actions (special issue 20):32–39

Ebissa G, Desta H (2022) Review of urban agriculture as a strategy for building a water resilient city. City Environ Interact 14(100081). https://doi.org/10.1016/j.cacint.2022.100081

El-Sayed A, Kamel M (2020) Climatic changes and their role in emergence and re-emergence of diseases. Environ Sci Pollut Res 27:22336–22352

Etana MB, Gonfa MT, Duresa CO (2019) Urban horticulture: the case of Finfinne Area Sebeta Town in Ethiopia. Int J Agric Agribus 4(2):133–138

Fanzo J, Davis C, McLaren R, Choufani J (2018) The effect of climate change across food systems: implications for nutrition outcomes. Glob Food Sec 18:12–19

FAO (2021) National agrifood systems and COVID-19 in Ethiopia: effects, policy responses and long-term implications. Rome. https://doi.org/10.4060/cb7850en

FAO, Rikolto, RUAF (2022) Urban and peri-urban agriculture sourcebook—from production to food systems. Rome, FAO and Rikolto. https://doi.org/10.4060/cb9722en

Farinati S, Betto A, Palumbo F, Scariolo F, Vannozzi A, Barcaccia G (2022) The new green challenge in urban planning: the right genetics in the right place. Horticulturae 8(761). https://doi.org/10.3390/horticulturae8090761

Farm Africa: Urban agriculture. https://www.farmafrica.org/ethiopia/urban-agriculture

Fekadu M, Lulekal E, Tesfaye S, Ruelle M, Asfaw N, Awas T, Balemie K, Asres K, Guenther S, Asfaw Z, Demissew S (2023) The potential of Ethiopian medicinal plants to treat emergent viral diseases. Phytotherapy Res

Feyissa G, Zeleke G, Bewket W, Gebremariam E (2018) Downscaling of future temperature and precipitation extremes in Addis Ababa under climate change. Climate 6(3):58

Fikeraddis M, Endeshaw L (2020) Influence of temperature and relative humidity on air pollution in Addis Ababa, Ethiopia. J Environ Earth Sci 2(2):19–25

Frontera A, Cianfanelli L, Vlachos K, Landoni G, Cremona G (2020) Severe air pollution links to higher mortality in COVID-19 patients: the "double-hit" hypothesis. J Infect 81(2):255–259

Further Africa (2020) Urban agriculture in Addis Ababa to ease COVID-19 impact. https://furtherafrica.com/2020/05/28/urban-agriculture-in-addis-ababa-to-ease-covid-19-impact/

Gietaneh W, Simieneh MM, Endalew B, Tarekegn S, Petrucka P, Eyayu D (2023) Traditional healers' roles, and the challenges they face in the prevention and control of local disease outbreaks and pandemics: the case of the East Gojjam Zone in northwestern Ethiopia. Front Trop Dis 4:978528

Gebremichael D, Gebremichael AT, Worku A, Abshare MW, Habtemariam YM, Balcha G, Gebremichael D (2014) Building urban resilience: assessing urban and peri-urban agriculture in Addis Ababa, Ethiopia. In: Padgham J, Jabbour J (eds) United Nations Environment Programme (UNEP), Nairobi, Kenya

Gebreyesus SH, Lunde TM, Mariam DH, Woldehanna T, Lindtjørn B (2014) Climate change, crop production and child under nutrition in Ethiopia; a longitudinal panel study. In: Spatial variations in child undernutrition in Ethiopia: implications for intervention strategies

Gupta S, Rouse BT, Sarangi PP (2021) Did climate change influence the emergence, transmission, and expression of the COVID-19 pandemic? Front Med 8:769208

Hagos S, Lunde T, Mariam DH, Woldehanna T, Lindtjørn B (2014) Climate change, crop production and child under nutrition in Ethiopia; a longitudinal panel study. BMC Public Health 14:1–9

Haile AA (2021) Important medicinal plants in ethiopia: a review in years 2015–2020. Herbs and Spices-New Processing Technologies

Haile B, Babege T, Hailu A (2016) Diseases and insect pests of onion (Allium cepa L.) in Masha District of Sheka Zone, South-West Ethiopia. Acad J Agric Res 4(10):629–632. https://doi.org/10.15413/ajar.2016.0180

Hayes K, Blashki G, Wiseman J, Burke S, Reifels L (2018) Climate change and mental health: risks, impacts and priority actions. Int J Ment Heal Syst 12:1–12

Hirvonen K, Abate GT, de Brauw A (2020) Food and nutrition security in Addis Ababa, Ethiopia during COVID-19 pandemic: May 2020 report. Intl Food Policy Res Inst 143

Hirvonen K, de Brauw A, Abate GT (2021) Food consumption and food security during the COVID-19 pandemic in Addis Ababa. Am J Agric Econ 103(3):772–789. https://doi.org/10.1111/ajae.12206

IPCC (2023) Summary for policymakers. In: Climate change 2023: synthesis report. Contribution of working Groups I, II and III to the sixth assessment report of the intergovernmental panel on climate change (Core Writing Team, Lee H and Romero J (eds)). IPCC, Geneva, Switzerland, pp 1–34, https://doi.org/10.59327/IPCC/AR6-9789291691647.001.

John M, Johan S (eds) (2022) COVID-19 and global food security: Two years later. International Food Policy Research Institute (IFPRI), Washington, DC, USA. pp 193. https://doi.org/10.2499/9780896294226

Ju JH, Yoon YH, Shin SH, Ju SY, Yeum KJ (2022) Recent trends in urban agriculture to improve bioactive content of plant foods. Horticulturae 8(767). https://doi.org/10.3390/horticulturae8090767

Kebebew M (2018) Diversity and management of useful homegardens plant species in Arba Minch Town, Southern Ethiopia: implication for plant diversity conservation and food security. Int J Econ Plants 5(3):137–148

Kennard NJ, Bamford RH (2020) Urban agriculture: opportunities and challenges for sustainable development. Zero Hunger 1–14

Khan A, Yasin M, Aqueel MA, Farooqi MA, Akram MI, Yousuf HMB, Noor M, Maqsood A (2023) Vector-borne disease and climate change. In: Arthropods-new advances and perspectives. IntechOpen

Laddha D (2021) Agriculture: Top 10 cities for urban AG. https://www.agritecture.com/blog/2021/2/23/top-10-cities-for-urban-ag

Lawrance E, Thompson R, Fontana G, Jennings N (2021) The impact of climate change on mental health and emotional wellbeing: current evidence and implications for policy and practice. Grantham Institute briefing paper, 36

Martelletti L, Martelletti P (2020) Air pollution and the novel Covid-19 disease: a putative disease risk factor. SN Compr Clin Med 2(4):383–387

McCann J, Sinno L, Ramadhan E, Assefa N, Berhane HY, Madzorera I, Fawzi W (2023) COVID-19 disruptions of food systems and nutrition services in Ethiopia: evidence of the impacts and policy responses. Ann Glob Health 89(1)

McCarney R, Kent J (2020) Forced displacement and climate change: time for global governance. Int J 75(4):652–661

Mengistu DT, Gebremariam E, Wang X, Zhao S (2022) Pandemic-resilient urban centres: a new way of thinking for industrial-oriented urbanization in Ethiopia. Urban Sci 6(26). https://doi.org/10.3390/urbansci6020026

Mengistu F, Alemayehu M (2017) Species assortment and biodiversity conservation in homegardens of Bahir Dar City, Ethiopia. Ethiopian J Agric Sci 27(2):31–48

Mensah JK (2023) Urban agriculture, local economic development and climate change: conceptual linkages. Int J Urban Sustain Dev 15(1):141–151

Mezgabu F, Tolossa D (2015) The contribution of urban agriculture to food security of individual urban farmers in Yeka Sub City, Addis Ababa. Ethiopian J Dev Res 37(1):177–207

Minten B, Habte Y, Tamru S, Tesfaye A (2020a) The transforming dairy sector in Ethiopia. PLoS ONE 15(8):e0237456. https://doi.org/10.1371/journal.pone.0237456

Minten B, Mohammed B, Tamru S (2020b) Emerging medium-scale tenant farming, gig economies, and the COVID-19 disruption: the case of commercial vegetable clusters in Ethiopia. Eur J Dev Res 32(5):1402–1429

Mirzabaev A, Olsson L, Kerr RB, Pradhan P, Ferre MGR, Lotze-Campen H (2023) Climate change and food systems. In: von Braun J, Afsana K, Fresco LO, Hassan MHA (eds) Science and innovations for food systems transformation. Springer, Cham. https://doi.org/10.1007/978-3-031-15703-5_27

Myers SS, Smith MR, Guth S, Golden CD, Vaitla B, Mueller ND, Dangour AD, Huybers P (2017) Climate change and global food systems: potential impacts on food security and undernutrition. Annu Rev Public Health 38:259–277

Newbury JB, Stewart R, Fisher HL, Beevers S, Dajnak D, Broadbent M, Pritchard M, Shiode N, Heslin M, Hammoud R, Hotopf M, Bakolis I (2021) Association between air pollution exposure and mental health service use among individuals with first presentations of psychotic and mood disorders: retrospective cohort study. Br J Psychiatry 219(6):678–685

Nigussie S, Liu L, Yeshitela K (2021) Towards improving food security in urban and peri-urban areas in Ethiopia through map analysis for planning. Urban Forestry Urban Greening 58. https://doi.org/10.1016/j.ufug.2020.126967

Ouda S, Zohry AEH, Noreldin T, Ouda S, Zohry AEH (2020) Water scarcity leads to food insecurity. In: Deficit irrigation: a remedy for water scarcity, pp 1–13

Palinkas LA, Wong M (2020) Global climate change and mental health. Curr Opin Psychol 32:12–16

Pauleit S, Pribadi DO, El Wafa HA (2019) Peri-urban agriculture: lessons learnt from Jakarta and Addis Ababa. Field Actions Sci Rep SI 20:18–25

Regassa R (2016) Useful plant species diversity in homegardens and its contribution to household food security in Hawassa city, Ethiopia. Afr J Plant Sci 10(10):211–233

Schnitter R, Berry P (2019) The climate change, food security and human health nexus in Canada: a framework to protect population health. Int J Environ Res Public Health 16(14):2531

Serbessa H, Makonnen T, Abi M (2023) The contribution of urban agriculture to household food security among urban vegetable producers in Addis Ababa, Ethiopia. J Sustain Dev Africa 25(1)

Simane B, Beyene H, Deressa W, Kumie A, Berhane K, Samet J (2016) Review of climate change and health in Ethiopia: status and gap analysis. Ethiop J Health Dev 30(1):28–41

Tadesse ST, Oenema O, van Beek C, Ocho FL (2018) Diversity and nutrient balances of urban and peri-urban farms in Ethiopia. Nutr Cycl Agroecosyst 111:1–18. https://doi.org/10.1007/s10705-018-9911-0

Tarekegn MM, Gulilat TY (2018) Trends of ambient air pollution and the corresponding respiratory diseases in Addis Ababa. Res Rep Toxi 2(5)

Tefera MM (2021) Food security impact of COVID-19 and policy responses in Ethiopia. In: Choongjae C (ed) Crisis and fragility: economic impact of C VID-19 and policy responses, KIEP Visiting Scholars Program, pp 357–368

Tefft J, Jonasova M, Adjao R, Morgan A (2017) Food systems for an urbanizing world: knowledge product. World Bank Group, New York, USA, and FAO, Rome, Italy. pp xvii+154

Tuasha N, Fekadu S, Deyno S (2023) Prevalence of herbal and traditional medicine in Ethiopia: a systematic review and meta-analysis of 20-year studies. Syst Rev 12(1):232

Turnšek M, Gangenes SL, Piirman M, Thorarinsdottir RI, Bavec M, Junge R (2022) Home gardening and food security concerns during the COVID-19 pandemic. Horticulturae 8(778). https://doi.org/10.3390/horticulturae8090778

UN-HABITAT (2008) Ethiopia: Addis Ababa urban profile. Participatory slum upgrading programme in the African, Caribbean and Pacific Countries, Nairobi, Kenya

Vandercasteelen J, Minten B, Tamru S (2019) Cities, value chains, and dairy production in Ethiopia. In: International food policy research institute—the Ethiopia strategy support program (IFPRI-ESSP), Working Paper 137, Addis Ababa, Ethiopia.pp 36

Vanhala L, Calliari E (2022) Governing people on the move in a warming world: framing climate change migration and the UNFCCC task force on displacement. Glob Environ Chang 76:102578

Woldeamanuel MM, Geda MK, Mohapatra S, Bastia TK, Rath P, Panda AK (2022) Ethnobotanical study of endemic and non-endemic medicinal plants used by indigenous people in environs of Gullele botanical garden Addis Ababa, central Ethiopia: a major focus on Asteraceae family. Front Pharmacol 13:1020097

Woldetsadik D, Drechse P, Keraita B, Itanna F, Gebrekidan H (2017) Heavy metal accumulation and health risk assessment in wastewater-irrigated urban vegetable farming sites of Addis Ababa, Ethiopia. Int J Food Contam 4(9). https://doi.org/10.1186/s40550-017-0053-y

Worku H (2017) Rethinking urban water management in Addis Ababa in the face of climate change: an urgent need to transform from traditional to sustainable system. Environ Qual Manage 27(1):103–119

Worku M, Sahela S (2018) Review on disease management practice of tomato wilt caused fusarium oxysporum in case of ethiopia. J Plant Pathol Microbiol 9:460. https://doi.org/10.4172/2157-7471.1000460

World Bank (2015) Poverty global practice Africa Region; Ethiopian poverty assessment. Report No: ACS12005, The World Bank, Washington DC, USA

Wright CY, Kapwata T, Naidoo N, Asante KP, Arku RE, Cissé G, Simane B, Atuyambe L, Berhane K (2024) Climate change and human health in Africa in relation to opportunities to strengthen mitigating potential and adaptive capacity: strategies to inform an African "Brains Trust". Ann Glob Health 90(1)

Zakar MZ, Zakar DR, Fischer F (2020) Climate change-induced water scarcity: a threat to human health. South Asian Stud 27(2)

Zegeye H (2018) Climate change in Ethiopia: impacts, mitigation and adaptation. Int J Res Environ Stud 5(1):18–35

Zhang W, Persoz L, Hakiza S, Biru L, Girmatsion L (2022) Impact of COVID-19 on food security in Ethiopia. Epidemiologia 3(2):161–178

# Chapter 18
# Impacts of Climatic Changes on Parasitic and Vector-Borne Diseases in Nigeria

**Etiosa Uyigue, Blessing Idon, and Dorothy Egbon**

**Abstract** The impacts of climatic changes may pose threats to public health. The breeding and lifestyle of disease vectors such as mosquitoes are highly sensitive to climatic conditions. Assessing vulnerability and adaptation strategies can reshape programmes for climate-sensitive infectious diseases. The objectives of this paper are to understand the likely impacts of climatic changes on the spread of parasitic infections and vector-borne diseases in Nigeria and propose strategies to combat the current and future health-related impacts of climate change. The analysis in this paper was carried out using secondary data derived from several sources, which include the Nigerian Meteorological Agency and the Federal Ministry of Health. Findings indicate that Nigeria's average atmospheric temperature rose by 1.11 °C, surpassing the global average of 0.74 °C. Rainfall increased by 245.72 mm, with regional variations. Northern Nigeria saw substantial rainfall increases compared to the south. Rising temperatures significantly correlated with increased malaria cases. Increase in rainfall had no major impact on the prevalence of malaria, but inversely affected the prevalence of filariasis and diarrhoea cases. Temperature did not significantly affect the prevalence of filariasis. The study concludes that mitigating the health impacts of climate change in Nigeria requires a multi-faceted approach. Key recommendations include increasing climate action to reduce health-related risks, investing in research to develop alternative strategies for parasite control, enhancing public education and awareness, and improving the capacity of local communities to prevent the breeding of disease vectors. Policymakers and stakeholders must prioritise these actions to safeguard public health in the face of ongoing and future climatic changes.

E. Uyigue (✉) · B. Idon · D. Egbon
Community Research and Development Centre, Abuja, Nigeria
e-mail: etiosa.uyigue@gmail.com

B. Idon
e-mail: blessingidon27@gmail.com

D. Egbon
e-mail: dorothyegbon@gmail.com



W. Leal Filho and F. Wolf (eds.), *Climate Change and Health: Perspectives from Developing Countries*, Climate Change Management,
https://doi.org/10.1007/978-3-031-93177-2_18

**Keywords** Environmental factor and diseases · Malaria and climate change · Vector borne diseases · Climate impact on health

## Introduction

The transmission patterns of many communicable diseases are influenced by many factors, and these may include climatic and ecological elements such as temperature and rainfall. Climatic changes have the potential to affect the spread of parasitic and non-parasitic infections, especially in countries in hot climates. The impacts of these changes may be favourable and in other cases may pose threats to public health. The breeding and lifestyle of disease vectors such as mosquitoes, sand flies, ticks, tsetse fly and others are highly sensitive to climatic conditions, mainly temperature and humidity (Fig. 18.1). Temperature increases facilitate the development of arthropod vectors and parasites themselves (Moore et al. 2012). The distribution pattern of these disease vectors is likely to shift due to climatic changes.

Arthropod vectors are ectothermic and are especially sensitive to climatic conditions. Climatic elements influence the survival and reproduction rates of vectors, in turn influencing habitat suitability, distribution and abundance. Moreover, climatic conditions influence the intensity and temporal pattern of vector activity (particularly biting rates) throughout the year as well as the rates of development, survival, and reproduction of pathogens within vectors (Rogers and Randolph 2006).

The Intergovernmental Panel on Climate Change (IPCC) asserted that global temperatures are rising (Fig. 18.2) due to human activities. The rising global temperature will increase the likelihood of many vector-borne diseases shifting to new areas. Health risks due to changes in climatic conditions will differ between countries. Countries that have developed sufficient health infrastructures will be able to cope better than those that do not have good and sufficient healthcare facilities (Andrew

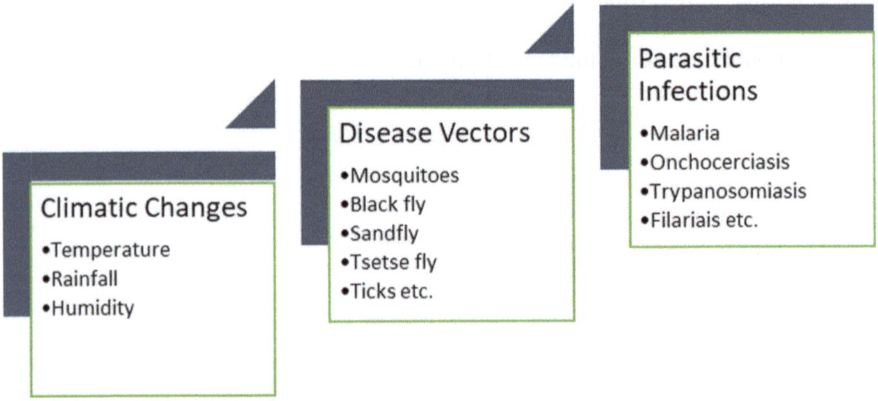


**Fig. 18.1** Climatic changes and the incidents of parasitic infections

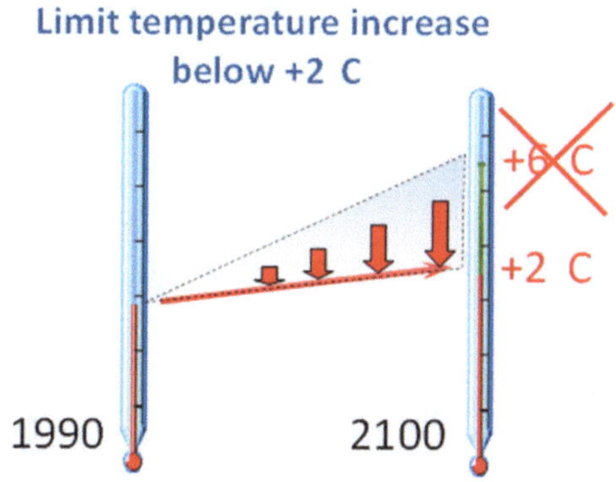


**Fig. 18.2** Expected global temperature rise by 2100 (Lebot 2009)

et al. 2000). Experts have asserted that the effects of climatic changes are felt everywhere in Nigeria, that the rising global temperatures will have a catastrophic effect on human health, and that the patterns of infection of many diseases will change, with insect-borne diseases such as malaria and dengue fever spreading more easily (The Guardian 2016).

It is important for policymakers and other stakeholders to have sufficient knowledge of the health risks that changes in climatic conditions pose to the populace. Empirical and systematic knowledge on the subject seems to be lacking among stakeholders in Nigeria. Moreover, an assessment of vulnerability and adaptation to changes in climatic conditions can provide modifications to current and planned programmes designed to reduce burdens of climate-sensitive infectious diseases. The objectives of this report are to empirically understand the likely impacts of changes in climatic conditions on the rate and spread of parasitic infections and vector-borne diseases in Nigeria and propose strategies to combat the current and future health-related impacts of climate change.

## Method of Data Collection

The analysis in this report was carried out using secondary data derived from several sources. Temperature and rainfall data were computed from data released by the Nigerian Meteorological Agency (NMA) representing a period of 15–60 years. Some of the meteorological data were sourced from published reports. Sixteen years of epidemiological data from the 36 states of Nigeria were sourced from data released by the Federal Ministry of Health and from existing literatures. Some of the data collected were subjected to inferential statistics such as correlation and regression analysis. To understand community vulnerability, primary data were collected from selected communities in the southern Nigeria using questionnaires and focused group discussions (FGD). Data collected were subjected to quantitative analysis using different statistical packages and qualitative analysis.

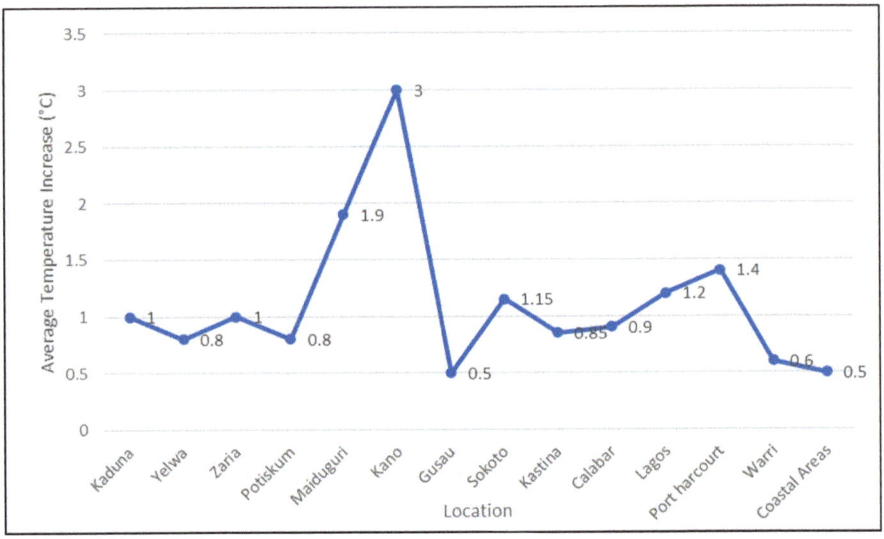


**Fig. 18.3** Average temperature increase in selected cities in Nigeria

## *Temperature*

The temperature data computed from 14 locations revealed that there has been general increase in atmospheric temperature within the study period. The atmospheric temperature has increased at an average of about 1.11 °C in Nigeria. This average temperature value computed in the current study is the same as the value reported in a similar study, where the mean air temperature was computed to increase by 1.1 °C for a period of 105 years (Odjugo 2010). The average temperature increase computed in the current study is higher than the global average of 0.74 °C recorded since 1860, when actual scientific temperature measurement started (IPCC 2007). The location with the highest average temperature increase was Kano, which experienced an average temperature increase of 3 °C, while the station with the lowest average temperature increase was the coastal region with an average increase of 0.5 °C. Many of the stations in the northern part of Nigeria experienced increases in temperature above the calculated general average (Fig. 18.3).

## *Rainfall*

The rainfall data collected from fifteen locations was computed and shows that rainfall has increased by an average of 245.72 mm within the study period. From the trend analysis, Sokoto (North–West) and Owerri (South–East) experienced a decrease

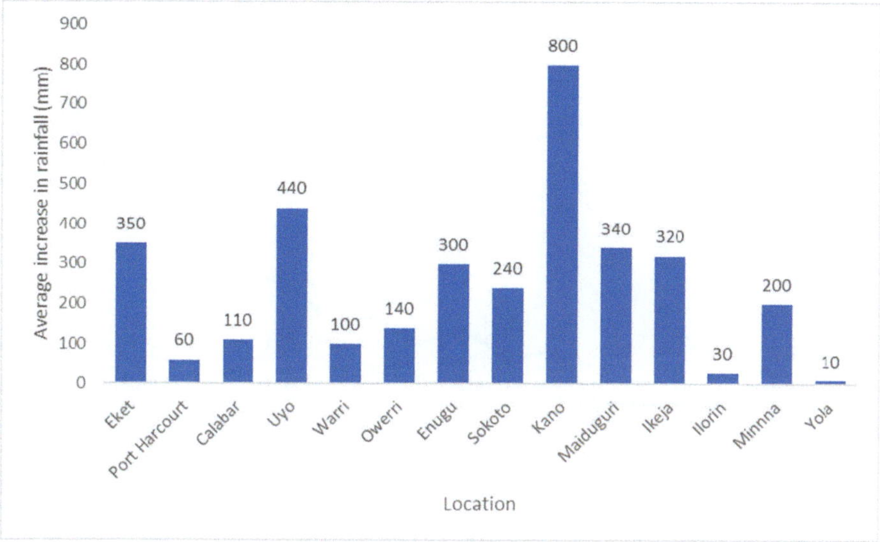


**Fig. 18.4** Average increase in rainfall in selected cities in Nigeria

in rainfall. The other stations, Eket, Port Harcourt, Calabar, Uyo, Warri (South–South), Enugu (South–East), Kano (North–West), Maiduguri, Yola (North–East), Ikeja (South–West), Ilorin and Minna (North-Central) experienced a general increase in rainfall within the period of the study. Yola experienced the lowest average rainfall increase of 10 mm in a period of 29 years, while Kano experienced the highest average rainfall increase by 800 mm in the same period.

An earlier report shows that the rainfall trend in Nigeria between 1901 and 2005 shows a general decline (Odjugo 2010). The same report shows that within 105 years, the rainfall amount dropped by 81 mm, and the declining rainfall became worse from the early 1970s. It is interesting to know from the current study that locations like Kano, Maiduguri, Minna, and Yola, located between the middle belt and the northern part of Nigeria, experienced a general increase in rainfall. From the current study, the increase in rainfall from the stations in northern Nigeria was generally high compared to the stations in southern Nigeria (Fig. 18.4).

## *Cases of Malaria Infection*

Recorded cases of Malaria from the 36 states of Nigeria show an increasing pattern for a period of sixteen years from the year 1994–2009 (Fig. 18.5). The study revealed that cases of malaria increased from 1,350,023 cases in 1994 to 7,307,155 cases in 2009. When the numbers of malaria cases were compared with temperature data using different statistical packages, it was discovered that temperature had played a

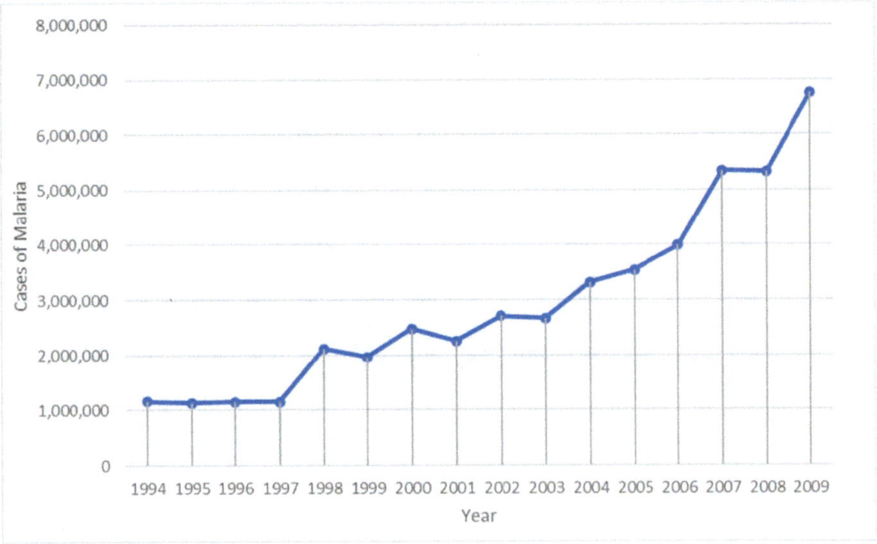


**Fig. 18.5** Trend of malaria infection in Nigeria—1994–2009

crucial role in the increase in the number of cases of malaria. At the 95% confidence limit, the regression and correlation between atmospheric temperature and the cases of malaria were significant. The correlation between the two variables was positive; in other words, as temperature increased, the number of cases of malaria also increased.

Many factors could be responsible for this direct relationship between cases of malaria and temperature. Malaria is a protozoa parasite transmitted through the bites of the female Anopheles mosquitoes. The findings from this study support a previous report, which shows that mosquito species such as Anopheles gambiae—responsible for the transmission of most vector-borne diseases—are sensitive to temperature changes during the immature stages in the aquatic environment and as adults (Andrew et al. 2000). Another study reported that if water temperature rises, the larvae of mosquitoes take a shorter time to mature and consequently there is a greater capacity to produce more offspring (Rueda et al. 1990). The period for the Plasmodium parasite to mature inside the mosquito is dependent on temperature (Smyth et al. 1996). In warmer climates, adult female mosquitoes digest blood faster and feed more frequently, thus increasing transmission intensity (Gillies et al. 1953). Similarly, malaria parasites complete extrinsic incubation within the female mosquito in a shorter time as temperature rises, thereby increasing the proportion of infective vectors (Turell 1989).

When rainfall data was compared statistically with cases of malaria, changes in rainfall did not significantly affect the cases of malaria. Although there has been a general increase in rainfall, it has not made any significant impact on the rising cases of malaria in Nigeria. Increasing precipitation has the potential to increase the number and quality of breeding sites for vectors such as mosquitoes (Andrew et al. 2000).

## *Cases of Filariasis*

Filariasis is a disease caused by a worm-like parasitic organism called Wuchereria bancrofti, found only in subtropical and tropical countries, and it is mainly transmitted by the Culex mosquitoes. The increase in temperature did not have a significant impact on the cases of filarial disease. When the number of filarial cases was statistically compared with the amount of rainfall, the correlation between the two variables was significant at 0.05 confidence limits. The relationship between them was an indirect relationship, which implied that as the amount of rainfall increased, the number of cases of filarial infection decreased. Excess rainfall may result in flooding, thereby affecting the habitat that is suitable for the breeding of the Culex mosquito (Fig. 18.6).

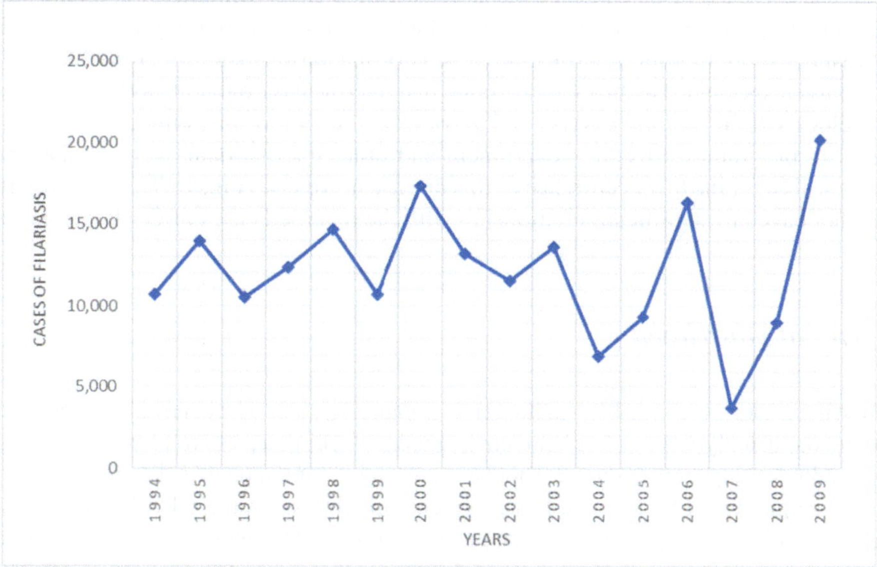


**Fig. 18.6** Case of filariasis in Nigeria from 1994 to 2009

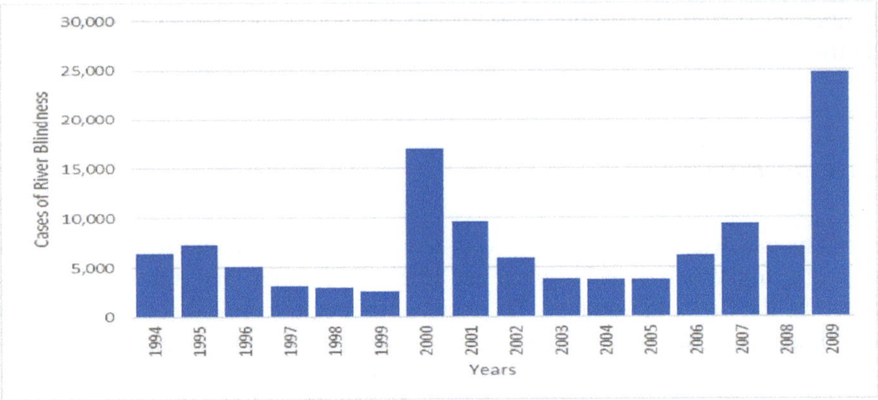


**Fig. 18.7** Case of onchocerciasis in Nigeria from 1994 to 2009

## *Cases of Onchocerciasis*

The causative agent of Onchocerciasis (also called River Blindness) is a parasitic organism called *Onchocerca volvulus*, which is similar in biology to the filarial parasite, except that the intermediate host is of the various species of the black fly (*Simulium* species). The black fly is found in a riverine environment, and the parasite develops within the black fly in 6–7 days at 27 $^{0}$C before infesting another host. Control programmes by the government and other international agencies may be responsible for the low occurrence of this disease in Nigeria (Fig. 18.7). Temperature did not have any significant relationship with the recorded cases of Onchocerciasis.

## *Cases of Diarrhoea*

Diarrhoea can be caused by pathogenic organisms such as bacteria, viruses, and protozoa, usually in poor hygienic environments lacking access to potable drinking water. Increasing temperature did not have any significant impact on diarrhoea, which could be prevented by keeping basic hygiene principles. On the other hand, rainfall had a significant impact on the cases of diarrhoea. The increase in rainfall was inversely proportional to the cases of diarrhoea, that is, as the amount of rainfall increased, the cases of diarrhoea decreased. Rainwater harvesting is a common practice in many communities in Nigeria; both rural and urban communities rely on rainwater as their main source of water. An increase in rainfall will mean more access to water for cooking, washing, and other domestic uses that will reduce the incidence of diarrhoea-related diseases (Fig. 18.8).

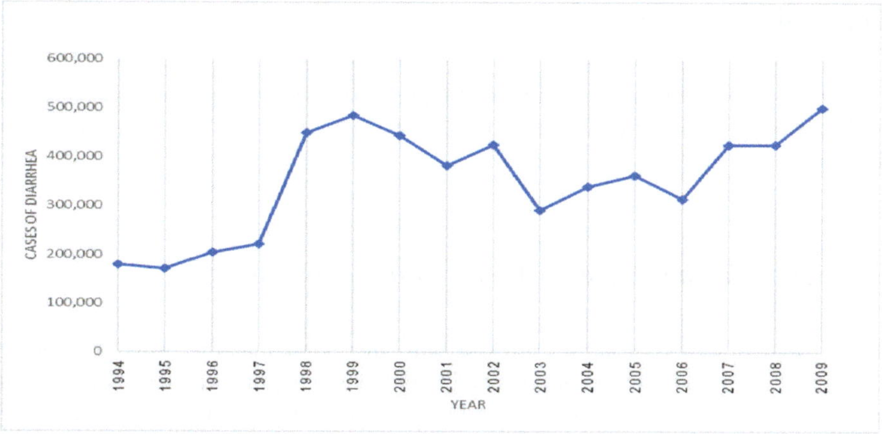


**Fig. 18.8** Cases of diarrhoea in Nigeria from 1994 to 2009

## *Vulnerability of Local Communities to Climatic Changes*

The IPCC defines vulnerability to climate change as the degree to which a system is susceptible to, or unable to cope with, the adverse effects of climate variability and change. Adaptive capacity describes the general ability of individuals, communities, and institutions to effectively prepare for and cope with the consequences of climate variability and change. The nature and magnitude of the impacts of climate change on infectious disease depend on local vulnerability, which in turn depends on local conditions, their resilience to change, and their capacity to adapt.

The current study has shown that rural communities in Nigeria are highly vulnerable to the health-related impacts of climate change. The health systems in the communities visited during the study were generally weak. In some of the communities, primary healthcare facilities were not available. In places where they were available, there was no sufficient and qualified staff to run the facilities. In some of the communities, people must travel several kilometres to access quality health care. All this makes rural communities in Nigeria highly vulnerable to health-related climate change impacts.

Socio-economic factors, health-seeking behaviour, geographical location, and population growth will determine the vulnerability of populations to climate change (Andrew et al. 2000). Weak health systems and infrastructural development in our communities will make Nigeria highly vulnerable to the impact of climate change.

## Discussion

The analysis of the impacts of climatic changes on parasitic and vector-borne diseases in Nigeria highlights critical interactions between environmental factors and public health.

### *Impact of Temperature on Vector-Borne Diseases*

Rising temperatures strongly correlate with increased malaria cases, as warmer conditions accelerate the development of Anopheles mosquitoes and the Plasmodium parasite. However, temperature has a lesser impact on diseases like filariasis and onchocerciasis, indicating other factors may influence their spread.

### *Rainfall's Mixed Impact on Disease Prevalence*

Rainfall showed an inverse relationship with filariasis and diarrhoea but had no significant impact on malaria cases. Excessive rain may disrupt Culex mosquito breeding sites, reducing filariasis, while improved water access reduces diarrhoea cases. However, rising temperatures in northern Nigeria could expand the range of vector-borne diseases like malaria into previously unaffected areas.

### *Community Vulnerability and Health Infrastructure*

The study reveals that rural communities in Nigeria are highly vulnerable to climate-related health impacts due to weak healthcare infrastructure and limited awareness. Climate change will worsen these vulnerabilities, especially in marginalised areas, unless healthcare services and community resilience are improved. Targeted interventions are needed to strengthen healthcare, disease surveillance, and community education on climate change impacts.

### *Policy Implications and Adaptation Strategies*

The findings of this study highlight the need for Nigeria's health sector to prioritise climate change adaptation by incorporating climate-sensitive disease control into national health plans. Policymakers should improve local infrastructure, strengthen public health systems, and promote climate action to mitigate health risks. Additionally, investment in research and climate-resistant vector control methods is essential, alongside collaboration with local and international organisations to ensure effective interventions.

### *Broader Global Context*

The impacts of climate change on vector-borne diseases in Nigeria mirror global trends, with rising temperatures increasing the spread and intensity of diseases, especially in vulnerable countries with weak health systems. Socio-economic factors like poverty and urbanisation complicate disease management, emphasising the need for integrated, cross-sectoral approaches. Addressing these challenges requires strategies that tackle both climate change and underlying social inequalities.

## Conclusion and Recommendations

The analysis of temperature data in this study has shown that there has been a general increase in temperature in Nigeria and atmospheric temperature has increased by 1.1 °C. Similarly, the analysis of rainfall data also revealed that there has been a general increase in rainfall. The study has shown that there has been a steady increase in the cases of malaria in Nigeria and that temperature has played a crucial role in the increase in the number of cases of malaria. The relationship between temperature and cases of malaria is a direct relationship; that is, as temperature increases, the number of cases of malaria also increases.

The number of cases of filariasis and Onchocerciasis were relatively low. The temperature rise did not contribute significantly to the cases of filariasis and Onchocerciasis. However, the amount of rainfall was indirectly proportional to the cases of filarial disease, which implied that as the amount of rainfall increased, the number of cases of filarial infection decreased. In the case of Onchocerciasis, we found out that temperature did not have any significant relationship with the recorded cases of Onchocerciasis. It was found that rainfall had a significant impact on the cases of diarrhoea. The increase in rainfall was inversely proportional to the cases of diarrhoea; that is, as the amount of rainfall increased, the cases of diarrhoea decreased.

The ability of any nation or community to identify and implement responses to address health risks posed by climate change will depend on many factors. First, it is of primary importance that policymakers and other stakeholders have sufficient knowledge of the health risks that climatic changes pose, and the range of responses needed to reduce current and projected impacts. Empirical and systematic knowledge on the likely impacts of climatic changes on health may be lacking among policymakers and other Nigerian stakeholders. Moreover, an assessment of climate change vulnerability and adaptation can provide modifications to current and planned programmes designed to reduce the burdens of climate-sensitive infectious diseases. Therefore, it is important to understand the likely impacts of climatic changes on the spread of parasitic infections to build capacities within public health and healthcare organisations to mitigate these impacts. This will help to identify options that can be implemented outside the health sector.

### *Recommendations*

- Government and other stakeholders should consistently increase climate actions to minimise health-related risks associated with climatic changes. Such action may include interventions such as widespread vaccination campaigns and improved healthcare delivery systems, especially in rural areas.
- There is a need to initiate and sustain research into alternative strategies for parasite control.
- There is a need to build the capacity of local communities and create awareness among vulnerable communities on ways to minimise the breeding of parasite vectors and contact with the intermediate host of these parasites.
- Improvement in the health care delivery systems is imperative for vulnerable countries to cope with the growing health risk associated with climatic changes.

## References

Andrew K, Githeko SW, Lindsay UE, Patz CA, Patz JA (2000) Climate change and vector-borne diseases: a regional analysis. Bull World Health Org 7(9):1136–1147

Gillies MT (1953) The duration of the Gonotrophic Cycle in *Anopheles gambiae* and *Anopheles funestus* with a note on the efficiency of hand catching. East Afr Med J 30:129–135

IPCC (2007) The physical science basis. Contribution of working Group I to the fourth assessment report of the IPCC. In: Solomon S, Qin D, Manning M, Chen Z, Marquis M, Averyt KB, Tignor M, Miller HL (eds) Cambridge University Press, Cambridge, United Kingdom and New York, NY, USA, 996 pp

Lebot B (2009) Energy efficiency and market transformation: a short overview of best practices. In: A paper presented during the inception workshop of the UNDP-GEF project to promote energy efficiency in residential and public building in Nigeria, 14th July 2009

Moore S, Shrestha S, Tomlinson KW, Vuong H (2012) Predicting the effect of climate change on African trypanosomiasis: integrating epidemiology with parasite and vector biology. J R Soc Interface 2012(9):817–830

Odjugo PAO (2010) General overview of climate change impacts in Nigeria. J Hum Ecol 29(1):47–55

Rogers DJ, Randolph SE (2006) Climate change and vector-borne diseases. Adv Parasitol 62:345–381

Rueda LM et al (1990) Temperature-dependent development and survey rates of *Culex quinquefasciatus* and *Aedes aegypti* (Diptera: Culicidae). J Med Entomol 27:892–898

Smyth JD (1996) Animal parasitology. Cambridge University Press, Cambridge, United Kingdom, p 549

The Guardian (2016) Climate change fuels spread of diseases in Nigeria. A publication of The Guardian Newspaper, 25th Feb 2016

Turell MJ (1989) Effects of environmental temperature on the vector competence of *Aedes fowleri* for rift valley fever virus. Res Virol 140:147–154

# Chapter 19
# Knowledge and Perception of Climate Change and Its Impact on Human Health in Haut Sassandra Region Communities in Cote d'Ivoire

**Wassa Dairrassouba and Edward Mendy**

**Abstract** Climate change (CC) poses significant threats to human health, particularly in vulnerable communities like those in the Haut Sassandra region of Côte d'Ivoire. This study aims to assess the knowledge and perceptions of climate change and its potential human health impacts among these communities. A cross-sectional survey was conducted in 2022 using a structured questionnaire administered to 340 participants from four regions in Haut Sassandra region, with data analysed using Epi Info and SPSS software. The study revealed that 51.5% of participants had some knowledge of climate change, primarily acquired through various sources such as radio (21.8%), social media (9.7%), and other means (51.8%). A few (7.9%) acknowledged that climate change impacts human health, with some recognising the increased incidence of illnesses during extreme weather events. Statistical analysis indicated significant associations between knowledge of climate change and variables such as age, sex, and education level. Men were 1.58% times more likely to have knowledge about climate change when compared to women. The findings underscore the need for targeted educational interventions to enhance climate change awareness and its health implications in vulnerable communities like the region of Haut Sassandra, focusing on demographic groups identified as having lower levels of knowledge on the impact of climate change on human health.

**Keywords** Climate change · Human health · Vulnerable communities · Knowledge and perception

W. Dairrassouba (✉)
West African Science Service Centre on Climate Change and Adapted Land Use (WASCAl), Faculty of Agriculture and Environmental Science, University of the Gambia, Kanifing, Gambia
e-mail: wassa9105@gmail.com; dw222215009@utg.edugm

E. Mendy
West African Science Service Centre on Climate Change and Adapted Land Use (WASCAL), Université de Lomé, Lomé, Togo
e-mail: mendy.e@edu.wascal.org



W. Leal Filho and F. Wolf (eds.), *Climate Change and Health: Perspectives from Developing Countries*, Climate Change Management,
https://doi.org/10.1007/978-3-031-93177-2_19

## Introduction

Climate change refers to notable long-term variations and patterns of the Earth's climate, which may be attributed directly or indirectly to human activities that alter the atmospheric composition (Pielke 2004; Don et al. 2024). The human health of vulnerable communities usually depends on the unsure availability of required resources and food, the occurrence of natural disasters, climate change, and the health impacts of variables. The biggest hurdle to considerable community health enhancement is the reliance of populations on farming, which suffers the most from the change in weather patterns. Cote d'Ivoire is very vulnerable to climate change impacts like extreme heat events, where children under 5 years of age are more affected (Dongo et al. 2018).

The frequency of vector-borne diseases, such as malaria, dengue, chikungunya, and leptospirosis, has risen (Ma et al. 2022). The World Health Organization (WHO) has described climate change as the most significant issue of the century (Greibe Andersen et al. 2023). The entire world population today is expected to suffer from the impact of climate change on human health; however, the most obvious victims are populations in vulnerable communities, as stated by Ncube and Tawodzera (2019). In most parts of the Global South, countries already suffer the challenges of facing inadequate resources amidst health issues, mainly triggered by climate change. Climate change impacts on human health are influenced by people's perceptions of the linkages between climate change and health implications. According to (Lybarger et al. 2023), community members will select and adapt to climate health-related impacts if they acknowledge and understand how climate change affects their health.

Research on people's perception of climate change is biased by cultural understanding, linguistic competence, and variations in communications and their impacts of climate change on human health. Various types of studies recognise the risks that place most vulnerable communities at greatest danger because of their economy, infrastructure quality and management, and mitigation approach. Africa is increasingly vulnerable to climate change due to its geographical position. Since the 1960s, few African regions have been experiencing persistent droughts over the past six decades (Spinoni et al. 2014). According to Naz et al. 2022, climate change is a global phenomenon that is affecting both developed and developing countries. Its impacts on human society and the earth are increasingly visible, and the links with public health are being addressed in public health research. It plays a major role in development, poverty, and wellbeing, as well as in environmental sustainability (Greibe Andersen et al. 2023).

According to a recent calculation, climate-related trends have caused 606,000 deaths between 1998 and 2017. The global effect of heat stress on agriculturally important crops was calculated with positive emission scenarios, i.e., leading to 300,000 deaths each year, with the current year being warmer than the past. One global scenario of $CO_2$ emissions, according to the Intergovernmental Panel on Climate Change, as a part of the continuation of warming rates, predicts that 250,000 people may die each year in Africa, and if the temperature rises to 3 °C, 150,000 people may die (Tripathi et al. 2024; Pathak 2022). The development of effective public health adaptation requires public health authorities to systematically consider how climate change and its associated health risks interact with the vulnerability of local communities. Social vulnerability and adaptive capacity are closely connected to aspects of health risks associated with environmental exposure. Various studies in the past have focused on the impact of climate change on communities. They suggest that social vulnerability and exposure to natural hazards are among the principal factors that influence the potential impact of climate change on human health (Li et al. 2022b). It is necessary to identify areas of particular social vulnerability that are connected to the perception of the people on climate health impacts (González and Sánchez 2022). In sub-Saharan Africa, especially West Africa, many people are poorly informed about health issues. Poverty, poor housing, communicable diseases, and healthcare shortages worsen the health impacts of climate change. It is uncertain if locals understand these climate-related health risks. A survey in a community without major cities tested whether inhabitants recognise these risks and their management and identified local perceptions of climate change's impact on health (Amadou and Diakarya 2021).

Climate change and its consequences are identified as major human health risks to the global population (Greibe Andersen et al. 2023). Climate change and human health interactions are especially pronounced in vulnerable populations such as those living in poverty, in substandard housing, or in marginal climatic conditions (Ragavan et al. 2020). Some essential elements of climate change and health risks are well established, but recognising these risks in low-income and less-resourced areas of developing nations presents significant challenges.

Social scientific research has determined that perception plays an important role in climate change risk evaluations. Individuals develop their perception about climate change based on available information and other events. The first step in establishing effective community-based, climate change-health adaptation initiatives is understanding the community members' knowledge and perceptions (Ncube and Tawodzera 2019). The findings of such inquiries can inform policies and programmes aimed at building community resilience to adapt to health impacts related to climate change (Li et al. 2022b). The public's perceptions and knowledge about climate change's impact on human health are believed to shape attitudes and behaviours, such as voting or supporting government policies that can mitigate current and future consequences (Salhi et al. 2021).

Studies in African countries, including Zimbabwe, show that communities rely on the environment and believe that climate change affects their health and livelihoods by expanding the range and season of some diseases (Ncube and Tawodzera 2019). Studies have concluded that climate change, specifically the increase in temperature, negatively impacts height growth (Parenti et al. 2020).

West Africa faces extreme weather events like heavy rains, wind gusts, heatwaves, and droughts, impacting the quality of life, health, and socio-economic development. Climate change exacerbates health issues such as malaria and exposure to hazardous substances, while also affecting water resources. Increasing temperatures are likely to worsen these impacts, depending on local vulnerabilities and responses (Amadou and Diakarya 2021). Climate change threatens 21st-century health by impacting air quality, water access, food supply, and shelter (Datoo et al. 2021). It increases exposure to vectors, pathogens, and pollutants, affecting food security, undernutrition, child health, mental health, and non-communicable diseases.

Côte d'Ivoire's health system is largely private, expensive, and extremely low-quality, as the healthcare demand of the multi-ethnic inhabitants and refugees shows a pattern of health disparities. Climate drives the incidence of famine primarily through drought, flood, and crop disease failures, and it affects food security directly and indirectly by declining access to services and from the burden of the disease (Ab Kader et al. 2022). Based on this background, this survey aims to describe determinants of local population perceptions and beliefs on climate change directly affecting human health.

Despite the growing global awareness of climate change and its significant impacts on human health, there remains a notable gap in region-specific data, particularly in vulnerable and under-researched areas such as the Haut Sassandra region of Côte d'Ivoire. Existing literature predominantly focuses on urban populations or well-studied rural regions, leaving a critical gap in understanding the knowledge, perceptions, and practices related to climate change and health in rural, agricultural communities. This gap is particularly concerning given the Haut Sassandra region's unique ecological and socio-economic context, characterised by its heavy reliance on agriculture and its susceptibility to climate-related health issues due to its abundant vegetation and favourable conditions for disease vectors like mosquitoes. Based on this background, this study addresses this gap by conducting a comprehensive cross-sectional survey to assess and characterise the knowledge and perceptions of climate change and its health impacts among residents of the Haut Sassandra region. By focusing on this specific population, the study aims to elucidate how local knowledge and perceptions influence awareness, attitudes, and practices related to disease prevention and management. This research is pivotal, as it provides baseline data that can inform targeted public health interventions and climate adaptation strategies tailored to the unique needs of vulnerable rural communities in Côte d'Ivoire.

## *The Study Area*

The Haut Sassandra region is located in the centre-west part of Côte d'Ivoire between latitude 6°12′ and 7°49′ North and longitude 5°55′ and 7°10′. This region is drained by the Sassandra river, from which it takes the name "high Sassandra" over an area of 15,200 $km^2$ and has an estimated population of 1,739,697 inhabitants. It is also home to Daloa, the third largest city in Côte d'Ivoire, which provides human resources and food distribution for livelihoods centring on rice cultivation and food crops (Giscard et al. 2022) (Fig. 19.1).

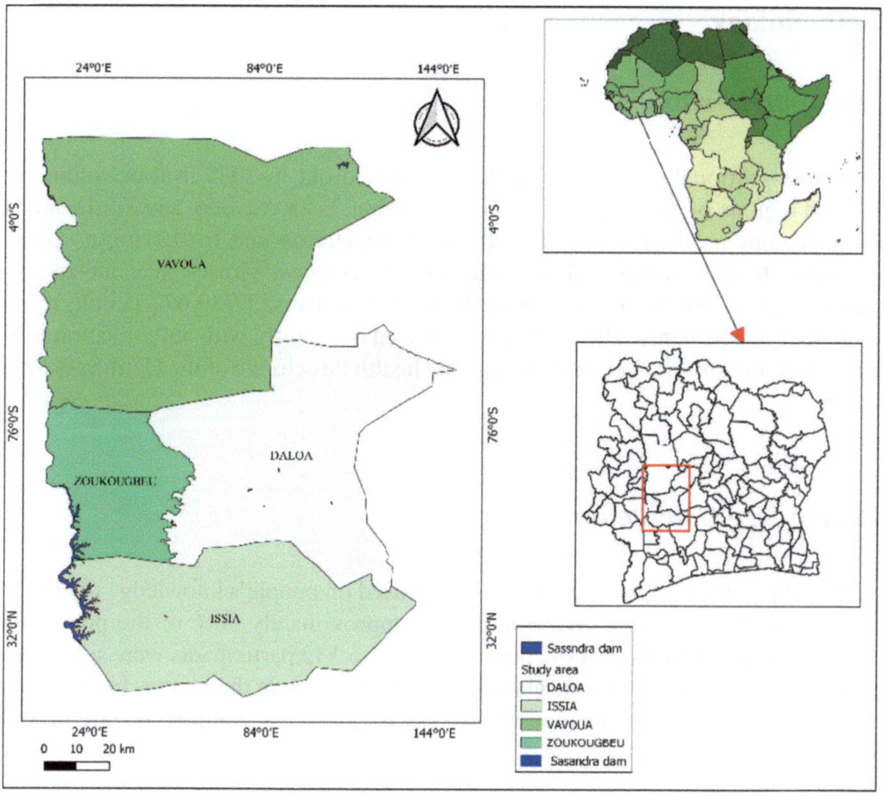


**Fig. 19.1** Map of the study area

## *General Objective*

Assess and characterise the community members' knowledge and perceptions of climate change and its potential impact on health.

## *Specific Objective*

To understand how community members' knowledge and perception can influence the awareness, attitudes, and practices related to disease prevention and management in the Haut-Sassandra region in Côte d'Ivoire.

## Methodology

### *Design and Setting*

This cross-sectional community study was carried out in 2022 in four voluntarily selected regions of Haut-Sassandra, namely Daloa, Issia, Vavoua, and Zoukougbeu. These regions are known for their abundant vegetation and for having a climate conducive to the growth and reproduction of parasites, viruses, and mosquitoes (Kouman et al. 2022). According to (INS 2021) census, 1,739,697 people reside in the four departments. This survey was carried out jointly with authorisation from the prefect, the sub-prefect, and the regional health directorate of the Haut-Sassandra region.

### *Sample Size Calculation*

We calculated the sample size for this study based on people's knowledge of climate change and its prevalence. We assume that approximately 50% of the population had knowledge of the impact of climate change, so 342 participants were selected in the four regions, within 34 households in the 10 villages in the region. In this study, knowledge about climate change was assessed only from available people in each household.

### *Development of Climate Change Knowledge-Perception Questionnaire*

A structured and pretested interview schedule, administered by the interviewers, constituted one of the main tools in this study. The development of questionnaires and measures of CC-related health knowledge, perception, attitude, and behaviours was based on the literature. We also conducted an exploration of the questionnaire by experts in the field to verify the availability of a valid tool for assessing CC knowledge. All perception questions about extreme CC, health, or CC events had responses using various response types such as "yes, no, or I don't know" or categorical options. The final interview included a set of 28 questions on various aspects of climate change and health issues.

### *Measurement of Different Variables*

Reported illnesses occurring after or during extreme weather events were collected from the preceding 10 years. The number of children infected with malaria and schistosomiasis during the previous 12 months was recorded. Participants who stated that they noticed evidence of climate change were asked to give their explanation of how this was perceived, followed by questions about the weather and its causes. The knowledge score was obtained from this information.

### *Data Collection*

A cross-sectional survey was carried out from November to April 2023 among communities in rural areas of central-west Côte d'Ivoire. Before conducting the survey, we carried out a pre-test in a comparable rural context, and this led to the development and finalisation of the questionnaire. Data on participants' sociodemographic characteristics, as well as their understanding and opinions on extreme conditions and recurrent illnesses, were collected. Two months were needed to collect data at the household level under the direct supervision of a field supervisor.

### *Statistical Analyses*

The questionnaires were developed using Epi info 7.2.5 statistical software. The completed questionnaires were then collected and entered into the computer and validated using the Statistical Package for the Social Sciences (IBM SPSS), Version 26. Summary statistics were reported as means with standard deviation (SD) for continuous variables, or as percentages with 95% confidence intervals (CI) for categorical variables. Sociodemographic variables and participant responses were summarised and presented using frequency tables. Evidence of association between sociodemographic variables and climate change knowledge was explored by cross-tabulation and measured using a chi-square test. Logistic regression models were applied to further explore significant knowledge predictors using bivariate analysis.

### *Ethical Approval and Consent*

The study protocol was approved by the Medical Research Council of Cote d'Ivoire and sub-prefects of the departments, followed by the ethics committee of Nangui Abrogoua University. Written informed consent (thumbprint if unable to write) was obtained from the family after reading the information note at the beginning of the interview.

## Results

For this study, we interviewed 340 participants in the four departments of the Haut Sassandra region. In this group, 220 (64%) were men compared to 120 (35%) women, with an average age of 46 years and a standard deviation (SD) of 12.79 years. The average length of residence in the locality was 21 years (median 20 SD 12). Subsequently, 312 or 91.8% of the participants claimed to have experienced extreme weather events during their time living there. Given that the survey took place in rural areas, most (91.09%) of the population had access to a water source in the village, with reported sources of boreholes (34.7%), backwaters (45.3%), wells (9.4%), and tap water (8.8%). Concerning toilet facilities, 56.2% had access to toilets in contrast with 43.8%, who did not. 55.6% had access to traditional toilets, 2.1% had access to model latrines for village officials, and 42.4% practiced open defecation (Table 19.1).

Table 19.2 presents the participants' responses to their knowledge of climate change. Indeed, 51.5% of participants said they have some knowledge about it, while 48.5% reported no prior knowledge about it. Moreover, the majority (51.8%) of the respondents said that they have heard about climate change through sources other than radio (21.8%), television (16.8%) and social media (9.7%).

The meteorological event mentioned most by the participants was drought (56.8%), while 23.8% confirmed a long rainy period had had an impact on crops, given that our study was carried out in a rural environment, 8.5% reported compromises caused by flooding, and 4.1% reported extreme heat.

According to the results, 66.47% of participants said that they know the causes of climate change, citing deforestation (58.82%), transportation emissions (28.83%), and population growth (4.11%).

According to the participants, 7.9% affirmed that illnesses are due to the external events of climate change. Participants felt that they are more exposed to diseases during the dry seasons (47.6%), during the rainy season (17%), at any time (35%), and other times (8.22%).

For relationship studies, we used cross-tabulations to examine the association between sociodemographic variables and participants' knowledge. Indeed, participants' knowledge was significantly correlated with age, sex, and education (Table 19.3).

After analysis, we note that our overall $p$ value $< 0.05$, which means that the dependent variables are significantly associated with knowledge of climate change. $R^2 = 11.5\%$ therefore $R^2 < 1$—this confirms that the probability of having knowledge about climate change can be attributed to age, sex, and education levels. According to the Table 19.4, according to the results, men are more likely to have knowledge about climate change than women, with $p = 0.05$. OR = 1.58—therefore, greater than 1—this means that men are 1.58 times more likely to have knowledge about climate change than women are (0.981 to 2.558), and the variables of primary school and secondary school are: $p$ value $< 0.05$ with OR, respectively 0.563 (0.380 to 1.435), 0.443 (0.244 to 0.806).

**Table 19.1** Sociodemographic and environmental characteristics of the participants

| Valid | Frequency (%) |
|---|---|
| *Genre* | |
| Man | 220(64) |
| Woman | 120(35) |
| Age (Mean; SD) | 46.37 (12.79) |
| *Education of respondent* | |
| No school | 133(39.1) |
| Primary schools | 128(37.6) |
| Secondary school | 76(22.4) |
| Higher school | 3(9) |
| Duration of stay in this locality (mean, media, (SD)) | 21.39, 20.1:(12.24) |
| *Presence of water source* | |
| Yes | 316 (92.09%) |
| No | 24(7.1) |
| *Type of water sources* | |
| Forage | 118(34.7) |
| Backwater | 154(45.3) |
| Well water | 32(9.4) |
| Tap water | 30(8.8) |
| *Present of latrine* | |
| Yes | 191(56.2%) |
| No | 149(43.8%) |
| *Type of latrine* | |
| Traditional latrine | 189(55.6%) |
| Model latrine | 7(2.1) |
| Latrine missing | 144(42.4%) |

## Discussion

Our present study aimed to investigate the knowledge and perceptions of populations on the relevance of climate change and the link between climate change and population health. For this reason, several analyses were carried out.

The socio-demographic analyses of the participants were studied, and the results indicate that the majority of the respondents were men, with a rate of 63% compared to 35%. This gender difference is explained by the fact that in most African societies, men are the heads of the family and women do not have the right to speak in the presence of their husbands. Likewise, the average age of the participants was 47 years, and the average residence in the locality was 21 years; spending more than ten years in a region can imply that these participants have experienced or noticed the

**Table 19.2** Population's knowledge about climate change and human health

| Variable N = 340 | Frequency (%) |
|---|---|
| *Knowledge about climate change?* | |
| Yes | 175(51.5) |
| No | 165(48.5) |
| *Where did you hear about it?* | |
| Radio | 74(21.8) |
| Television | 57(16.8) |
| Social media | 33(9.7) |
| Other | 176(51.8) |
| *What kind of changes have you noticed?* | |
| Long dry season | 193(56.8) |
| Long rainy season | 81(23.8) |
| Long cold periods | 10(2.9) |
| Long hot periods | 14(4.1) |
| Flooding | 29(8.5) |
| I don't know | 13(3.8) |
| *Do you know the causes of CC?* | |
| Yes | 226(66.47) |
| No | 114(33.52) |
| *Which ones do you know?* | |
| Population growth | 14(4.11) |
| Transportation vehicles | 98(28.83) |
| Deforestation | 200(58.82) |
| *Do you think you are exposed to disease during climate change?* | |
| Non | 313(92.1) |
| Yes | 27(7.9) |
| *When are you exposed to diseases?* | |
| Dry season | 162(47.6) |
| Rainy | 59(17.4) |
| Anytime | 119(35) |
| Other | 28(8.22) |

different environmental changes that occurred in this region. According to the PNCC, Côte d'Ivoire has observed several external events in recent years due to an average increase in temperatures from 1957 to the present.

Our survey indicates that most participants were aware of climate change, but the depth of their knowledge varied. In fact, 51.5% of the population was aware of climate change. This can serve as a starting point to encourage participants to tackle climate change impacts and broader environmental issues related to public health.

**Table 19.3** Association test of community knowledge about climate variability with sociodemographic factors

| Variable | Knowledge of climate change | | | Statistics |
|---|---|---|---|---|
| | Yes (%) | No (%) | Total (%) | |
| *Age of people* | | | | |
| [19–30] | 8(4.90) | 20(11.40) | 28(8.30) | $X_2 = 6.11$ |
| [31–60] | 123(75.50) | 131(74.90) | 254(75.10) | df = 2 |
| [61–80] | 32(19.60) | 24(13.70) | 56(16.60) | $p < 0.04$ |
| Total | 163(1001) | 175(100) | 340(100) | |
| *Genre* | | | | |
| Woman | 43(28.10 | 77(41.20) | 120(35.30) | $X_2 = 6.26$ |
| Man | 110(71.90) | 110(58.80) | 220(64.70) | df = 1 |
| Total | 153(100) | 187(100) | 340(100) | $p < 0.01$ |
| *Education level* | | | | |
| Primary school | 56(36.40) | 72(38.70) | 128(37.60) | $X_2 = 21.18$ |
| Secondary school | 49(31.80) | 27(14.50) | 76(22.40) | df = 3 |
| High school | 3(0) | 0 | 3(0.90) | $p < 0.001$ |
| On school | 46(29.90) | 46(29.90) | 133(39.10) | |
| Total | 154(100) | 186(100) | 340(100) | |

**Table 19.4** Predictor of knowledge of climate change from logistic regression analyses

| Variable | *P* value | OR (95%CI) |
|---|---|---|
| Sex (1) | **0.059** | 0.982 (0.964 to 1.001) |
| Age | 0.06 | 1.584 (0.981 to 2.558) |
| *Education level* | | |
| Primary school | **0.002** | 1 |
| Secondary school | **0.008** | 0.443 (0.244 to 0.806) |
| High school | 0.999 | 0 |
| No school | 0.129 | 1.486 (0.892 to 2.478) |
| Time | 0.667 | 1.004 (0.985 to 1.024) |

This level of awareness was low compared to those reported in adults by the two studies conducted in (Korir 2019) (Banstola et al. 2014); the differences between our study and the other studies may be largely due to differences in the populations studied.

Moreover, the majority of the population had an idea about the causes of climate change. In fact, several participants cited deforestation and excessive vehicle transportation as the main origins of climate change, along with population growth. Having completed higher levels of education can play an important role in raising awareness among participants. These results are also supported by other studies (Uralovich et al.

2023) (Asekun-Olarinmoye et al. 2014). Most of our participants said they had heard about climate change from television, radio, social media, and other sources. Indeed, after the analyses, discussions between friends are considered to be the main source of communication on climate change in this locality. This study differs from that of (Nigatu et al. 2014) in Ethiopia, where several female students cited electronic media as the main source of communication. This difference in information is certainly due to differences in the area and population studied or to the absence of electricity in this locality, which explains why radio transmissions and discussions between friends are the main sources of information (Madaki et al. 2023) (Sacolo et al. 2018).

After the analysis of the cross-tabulation, it appears that the variables of age, sex, and education are significantly correlated with knowledge of climate change, which means that as the participants have an advanced age, they are more likely to know about climate change. Then, the gender variable was correlated with CC, which suggests that men in this locality are more aware of CC than women. This is contrary to the study carried out in Ghana by (Partey et al. 2020), which found that men and women had similar perceptions about climate change. These gender differences can be attributed to the social responsibility of men as heads of families in Africa. Similarly, education plays an important role: the higher the level of education achieved, the more likely you are to be informed about climate change and its impacts. This result confirms that of numerous studies (Wijk et al. 2020; Rousell and Cutter-Mackenzie-Knowles 2020) carried out in the field of climate change and education. Our study recruited participants in four departments of the Haut Sassandra region, which constitutes one of the strengths of our study. In addition, by collecting data on the local context of the extreme weather events they experienced, these results will be useful for creating plans to encourage behavioural change among participants and in localities so that they can adapt to minimise public health problems and better manage CC. Also, our study presents limits, particularly on the number of participants in the households studied, due to lack of budget and study time. Future studies could be carried out on several population groups in other regions of Côte d'Ivoire with the commitment of its government and national and international organisations.

## Conclusion

The findings of this study highlight the critical gaps in knowledge and perception of climate change and its impacts on human health among the residents of the Haut Sassandra region in Côte d'Ivoire. Despite 51.5% of participants having some awareness of climate change, only a small fraction recognised its direct impact on health. The significant associations between knowledge of climate change and variables such as age, sex, and education level underscore the importance of demographic factors in shaping understanding and awareness. These insights are pivotal for developing targeted interventions aimed at enhancing climate change literacy in vulnerable communities.

The study underscores the necessity for tailored educational initiatives that focus on improving the knowledge of climate change and its health impacts, especially among women and less-educated individuals. The reliance on informal sources such as radio and social media for information dissemination highlights the need for leveraging these channels for educational campaigns. Additionally, the finding that men are more likely to be knowledgeable about climate change than women points to a gender disparity that must be addressed through inclusive and accessible educational programmes.

To effectively mitigate the adverse health impacts of climate change in the Haut Sassandra region, it is crucial to implement community-based adaptation strategies that are informed by the local context. These strategies should include enhancing infrastructure to withstand extreme weather events, promoting sustainable agricultural practices, and improving access to healthcare services. By focusing on these areas, it is possible to build resilience within these communities, ensuring that they are better equipped to handle the health challenges posed by climate change.

## Recommendations

Based on the findings of this study, we propose some recommendations to enhance climate change awareness and health resilience in the Haut Sassandra region. By implementing these recommendations, the Haut Sassandra region can move towards a more informed, resilient, and health-secure future in the face of climate change.

Firstly, educational programmes should be designed and implemented with a major focus on the most vulnerable groups, particularly women and individuals with lower levels of education. These programmes should utilise local communication channels (community groups), including radio and social media, to effectively reach and engage the community. Furthermore, there is a need for policy interventions that support the integration of climate change education into the school curriculum and community workshops. This could help foster a deeper understanding of the environmental and health impacts of climate change from an early age. Additionally, training healthcare providers to recognise and address climate-related health issues can improve community health outcomes in any given region. Finally, enhancing local infrastructure and promoting sustainable agricultural practices are essential for building community resilience. Investment in water and sanitation facilities, as well as initiatives to prevent deforestation and promote reforestation, can help mitigate the environmental impacts of climate change. Furthermore, establishing early warning systems for extreme weather events and creating emergency response plans can significantly reduce the health risks associated with climate-related disasters.

## References

Amadou D, Diakarya B (2021) Analysis of the evolution of agroclimatic risks in a context of climate variability in the region of Segou in Mali 1–25. http://arxiv.org/abs/2106.12571

Asekun-Olarinmoye EO, Odu OO, Adeomi AA (2014) Public perception of climate change and its impact on health and environment in rural Southwestern Nigeria 1–10

Banstola A, Stebbing M, Banstola A (2014) Willingness: knowledge related to climate change and willingness to act for adaptation and mitigation practices in rural Nepal. Vietnam J Public Health 23

Datoo MS, Natama MH, Somé A, Traoré O, Rouamba T, Bellamy D, Yameogo P et al (2021) Efficacy of a low-dose candidate Malaria Vaccine, R21 in adjuvant matrix-M, with seasonal administration to children in Burkina Faso: a randomised controlled trial. Lancet 397(10287):1809–1818. https://doi.org/10.1016/S0140-6736(21)00943-0

Don A, Seidel F, Leifeld J, Kätterer T, Martin M, Pellerin S, Emde D, Seitz D, Chenu C (2024) Carbon sequestration in soils and climate change mitigation—definitions and pitfalls. Glob Change Biol 30(1). https://doi.org/10.1111/gcb.16983

Dongo K, Kablan AKM, Kouamé FK (2018) Mapping urban residents' vulnerability to heat in Abidjan, Côte d'Ivoire. Climate Dev 10(7):600–613. https://doi.org/10.1080/17565529.2018.1450213

Giscard AA, Lorougnon Guédé, Christophe NK, Rodolphe AAR (2022) Impact of climate variability on agricultural calendars in the vavoua sub-perfecture (Center-West Ivorian), 255–74. https://doi.org/10.19044/esj.2022.v18n27p255

González JB, Sánchez A (2022) Multilevel predictors of climate change beliefs in Africa. PLoS ONE 17(4):e0266387. https://doi.org/10.1371/journal.pone.0266387

Greibe Andersen J, Kallestrup P, Karekezi C, Yonga G, Kraef C (2023) Climate change and health risks in Mukuru informal settlement in Nairobi, Kenya—knowledge, attitudes and practices among residents. BMC Public Health 23(1). https://doi.org/10.1186/s12889-023-15281-y

INS (2021) Institut National de La Statistique. Recensement Général de La Population et de l'habitat (R.G.P.H.) 1–37

Ab Kader NI, Yusof UK, Khalid MNA, Nik Husain NR (2022) Recent techniques in determining the effects of climate change on depressive patients: a systematic review. J Environ Public Health. https://doi.org/10.1155/2022/1803401

Korir JC (2019) Level of awareness about climate change among the pastoral community. Environ Ecol Res 7(4):197–207. https://doi.org/10.13189/eer.2019.070401

Kouman KJ, Marc AT, Kouakou M, Kpangui KB, Bamba I, Barima YSS, Bogaert J (2022) Dynamics of the natural regeneration of vegetation in an anthropized forest in Côte d'Ivoire, West Africa. Ecologies 3(2):66–77. https://doi.org/10.3390/ecologies3020007

Li B, Deng A, Li K, Yao H, Li Z, Shi Y, Xiong Q et al (2022a) Viral infection and transmission in a large, well-traced outbreak caused by the SARS-CoV-2 delta variant. Nat Commun 13(1):1–9. https://doi.org/10.1038/s41467-022-28089-y

Li X, Hoffmann R, Muttarak R (2022) Measuring global social vulnerability to natural hazards at the subnational level 1–23. www.iiasa.ac.at

Lybarger K, Yetisgen M, Uzuner Ö (2023) The 2022 N2c2/UW shared task on extracting social determinants of health. J Am Med Inf Assoc 30(8):1367–1378. https://doi.org/10.1093/jamia/ocad012

Ma J, Guo Y, Gao J, Tang H, Keqiang X, Liu Q, Lei X (2022) Climate change drives the transmission and spread of vector-borne diseases: an ecological perspective. Biology 11(11):1–12. https://doi.org/10.3390/biology11111628

Madaki MY, Muench S, Kaechele H, Bavorova M (2023) Climate change knowledge and perception among farming households in Nigeria. Climate 11(6). https://doi.org/10.3390/cli11060115

Naz S, Fatima Z, Iqbal P, Khan A, Zakir I, Ullah H, Abbas G et al (2022) An introduction to climate change phenomenon BT—building climate resilience in agriculture: theory, practice and future perspective. In: Jatoi WN, Mubeen M, Ahmad A, Cheema MA, Lin Z, Hashmi MZ (eds). Springer, Cham, pp 3–16. https://doi.org/10.1007/978-3-030-79408-8_1

Ncube A, Tawodzera M (2019) Communities' perceptions of health hazards induced by climate change in mount Darwin District, Zimbabwe. JAMBA J Disas Risk Stud 11(1):1–11. https://doi.org/10.4102/JAMBA.V11I1.748

Nigatu AS, Asamoah BO, Kloos H (2014) Knowledge and perceptions about the health impact of climate change among health sciences students in Ethiopia: a cross-sectional study. BMC Public Health 14(1). https://doi.org/10.1186/1471-2458-14-587

Parenti I, Rabaneda LG, Schoen H, Novarino G (2020) Neurodevelopmental Disorders: from genetics to functional pathways. Trends Neurosci 43(8):608–621. https://doi.org/10.1016/j.tins.2020.05.004

Partey ST, Dakorah AD, Zougmoré RB, Ouédraogo M, Nyasimi M, Nikoi GK, Huyer S (2020) Gender and climate risk management: evidence of climate information use in Ghana. Clim Change 158(1):61–75. https://doi.org/10.1007/s10584-018-2239-6

Pathak H (2022) Impact, adaptation, and mitigation of climate change in Indian agriculture. Environ Monit Assess 195(1):52. https://doi.org/10.1007/s10661-022-10537-3

Pielke RA (2004) What is climate change? Energy & Environment 15(3):515–520. https://doi.org/10.1260/0958305041494576

Ragavan MI, Marcil LE, Garg A (2020) Climate change as a social determinant of health. Pediatrics 145(5):1–3. https://doi.org/10.1542/PEDS.2019-3169

Rousell D, Cutter-Mackenzie-Knowles A (2020) A systematic review of climate change education: giving children and young people a 'voice' and a 'hand' in redressing climate change. Children's Geographies 18(2):191–208. https://doi.org/10.1080/14733285.2019.1614532

Sacolo H, Chimbari M, Kalinda C (2018) Knowledge, attitudes and practices on schistosomiasis in Sub-Saharan Africa: a systematic review. https://doi.org/10.1186/s12879-017-2923-6

Salhi A, Benabdelouahab S, Bouayad EO, Benabdelouahab T, Larifi I, El Mousaoui M, Acharrat N, Himi M, Ponsati AC (2021) Impacts and social implications of landuse-environment conflicts in a typical mediterranean watershed. Sci Total Environ 764. https://doi.org/10.1016/j.scitotenv.2020.142853

Spinoni J, Naumann G, Carrao H, Barbosa P, Vogt J (2014) World drought frequency, duration, and severity for 1951–2010. Int J Climatol 34(8):2792–2804. https://doi.org/10.1002/joc.3875

Tripathi MK, Kumar H, Tyagi PK (2024) Climate change mitigation potential of forestry sector for sustainability of agro-ecosystem: a review. Curr World Environ 18(3):914–32. https://doi.org/10.12944/cwe.18.3.02

Uralovich KS, Toshmamatovich TU, Kubayevich KF, Sapaev IB, Saylaubaevna SS, Beknazarova ZF, Khurramov A (2023) A primary factor in sustainable development and environmental sustainability is environmental education. Caspian J Environ Sci 21(4):965–975. https://doi.org/10.22124/cjes.2023.7155

Van Wijk M, Naing SY, Diaz Franchy S, Heslop RT, Novoa Lozano I, Vila J, Balleste C (2020) Perception and knowledge of the effect of climate change on infectious diseases within the general public: a multinational cross- sectional survey-based study 1–14. https://doi.org/10.1371/journal.pone.0241579

# Chapter 20
# Uses of Urban Vacant Land, Green Infrastructure and Its Impact on Climate Change

**José G. Vargas-Hernández, Absalón J. Salmerón-Zapata, and M. C. Omar C. Vargas-González**

**Abstract** This paper aims to analyse the implications between the declining and shrinking cities, urban vacant lands, urban land uses, green infrastructures, urban green areas, and their impacts on climate hazard change. The analysis departs from the basic assumption that urban vacant land sites and spaces have a negative connotation, but when supported by the appropriate policies and programmes of incentives, they can turn around and develop the essential green infrastructure to enable the mitigation of climate change hazards, economic growth, and socio-ecological development. The method used is the analytical-descriptive based on the theoretical and empirical literature review. It is concluded that land uses from vacant land to urban green innovation infrastructures and forest areas contribute to mitigating the dangers of climate change, potentially creating green environments in support of human health, individual wellbeing and improved quality of life for residents in the respective community.

**Keywords** Climate change · Hazards · Declining cities · Forest areas · Green infrastructure · Land use · Shrinking cities · Urban vacant land

---

J. G. Vargas-Hernández (✉) · A. J. Salmerón-Zapata · M. C. O. C. Vargas-González
Financial Literacy Research Center, Universidad Cristòbal Colón., Boca del Rio, Mexico
e-mail: jvargas2006@gmail.com

A. J. Salmerón-Zapata
e-mail: absajose@ucc.mx

M. C. O. C. Vargas-González
e-mail: omar.vg@cdguzman.tecnm.mx

A. J. Salmerón-Zapata
Tecnológico Nacional de México, Cd. Guzmán, Jalisco, México

M. C. O. C. Vargas-González
Universidad Cristóbal Colón, Boca del Rio, Veracruz, México



W. Leal Filho and F. Wolf (eds.), *Climate Change and Health: Perspectives from Developing Countries*, Climate Change Management,
https://doi.org/10.1007/978-3-031-93177-2_20

## Introduction

Urban processes linked to shrinking cities feature characteristics, including economic decline associated with urban land uses and other socio-economic dimensions, that may lead to adaptive capacity building. This, in turn, increases urban green infrastructure and moderates the impact of climate hazard change. The vacant land is used as green infrastructure to improve the health of urban ecosystems.

City regions experiencing economic growth and increases in urban population are subject to the adversity of losing green infrastructure resources, leading to a disequilibrium in adaptive capacity to climate change hazards and quality-of-life goals.

The importance of urban vacant land lies in its potential to transform underutilised spaces into valuable green infrastructure that can improve urban ecosystems and mitigate the risks of climate change. Historically, vacant urban land has been viewed negatively because of its association with disinvestment and degradation. However, if properly managed and restored, these spaces can provide essential ecosystem services such as stormwater management, pollutant dispersal, and biodiversity habitat. By transforming vacant land into green space, cities can improve the quality of life of their residents, drive economic growth, and promote social and environmental development.

Today, many cities struggle with inadequate policies and economic incentives to support the maintenance and redevelopment of urban open space. Successful examples, such as the Green Belt project in Vitoria-Gasteiz, Spain, demonstrate the potential benefits of greening open space, including reducing the urban heat island effect, improving air quality, and increasing community engagement or the practices applied to maximizing ecosystem services and habitat in vacant lots in Chicago, Illinois (USA). Despite these examples, the problem remains severe as urban sprawl and economic downturn create vacant land faster than it can be reclaimed. Integrated strategies involving local authorities, communities, and private investment are essential to exploiting the full potential of vacant urban land for a sustainable and adaptive urban development.

This analysis aims to find the relationships between the implications of declining and shrinking cities, urban vacant lands, urban land uses, green infrastructures, and urban green areas and their impact on climate hazard change.

## Background

### *Declining and Shrinking Cities*

Urban decline comprises urban environments subject to population displacements, where the main economic activity becomes obsolete, generating inhabitants' migration to more competitive environments; this term, described by Oswalt, is described as a problem of multiplicity and different scales (Cutieru 2021). Urban vacant land in shrinking cities has long been associated with high crime rates and economic decline (Zhang and Park 2023). Urban growth and shrinkage of cities constitute a pattern of growing urbanization across the globe yet studies on urban vacant land are few and have proved to be rudimentary and subjective (Song et al. 2020). Adequate urban planning should consider demographic and economic growth to avoid space contraction and generate strategies to understand the urban decline caused by deindustrialization, internal migration, or resource depletion (Cutieru 2021).

In 2018, the Inter-American Development Bank established cities as the main social, economic, and territorial development engine. The disproportionate growth of cities towards the peripheries has occurred due to neoliberal urbanization policies, which have facilitated the movement of people over greater distances and in less time (Herce and Vallejo 2009). The economic growth processes of cities that benefited from the development of some comparative and competitive advantages in the past have led to the corresponding deindustrialization processes while unplugged from the global value chain and resulting in urban population and economic decline (Pike et al. 2016; Martinez-Fernandez et al. 2012: 220). Patterns of economic decline and depopulation in communities caused by contextual conditions result in vacant land with implications in institutional, fiscal, and programmatic responses that may bring potential benefits.

City regions and urban areas develop, expand, or decline, remaining sensitive to environmental and climate change agendas guided by spatial planning regulations, but cities that are experiencing shrinkage and decline face different challenges. Around the globe, cities are either growing and expanding or shrinking and declining because of dynamic processes such as deindustrialization, globalization, and demographic change (Martinez-Fernandez et al. 2012; McKinsey and Company 2016; Sassen 2001). Besides the complexity of these challenges, global warming and climate change are affecting urban settings (Ruth and Coelho 2007).

Economic and urban decline is linked to limited regulation over land use in the different locations and forms of new urban development supported by deficient urban spatial planning, small changes in green infrastructure cover, and a lack of strategies to mitigate socioeconomic inequality and environmental degradation. All these factors impact the implications for adaptive capacity to cope with the risks of urban weather and climate change. Low-income communities lack green infrastructure compared with the rest of the city and are exposed to more urban vacancies in their neighborhoods (Kim et al. 2023).

Finally, the concept of shrinking cities is widely accepted as a global phenomenon, which dates to the urban sociology theories of cyclical growth and decline (Martinez-Fernandez et al. 2012). Shrinking cities are places characterized by economic and population decline that harbor vacant properties and land parcels (Dietersdorfer et al. 2012; Burkholder 2012; Schilling and Logan 2008). Urban expansion, decline, and shrinkage processes will continue influencing the global landscape of city spaces, urban land uses, and green infrastructure. Declining urban land uses, derelict and disused land experiencing declines in surface area may replace informal open green spaces and can be considered prospects for urban green infrastructure. Developing planning Tools to Revitalize will improve the quality of life of the urban inhabitants, attract new investments, and ultimately encourage the repopulation of declining cities (Mohamad et al. 2023).

## *Urban Vacant Land*

Many cities are dominated by abandoned and vacant land. Cities that have experienced land abandonment use vacant land to create open green spaces and sites for productive purposes (Alexander 2005; U.S. Department of Housing and Urban Development 2009; Wachter et al. 2010). Urban vacancies have been a concern for neighborhood distress and economic decline and have gained more recent attention as potential green infrastructure benefits communities in diverse ways (Kim et al. 2023). The average vacancy rate of residential land amounts to 11.48% in 65 U.S. cities (Song et al. 2020). Rapid deindustrialisation, followed by population loss, disinvestment, and decentralization, has left many industrial legacy cities with abandoned factory structures and extensive urban blight and vacancy (Zhang and Park 2023).

In urban context, vacant land often has a negative connotation and is considered a problem (Jakle and Wilson 1992). Vacant land in physical blight conveys negative images that contribute to disinvestment, reducing reinvestment and development and increasing urban vacancy rates (Neukrug 2009). Vacant land deters and discourages investments in the community, imposing the costs of maintaining vacant properties. Urban vacant land, empty urban spaces, and abandoned buildings have become a wasteland undervalued as a normal part of the urban fabric despite the lack of public policy and economic investment for renewal. While some efforts have been made to repurpose vacant land for tax revenue generation and temporary environmental restoration, a comprehensive framework for city-scale assessment and reprogramming remains lacking (Zhang and Park 2023).

Urban vacant lands are thus structural assets with economic and socioecological value relative to other urban land uses. They are ecological habitats for a wide range of biodiversity, including wildlife, plants, and animals, and provide ecological corridors and green network systems, wetland and stormwater management, and regulation services to disperse pollutants (Kamvasinou 2011). Nowadays, some research uses multi-spectral remote sensing images and apply post-processing operations to improve the segmentation performance of different types of urban vacant land (Hu and Zhuang 2024). Most cities do not have adequate programs of economic incentives to support the maintenance and improvements of abandoned buildings and vacant land. Vacant land and abandoned buildings are not maintained and negatively impact the community's quality of life and property values.

The emergence of open vacant land spaces provides some benefits and opportunities associated with securing adaptive capacity with new land uses and green infrastructure. Revitalising urban vacant land and reusing urban wasted land is critical to urban land use in urban spaces. The capacity of urban vacant land to design, plan, and manage promotes the public socio-ecological and also human health related value of urban landscape as a natural resource and community asset to improve the quality of life in the surrounding neighborhood. Vacant urban land covers a significant spatial territory of the urban landscape that provides green infrastructure for ecosystem services used to enhance residents' health and quality of life (Pagano and Bowman 2000; Kim et al. 2015). In addition, this urban landscape resembles a valuable source of ecosystem services that benefit urban communities and provide transformation opportunities for community development (Burkholder 2012; Little 2008).

Urban vacant land therefore possesses a significant public value in terms of the potential to enhance socio-ecological benefits if redeveloped and integrated with other urban green infrastructure (Kim 2016). Integrated planning leads to a better appreciation of public value and design strategies to optimize the utilization of these spaces in landscape design, urban environmental regeneration, and renewal. Socio-ecosystem services may be provided in the form of open public spaces, urban green spaces, etc. Strategic use and reuse of land in the urban landscape resemble relevant enablers of ecological opportunities and benefits to create more prosperous, livable, healthy, and sustainable cities.

Different types of urban vacant land sites can be identified and explored by a wide range of typologies (Blakeman et al. 2008). These are different types of urban vacant land habitats, including vacant lots and areas alongside parking lots and roads, abandoned buildings in industrial areas, wastelands, old town centers, railway sites, etc. All of these sustain a diversity of urban plant communities and animal populations depending on the environmental conditions, natural habitats and management practices (Robinson and Lundholm 2012; Klimeš 1987; Sukopp et al. 1979; Brandes 1995; Hope et al. 2003; Burkholder 2012). Finally, urban vacant land can be defined as unused land within urban areas. It is distinguished by the key features of being unused with no current human activity and abandoned or deserted land with buildings, structures, and ancillary facilities (Song et al. 2020).

When properly managed, urban vacant land provides beneficial socioecological services that positively impact the surrounding communities by creating and developing infrastructure services, community and wildlife gardens, public plantings, and recreational areas (Bonham et al. 2002). Socio-ecological values and benefits provided by urban vacant land as a resource improve urban policies through urban planning, development, and management aimed to enhance environmental quality, human health, and quality of life for residents in the community, all urban greening projects, planting methods must be selected with specific species-, community-, and ecosystem-level goals in mind (Anderson and Minor 2021).

Clusters of urban vacant lots with overlapping urban areas of low concentration of green spaces and ecological value require a socio-ecological function associated with ecosystem services. Vacant land can be reused because of its socio-ecological value, but its financial value is an obstacle to redevelopment. Land vacancy rates in developed urban areas tend to increase due to deindustrialization processes, declining city population, urban sprawl, and preferences for new residential choices (Kremer et al. 2013).

The transformation of vacant land includes projects that may be components of urban community development, including comprehensive plans for greenways, waterfront parks, and pathways development (City of Pittsburgh, Greening Resources 2012; The Greening of Detroit 2012). The transformation of different types of vacant land and abandoned buildings into green infrastructure can promote community stabilization, economic growth, socio-ecological well-being, and quality of life. Investment policies, incentive programs, and regulatory strategies for rehabilitating vacant land and abandoned buildings, such as enforcement codes on maintenance requirements and increasing ownership rates, may encourage owners to improve their vacant properties and neighborhood stability. Local authorities can maintain enforcement codes if the owners fail to improve their vacant land and abandoned buildings (Accordino and Johnson 2000). Therefore, it is proposed to promote a positive incentive program and a graduated property tax system on the requirement of vacant land, vacant spaces, and abandoned buildings as primary assets for productive speculation and restructuring by encouraging redevelopment (Kim et al. 2015; Bowman and Pagano 2004).

## Methodology

This paper presents the definition of urban vacant land worldwide. It is based on bibliometric analysis and addresses three fundamental research questions:

(1) What is the impact of repurposing urban vacant lands on mitigating the hazards of climate change?
(2) How does the green infrastructure developed on these lands contribute to improving the quality of life for urban residents?
(3) What are the most effective policies and strategies for transforming vacant lands into functional and sustainable green spaces?

This study explores how converting these lands into green infrastructure can tackle ecological challenges and promote socioeconomic development and community cohesion.

The method employed in this study is meta-analytical and reflective based on conceptual, theoretical and empirical literature. A bibliometric review was used, through which a quantitative analysis of publications related to declining cities, urban vacant land uses, green infrastructure, and its impact on climate change risks was carried out. The bibliometric indicators taken as inclusion factors were the quality of the publications where the research is presented, the context covered by the publication, the relevance of the topic in the research (urban vacant land, green infrastructure, climate change, declining cities, forest areas, land use, shrinking cities) and exclusion, in addition to non-compliance with the above points, was based on the date of publication.

This bibliometric review was based on literary and scientific sources on declining cities, vacant urban land uses, green infrastructure and its impact on climate change risks. Ninety-seven documents were used for the research. It was conducted in two stages. In the first stage, the search criteria used in the different databases were defined, including temporality, date of consultation, type of document and journal, topic and keywords. The second stage consisted of classifying and analyzing the documents based on the identification criteria; these were grouped to identify the connections between the documents, selecting the most relevant ones.

With the bibliometric review of 88 documents, urban green space and climate change literature research clusters were identified:

- Green Infrastructure: Presenting the network of interconnected green spaces and the benefits of green infrastructure.
- Climate Change: Presenting the impact of climate change in urban environments and the loss of ecosystem services due to changes in land use.
- Declining Cities: Presenting the urban inequality and demographic transformation of shrinking cities and the global phenomenon of declining cities.
- Urban Land Uses: Presenting methods for establishing native plants on urban vacant land and the contribution of ecosystem services to air quality and climate change mitigation policies.

## Results and Discussion

### *Urban Land Uses*

Urban vacant land availability offers revitalization opportunities by improving ecological functions. However, less is known about the available planning tools to mobilize this effort (Mohamad et al. 2023). The socio-ecological values of urban green vacant sites can be developed and protected for various types of land uses, and those that are wasted, underused, and undervalued can be revitalized compared to other types. Transformation of expanding uncontrolled urban land uses leads to the loss of natural forest and vegetation followed by open spaces, a decline of ecosystem services provision, and environmental and climate changes (Sackey 2008; Weng 2001), With little or no intervention/management, natural succession occurs on urban vacant land, spontaneously bringing flora and even fauna species and creating informal green spaces in cities (Hu and Zhuang 2024).

Transdisciplinary approaches can be applied to address vacant urban land uses, urban design, and planning to implement strategies to improve urban green areas and urban green infrastructure, which may mitigate climate change hazards. Urban land use planning must be based on a comprehensive benchmark for ecosystem services productivity, considering the benefits of other land uses and management of vacant land. Urban land use planning and development, urban forestry, and green spaces can help to mitigate the problems associated with climate change hazards. Vacant land policies and programs may alter the land use structure and nature to achieve higher redevelopment potential utility (Blakeman et al. 2008).

Local authorities, communities, land trusts, and greening organizations should support the maintenance of green programs and crop initiatives to enhance the socio-economic ecosystem by using different types of vacant land. Social organizations may manage special programs to improve and maintain different types of vacant land using project funding from tax levies, foundations, endowments from private donations, and personal funds.

## *Green Infrastructure and Forest Structure*

According to Benedict and McMahon (2012), Green infrastructure is an interconnected network of green spaces providing environmental functions and services such as improving water supply, air quality, recreation, scenic beauty, and protection against disasters. Hence, it includes generating strategies for conserving the landscape and its components of natural and cultural value. Urban vacant land sites can be considered a component of green infrastructure systems that improve socio-ecosystem health and have an enduring value for the community. They are a vital source of green infrastructure and create opportunities for urban green open spaces and creative landscapes. Greening vacant parcels can transform them into public green spaces, supporting economic development and local neighborhoods' social and environmental well-being by encouraging reinvestment and decreasing gun violence (Branas et al. 2011). Finally, some cities have begun greening their urban vacant land and public open spaces, supported by management tools incorporating the management strategy on urban stormwater to reduce rainfall drains.

Urban vacant land providing ecosystem services from urban forest structure and green infrastructure may differ from those provided among the different types of vacant land with different land uses. To create a more sustainable, healthy, and livable city, urban vacant land can be used for urban forests, including park systems, street trees, natural resources, and vegetation. Some functional benefits of urban vacant land used for forest structures are trees' structural value, air pollution removal, carbon sequestration and storage, energy savings, avoided runoff, etc. (Nowak et al. 2002). Forest infrastructures in urban vacant land sites help to counteract climate change by removing carbon dioxide through photosynthesis and altering energy consumption by reducing the emissions from the fossil fuels burned by power plants (Abdollahi et al. 2000).

Concerning adaptation related benefits, vacant land vegetation is a cost-effective way of reducing the management infrastructure of stormwater able to mitigate urban runoff (Crauderueff et al. 2012). Urban vacant land cover vegetation and bare soil is a major driver to regulating services and climate regulation, provisioning socio-ecosystem services among others, food production, air pollution removals, water supply, and support services, such as nutrient cycling and soil building, etc.

Urban green and forest structures have the structural value of providing substantially greater environmental and socio-ecosystem services, including, among others, decreased urban heat island effect, energy saving, air pollution removal, carbon sequestration and storage, rainfall interception, and other benefits (Rosenthal et al. 2008). Determining the characteristics of the different forest structures of urban vacant land is vital for a detailed evaluation and an accurate estimate of the ecosystem services productivity and value of the urban green infrastructure.

Green infrastructure projects on urban vacant land impact urban design properties, focusing more on functions of the socio-ecosystem services and considering their impact on sustainability, social responsibility, and health in socio-ecologically underserved communities. Many cities are by now greening urban vacant land sites as a strategy for managing stormwater infrastructure and potentially becoming a helpful component. Installing native plants in large vacant spaces via broadcast seeding can be an inexpensive and ecologically viable method for urban greening (Anderson and minor 2021).

In the long term, urban infrastructure as an urban green innovation can contribute to capacity building to adapt to global warming and climate change in urban areas, given their severity, increased risks, and related impacts. Urban green landscapes have an influence either to intensify or to moderate climate change impacts. Additional studies are needed to make studies from the perspective of growing cities provide insights into urban vacant land revitalization planning, considering the competing objectives of economic prosperity and green space preservation (Mohamad et al. 2023).

As a prerequisite for seizing any benefits, urban vacant land requires designing and implementing policies, planning and managing urban open green spaces, including urban forests, street trees, and any other vegetation as valuable natural and ecological resources to provide ecosystem services. Urban forest structures, green roofs, and other green innovation areas have a dual purpose of providing benefits but also some catastrophic events, disease spreading, and other adverse effects critical for urban planning, design, and development. This can be prevented by implementing an integrated approach for urban forest structure management of vacant land sites as part of the urban green infrastructure, aiming to achieve an optimum balance and maximize benefits and advantages (Baró et al. 2014). Spatial planning policies may enhance urban green infrastructure, which is interrupted and lost due to changes driven by pressures in economic development, population growth, and urban expansion (Lennon and Scott 2014; Angel et al. 2005). The design and implementation of policy mechanisms aimed at greening urban vacant lots improve the environmental benefits to the community, such as promoting urban agriculture to grow food crops for self-consumption and sale Hodgson et al. (2011) and Ackerman (2012).

An illustrative example of successful green infrastructure is the case study of the Temuco—Padre de las Casas intercommunal, Chile, which has territorial qualities such as protected spaces of high ecological and landscape value, recovered or restored habitats, multifunctional areas, facades, and green roofs that house biodiversity and favor the adaptation and mitigation of climate change. (Moreno et al. 2014). Another example is the Green Belt project of Vitoria Gastrin, the capital of the province of Alba in the Basque Country of Spain. This city is a strategic military, commercial, cultural, and agricultural point. It was named European Green Capital in 2012, and after ten years, environmental benefits, social, economic, ethnic, or other, have been recognized (Moreno et al. 2014). Altena in Germany is another success story that faced depopulation but invested in the residents' quality of life who decided to stay, such as the expropriation of neglected assets and investment in improving other services. Finally, the most current example is the Urbact project, which fosters sustainable urban development by training 550 cities in 29 countries of the European Union, based on a participatory approach through the creation of city networks and citizen participation. The purpose is this project, funded by the European Union, is to promote integrated sustainable development in cities, improve city policies, and improve the effectiveness of Cohesion policy in cities.

## *Socio-ecosystem Services of Urban Vacant Land*

Vacant land contributes to ecosystem services, benefiting local communities (Schilling and Mallach 2012). Ecosystem services are the benefits obtained from ecosystems as provisioning services, regulating services such as climate regulation, supporting services, recreation, and education (Millennium Ecosystem Assessment 2003). Vacant land supports ecosystem services, obtaining benefits depending on the natural systems, environmental conditions, and management practices. Vacant land has species diversity in the ecosystem relative to other land uses, providing natural sustainability and resilience to recover from natural hazards (Shah 2015).

Urban vacant land provides ecosystem services than differ from other urban land uses, such as biocontrol in gardens and lawns, soil food web, rainwater retention, habitat provision services, carbon capture and climate regulation, among others (Yadav et al. 2012; Grewal et al. 2011; Shuster et al. 2014; Robinson and Lundholm 2012). Rainwater retention is vital for the potential benefit of stormwater management, synthesizing qualitative and quantitative data and function (infiltration rates), measured parcel infiltration rates, and previous lot area into the application of a simple infiltration-excess model (Kelleher et al. 2020). To assess ecosystem services productivity, the benefits from forest structures in urban vacant land use must be considered, as they may have different results and outcomes depending on the type and extent of residential, industrial, commercial, etc., and characteristics of the surrounding urban areas. The relationship between land cover, the natural environment, and urban vacancies may be complex and green infrastructure can potentially interact with residential and business vacancies differently (Kim et al. 2023).

Al in all, urban vacant land has become a diffuse social, economic, and ecological field of research. However, it can also be considered a potential resource for enhancing urban sustainability (Song et al. 2020); more research is needed to prepare the policies to shed more light on the subject matter.

## *Climate Change Hazards Linked to Urban Vacant Land*

Rapid urbanisation processes have increased the socioecological and environmental threats resulting in climate change and its consequences in alterations of biochemical cycles, ozone depletion, chemical pollutants, flooding, modifications of the water balance, etc. (Fohrer et al. 2001; Rosa et al. 2004). The new context of urban areas requires building and maintaining an adaptive capacity and the ability to anticipate and influence the urbanization features of their landscapes, aimed at moderating and managing climate-related risks and impacts (Sun et al. 2013). Risk is a function of climate hazards and the vulnerability of the urban area. In urban areas, climate change patterns of vulnerability and risk are uneven (Carter et al. 2015).

Projections of people expected to live in urban areas highly exposed to climate change impacts have increased (IPCC 2022). The surface temperatures of land use exchange moisture and energy with the environment, with rapid variations depending on time and space, which impact climate change. These land surface temperatures impact heat and radiation balance in global climate changes (Wubet 2003; Ifatimehin 2007; Weng 2001).

The long-term relationship between urban green infrastructure and urban land use change has implications for the adaptive capacity of urban spaces to enhance the spatial context of climate change impacts on related biophysical processes (Blanco et al. 2011). Temporal and spatial changes in land uses, vacant land, and built-up spaces increase the land surface temperature and encourage environmental hazards associated with climate change.

Also, changes in land use affect urban spaces' capacity to address climate-related risks. A gradual decline in urban land use is reflected in the challenges of declining economic activities related to the slowing of development activities, other socioeconomic factors, and the rising environmental issues of urban adaptive capacity in the context of climate change. This could increase the risks linked to climate change hazards such as increased evapotranspiration of land cover, heat waves, flooding in urbanized spaces lacking green infrastructure, etc.

Many plans submitted in previous years focused strictly on climate risk reduction, missing opportunities to promote the co-benefits of climate change mitigation and sustainable development, exacerbating inequality and reducing welfare (IPCC 2022). Urban green spaces provide several benefits, including climate change adaptation (Rosenthal et al. 2008). There are several applications based on scenario approaches to analyze changes in the urban climate as an impact of land use change, although there is a gap in the research on the relationship between urban land use and climate change adaptation (Storch and Downes 2011; Tong et al. 2012).

Climate change adaptation in declining and expanding cities requires support from the strategic planning of green infrastructure. Cities must develop urban strategies to grow and expand by enhancing urban green infrastructure to moderate climate change risks, designing and implementing adaptation strategies on climate change, supporting decision-making to respond to the implications of shrinking and declining urban spaces and land uses, and enhancing the potential value of adaptive capacity to foster urban green innovation.

## *Challenges of Urban Vacant Land*

The dizzying changes faced by today's society affect the development of cities, causing them to expand or contract. This results in vacant land, which can be used to create green spaces. Now, as part of the bibliometric study, the literature review allowed us to recognize different relevant success cases that allow us to glimpse the uses of urban vacant land, green infrastructure, and its impact on the risks of climate change. With these challenges, we can identify that urban vacant lands can cooperate with the development of the communities in the following ways:

1. Improve air quality and reduction of the heat island effect.
2. Helping rainwater management and flood reduction.
3. Assist in the promotion of physical activity and mental well-being.
4. Generate habitat for biodiversity and ecosystem enhancement.
5. Help foster community engagement and social cohesion.
6. Drive economic growth and sustainable development.

The importance of using urban vacant land and how governments plan, design and maintain settlements and critical infrastructure determines the exposure patterns, social and physical vulnerability and resilience of cities (IPCC 2022). Concerning the above, the adaptation of abandoned land into new green areas provides a better image to cities, in addition to the fact that these recreational places encourage community participation, thus improving the quality of life of the inhabitants, as described by Bonham et al. (2002). Similarly, as described by Branas et al. (2011), greening vacant plots of land, transformed into public green spaces, helps social and environmental well-being, thereby reducing gun violence.

It should be noted that urban brownfield forestry infrastructure contributes to climate change mitigation by removing carbon dioxide, as noted by Abdollahi et al. (2000) and Nowak et al. (2002). Likewise, brownfield vegetation reduces stormwater management infrastructure, as Crauderueff et al. (2012) described. According to Schilling and Mallach (2012), Yadav et al. (2012), Grewal et al. (2011), Shuster et al. (2014), and Robinson and Lundholm (2012), therefore, brownfields contribute to systemic services that, in turn, have a beneficial impact on the community. This highlights the need for urban planners and policymakers to take a more holistic and forward-looking approach to managing urban wastelands, viewing them not as a burden but as a valuable asset for sustainable urban development. However, successfully converting urban vacant land into functional green infrastructure is challenging. The research reveals a significant gap in the policy and economic incentive frameworks to support the retention and redevelopment of these spaces. Many urban areas still face serious obstacles, including a lack of financial resources, a lack of public awareness, and inadequate governance structures.

Addressing these challenges requires a multifaceted approach, including solid policy reforms, increased financial investment, and active community engagement to ensure the long-term sustainability and viability of green urban projects. In summary, the adaptive reuse of vacant urban land represents a significant opportunity for cities to enhance their resilience to climate change while improving urban livability.

This study recommends developing comprehensive urban planning strategies prioritizing converting vacant land into green infrastructure. By fostering collaboration between local governments, private sector stakeholders, and community organizations, cities can unlock the full potential of these underutilized spaces. As previous studies show, some cities are making their first attempts to show the results of using urban vacant land to regenerate green infrastructure and thereby regenerate abandoned spaces. Future research should focus on developing detailed case studies and empirical evidence to illuminate better the most effective practices and policies for transforming vacant urban land into vibrant green infrastructure, promoting more sustainable and resilient urban environments.

## Conclusion

The changes in land surface temperatures of the land uses lead to an increase in the surface temperature of urban vacant land, urban built-up spaces, and streams. Poor urban planning of uncontrolled land uses and inadequate environmental management lead to socioecological and environmental problems that alter urban climate (UNCHS 1996; Zhau et al. 2006; Kalnay and Cai 2003). The consequences of urban climate change call for immediate action to avert negative environmental implications.

Revitalization of vacant land can be improved through the design and implementation of strategies addressing tax incentive structures, low rates of infill development of land forest structures, and rehabilitation abatement to increase the value. Therefore, the methodology used in this study and the examples reviewed in the bibliographic review make it possible to identify the significance of brownfield rehabilitation and its impact on climate change abatement.

The results of this study highlight the transformative potential of vacant urban land when integrated with green infrastructure initiatives. The analysis shows that despite traditional negative perceptions of vacant land as a symbol of urban decay, these spaces provide significant socio-ecological benefits when repurposed. Converting vacant land to green space mitigates climate change through carbon capture and stormwater management and improves urban biodiversity and community well-being.

Adaptive capacity is needed to design and implement urban planning-related strategies to adapt and deliver benefits linked to climate hazard change, including land use banking and vacant sites to create urban green spaces to develop green infrastructure projects (Schilling and Logan 2008). Urban areas must maintain and enhance their adaptive capacity to the changing climate to determine their vulnerability to the severity of climate hazards and related risks. High adaptive capacity in urban areas tends to moderate the vulnerability and severity of climate-related risks. Urban planning may incorporate vacant lots in its urban management strategy to promote cost-effective greening of vacant lots, as in the case of establishing the use of vacant lots in the stormwater management strategy. All of these aimed to reduce the drains of the rainfalls, protect the supply of drinking water, improve the safety of waterways, and mitigate the impact of hazards such as flooding. It is vital to quantify urban vacant land infiltration rates and translate these rates into rainfall detention capacities at storm and seasonal scales (Kelleher et al. 2020). In summary, taking advantage of vacant urban land spaces is essential to improve public health and community well-being.

Finally, one of the main limitations of this study is its reliance on theoretical and empirical literature without field research or primary data collection. While this analysis provides a comprehensive overview of existing research and successful case studies, it lacks direct observational data that could provide deeper insight into specific local conditions and practical challenges in converting vacant urban land. In addition, the study focused mainly on cities in developed regions, as these are the most studied cities at present and may not fully reflect the diverse socioeconomic and environmental contexts of cities in developing countries, where the dynamics of integration of urban vacant land and green infrastructure may differ significantly, but which on the other hand may yield equal or better results according to the studies. Another limitation is generalizing the results to different urban contexts without considering each city's unique characteristics. As seen in previous research, factors such as local climate, governance structures, economic conditions, and levels of community engagement vary widely. They can significantly influence the success of green infrastructure projects in greenfield areas.

## References

Abdollahi KK, Ning ZH, Appeaning A (2000) Gulf coast regional climate change council. In: Global climate change & the urban forest. Franklin Press, Baton Rouge, LA, USA

Accordino J, Johnson GT (2000) Addressing the vacant and abandoned property problem. J Urban Aff 22:301–315

Ackerman K (2012) The potential for urban agriculture in New York City: growing capacity, food security, & green infrastructure. Urban Design Lab at the Columbia Earth Institute, New York, NY, USA

Alexander FS (2005) Land bank authorities: a guide for the creation and operation of local land banks. Local initiatives support corporation, New York, NY, USA. Available online: http://www.nhc.org/media/files/Land_Bank_Authorities-Alexander.pdf. Accessed on 18 Dec 2015

Anderson E, Minor E (2021) Assessing four methods for establishing native plants on urban vacant land. Ambio 2021(50):695–705. https://doi.org/10.1007/s13280-020-01383-z)

Angel S, Sheppard S, Civco D (2005) The dynamics of global urban expansion. Transport and Urban Development Department, World Bank, Washington DC

Baró F, Chaparro L, Gómez-Baggethun E, Langemeyer J, Nowak DJ, Terradas J (2014) Contribution of ecosystem services to air quality and climate change mitigation policies: the case of urban forests in Barcelona, Spain. Ambio 43:466–479

Benedict MA, McMahon ET (2012) Green infrastructure: linking landscapes and communities. Island press

Blakeman C, Brown B, Fitzpatrick BT, Shaw I, Williamson A (2008) City-wide brownfield redevelopment plan. Report Prepared for the City of Roanoke, VA, USA. City Council, Roanoke, USA

Blanco H, McCarney P, Parnell S, Schmidt M, Seto K (2011) The role of urban land in climate change. In: Rozensweig C, Solecki W, Hammer S (vol eds) First urban climate change research network (UCCRN) assessment report on climate change in cities. Cambridge University Press, Cambridge

Bonham JB, Spilka G, Rastorfer D (2002) Old cities/green cities: planning advisory service report Number 506/507. American Planning Association, Chicago, IL, USA

Bowman AOM, Pagano MA (2004) Terra incognita: vacant land and urban strategies. Georgetown University Press, Washington, DC, USA

Branas CC, Cheney RA, MacDonald JM, Tam VW, Jackson TD, Ten Have TR (2011) A difference in differences analysis of health, safety, and greening vacant urban space. Am J Epidemiol 174:1296–1306

Brandes D (1995) The flora of old town centers in Europe. In: Sukoop H Numata M Huber A (eds) Urban ecology as the basis for urban planning. SPB Academic Publishing, Amsterdam, The Netherlands

Burkholder S (2012) The new ecology of vacancy: Rethinking land use in shrinking cities. Sustainability 4:1154–1172

Carter JG, Cavan G, Connelly A, Guy S, Handley J, Kazmierczak A (2015) Climate change and the city: building capacity for urban adaptation. Prog Plan 95:1–66

City of Pittsburgh, Greening Resources (2012) Available online: http://www.pittsburghpa.gov/neighborhoodinitiatives/greenup/resources.htm. Accessed on 19 June 2015

Crauderueff R, Margolis S, Tanikawa S (2012) Greening vacant lots: planning and implementation strategies. The Nature Conservancy, New York, NY, USA

Cutieru A (2021) Ciudades que se encogen: el auge y la caída de los entornos urbanos (Shrinking cities: the rise and fall of urban environments) 31 July 2021. ArchDaily México (Trad. Caballero, Pilar) Accedido el 27 Mar 2024. https://www.archdaily.mx/mx/965618/ciudades-que-se-encogen-el-auge-y-la-caida-de-los-entornos-urbanos. ISSN: 0719-8914

Dietersdorfer L, Efremova V, Fernandez Agueda B, Fleschurz R, Mangialardi G, Piscitelli C, Schmitz S, Scurrell B, Sosinski P, Willi C, Wolff M (2012) Urban shrinkage and chances for adaptation to climate change. Final report EU-COST action: TU0803

Fohrer N, Haverkamp S, Eckhardt K, Frede HG (2001) Hydrologic response to land use changes on the watershed scale. Hydrol Oceans Atmos 26(7–8):577–582

Grewal SS, Cheng Z, Masih S, Wolboldt M, Huda N, Knight A, Grewal PS (2011) An assessment of soil nematode food webs and nutrient pools in community gardens and vacant lands in two post-industrial American cities. Urban Ecosyst 14:181–194

Herce M, Vallejo MH (2009) Sobre la movilidad en la ciudad: propuestas para recuperar un derecho ciudadano 18. Reverté

Hodgson K, Campbell MC, Bailkey M (2011) Urban agriculture: growing healthy, sustainable places. American Planning Association, Chicago, IL, USA

Hope D, Gries C, Zhu W, Fagan WF, Redman CL, Grimm NB, Kinzig A (2003) Socioeconomics drive urban plant diversity. Proc Natl Acad Sci USA 100:8788–8792

Hu X, Zhuang S (2024) Large-Scale spatial–temporal identification of urban vacant land and informal green spaces using semantic segmentation. Remote Sens 16:216. https://doi.org/10.3390/rs16020216

Ifatimehin OO (2007) An assessment of urban heat Island of Lokoja town and surroundings using Landsat ETM data. FUTY J Environ 2(1):100–108

IPCC (2022) Chapter 6: Cities, settlements and key infrastructure. Intergovernmental Panel on Climate Change. Climate change 2022: impacts, adaptation and vulnerability. https://www.ipcc.ch/report/ar6/wg2/chapter/chapter-6/

Jakle JA, Wilson D (1992) Derelict landscapes: the wasting of America's built environment. Rowman & Littlefield, Savage, MD, USA
Kalnay E, Cai M (2003) Impact of urbanization and land use change on climate. Nature 423:528–531. https://doi.org/10.1038/nature01675
Kamvasinou K (2011) The public value of urban vacant land. Proc ICE Munic Eng 164:157–166
Kelleher C, Golden H, Burkholder S, Shuster W (2020) Urban vacant lands impart hydrological benefits across city landscapes. Nat Commun 11:1563. https://doi.org/10.1038/s41467-020-15376-9
Kim G (2016) The public value of urban vacant land: social responses and ecological value. Sustainability 8(5):486. https://doi.org/10.3390/su8050486
Kim G, Miller PA, Nowak DJ (2016) The value of green infrastructure on vacant and residential land in Roanoke, Virginia. Sustainability 8:296
Kim G, Miller PA, Nowak DJ (2015) Assessing urban vacant land ecosystem services: urban vacant land as green infrastructure in the City of Roanoke, Virginia. Urban for Urban Green 14:519–526
Kim Y, Lee R, Lee T, Shin Y (2023) Green infrastructure and urban vacancies: land cover and natural environment as predictors of vacant land in Austin, Texas. Land 12:2031. https://doi.org/10.3390/land12112031
Klimeš L (1987) Succession in road bank vegetation. Folia Geobot Phytotaxon 22:435–440
Kremer P, Hamstead ZA, McPhearson TA (2013) Social-ecological assessment of vacant lots in New York City. Landsc Urban Plan 2013(120):218–233
Lennon M, Scott M (2014) Delivering ecosystems services via spatial planning: reviewing the possibilities and implications of a green infrastructure approach. Town Plann Rev 85(5):563–587
Little GM (2008) 629 Main street—a revisioning of an urban void: an exploration into challenging cultural perceptions of an urban vacant lot. The University of Manitoba, Manitoba, Canada
Martinez-Fernandez C, Audirac I, Fol S, Cunningham-Sabot E (2012) Shrinking cities: urban challenges of globalization. Int J Urban Reg Res 36(2):213–225
McKinsey and Company (2016) Urban world: meeting the demographic challenge. McKinsey and Company
Millennium Ecosystem Assessment (2003) Ecosystems and human well-being: synthesis. World Health; Island Press, Washington, DC, USA
Mohamad I, Maruthaveeran S, Mohd M, Shahidan M (2023) Planning tools to revitalise urban vacant land from ecological perspectives: a systematic review. Urban Sci 7:58. https://doi.org/10.3390/urbansci7020058
Moreno O, Lillo C, Gárate V (2014) La infraestructura verde como espacio de integración. In: XI Simposio de la Asociación Internacional de Planificación Urbana y Ambiente (UPE 11) (La Plata)
Neukrug HM (2009) Clean water, green city: blending interests of land and water. Office of Watersheds, Philadelphia, PA, USA
Nowak DJ, Crane DE, Dwyer JF (2002) Compensatory value of urban trees in the United States. J Arboric 28:194–199
Pagano MA, Bowman AOM (2000) Vacant land in cities: an urban resource; brookings Institution. Center on Urban and Metropolitan Policy, Washington, DC, USA
Pike A, MacKinnon D, Coombes M, Champion T, Bradley D, Cumbers A, Robson L, Wymer C (2016) Uneven growth: tackling city decline. Joseph Rowntree Foundation, York
Robinson SL, Lundholm JT (2012) Ecosystem services provided by urban spontaneous vegetation. Urban Ecosyst 15:545–557
Rosa EA, York R, Dietz T (2004) Tracking the antropogenic drivers of ecological impacts. J Hum Environ 33(8):509–512
Rosenthal JK, Crauderueff R, Carter M (2008) Urban heat Island mitigation can improve New York City's environment. Sustainable South Bronx, New York, NY, USA
Ruth M, Coelho D (2007) Understanding and managing the complexity of urban systems under climate change. Clim Policy 7(4):317–336

Sassen S (2001) The global city: New York, London, Tokyo. Princeton University Press, Princeton, NJ

Schilling J, Logan J (2008) Greening the rust belt: a green infrastructure model for right sizing America's shrinking cities. J Am Plann Assoc 74:451–466. https://doi.org/10.1080/01944360802354956

Sackey ANA (2008) Environmental impacts of rapid urbanization of Ethiopia's cities. In: van Dijk MP, Fransen J (eds) Managing Ethiopian cities in an era of rapid urbanization. Eburon Delft 119–131

Schilling JM, Mallach A (2012) Cities in transition: a guide for practicing planners. American Planning Association, Washington, DC, USA

Shah A (2015) Why is biodiversity important? Who cares? Available online: http://www.globalissues.org/article/170/why-is-biodiversity-important-who-cares. Accessed on 12 Sept 2015

Shuster WD, Dadio S, Drohan P, Losco R, Shaffer J (2014) Residential demolition and its impact on vacant lot hydrology: Implications for the management of stormwater and sewer system overflows. Landsc Urban Plan 125:48–59

Song X, Wen M, Shen Y, Feng Q, Xiang J, Zhang W, Zhao G, Wu Z (2020) Urban vacant land in growing urbanization: an international review. J Geograph Sci 669–687. https://doi.org/10.1007/s11442-020-1749-0

Storch H, Downes N (2011) A scenario-based approach to assess Ho Chi Minh City's urban development strategies against the impact of climate change. Cities 28:517–526

Sukopp H, Blume HP, Kunick W (1979) The soil, flora, and vegetation of Berlin's waste lands. In: Nature in cities: the natural environment in the design and development of urban green space. Wiley, New Jersy, NY, USA, pp 115–132

Sun Z, Li X, Fu W, Li Y, Tang D (2013) Long-term effects of land use/land cover change on surface runoff in urban areas of Beijing, China. J Appl Remote Sens 8(1):https://doi.org/10.1117/1.JRS.8.084596

The Greening of Detroit. Urban Agriculture and Open Space (2012) Available online: http://detroitagriculture.met/open-space-stories/. Accessed on 7 May 2016

Tong S, Sun Y, Ranatunga T, He J, Yang Y (2012) Predicting plausible impacts of sets of climate and land use change scenarios on water resources. Appl Geogr 32:477–489

UNCHS (1996) An urbanizing world. World global report on human settlements. Oxford University Press, New York

U.S. Department of Housing and Urban Development (2009) Revitalizing foreclosed properties with land banks. Diane Publishing, Collingdale PA USA

Wachter S, Scruggs G, Voith R, Huang L (2010) Redevelopment authority of the city of Philadelphia land use and policy study. Penn Institute for Urban Research, Econsult Corporation, Philadelphia PA USA

Weng Q (2001) A remote sensing-GIS evaluation of urban expansion and its impact on surface temperature in the Zhujiang Delta, China. Int J Remote Sens 22(10):1999–2014

Wubet MT (2003) Estimation of absolute surface temperature by satellite remote sensing. Unpublished M. Sc thesis, International Institute for Geoinformation Science and Earth Observation, Netherlands

Yadav P, Duckworth K, Grewal PS (2012) Habitat structure influences below ground biocontrol services: a comparison between urban gardens and vacant lands. Landsc Urban Plan 104:238–244

Zhang P, Park S (2023) VLAS: vacant land assessment system for urban renewal and greenspace planning in legacy cities. Sustainability 15:9525. https://doi.org/10.3390/su15129525

Zhau S, Da L, Tang Z, Fang H, Song K, Fang J (2006) Ecological consequences of rapid urban expansion: Shanghai, China. Front Ecol Environ 4(7):341–346

# Chapter 21
# Mental Health and Climate Change from a Small Island Developing State Perspective-Research Gaps and Needs

**Rachael E. Williams**

**Abstract** Small Island Developing States (SIDS) experience some of the most devastating impacts of health effects of climate change, especially on mental health. Acute climate change events like hurricanes can cause displacement due to loss of habitable land and result in eco-anxiety, helplessness, worry and fear, which can significantly affect a population's mental health. Despite these severe mental health effects, major research gaps exist in research on the relationship between mental health and climate change in SIDS. To determine the current state of research conducted on mental health and climate change in SIDS an traditional literature review was undertaken. Five factors were identified which are linked to a climate change event's impact on mental health: (1) the local context of culture, society, economy, and development; (2) the exposure's spatial distribution; (3) the kind of meteorological event; (4) the length and intensity of the event; and (5) the expected severity and acuity of the associated effects on physical health and community well-being. This paper recommends greater use of multivariate analysis to address identified research gaps due to climate change in SIDS. This will allow for major gains in mental health by accounting for the myriad factors that vary within and between each SIDS to properly consider distinctive local contexts.

**Keywords** Mental health · Climate change · Small island developing states · Research gaps

R. E. Williams (✉)
Data for Development and Management, 33 Christina Gardens, Arima, Trinidad and Tobago
e-mail: rachaew12@gmail.com; rachaew@datadevmgt.com

 
W. Leal Filho and F. Wolf (eds.), *Climate Change and Health: Perspectives from Developing Countries*, Climate Change Management,
https://doi.org/10.1007/978-3-031-93177-2_21

# Introduction

## *Small Island Developing States*

Small island developing states or SIDS, can be defined as a unique collection of developing countries facing specific environmental, social, and economic hazards (Association of Caribbean States 2015). SIDS exist in three regions, the Caribbean; the Pacific; and the Atlantic, Indian Ocean, South China Sea (AIS) Region. There are thirty-eight which belong to the United Nations (UN), of which sixteen are in the Caribbean, thirteen in the Pacific and the rest in the Atlantic, Indian Ocean and South China Sea (*AIS*) region. There are also twenty non-UN/Associate members of Regional Commissions.

In 1992, at the United Nations Conference on Environment and Development, the International Community formally declared SIDS a special case for environment and development and committed to assist them to meet their sustainable development objectives (Agenda 21, n.d.). Apart from the shared problems that developing countries are facing, SIDS face several other challenges and need special assistance from the International Community to address them. These are climate change, environmental disasters and climatic variations, risks to freshwater resources, marine ecosystems, particularly high dependence on imports of oil, as well as tourism's threat (Association of Caribbean States 2015). All of this can contribute to increasing the health risks of SIDS populations.

Health can be defined as "a state of complete physical, mental, and social well-being and not merely the absence of disease or infirmity". It is suggested to include spiritual health to make the definition complete. Since 1992, Agenda 21 recognized the right of individuals to 'healthy physical, mental, and spiritual development' (Agenda 21, Chapter 6.23, as cited in Vader 2006) yet the use of spiritual health to improve mental health problems due to climate change effects is seldom mentioned in 'mainstream' research or research in SIDS.

Spiritual health can be defined as an intense degree of faith, hope, and dedication to a clearly defined worldview or belief system that offers an ethical route to personal fulfillment that involves a sense of connectedness with oneself, others, a higher power, or the larger reality (Hawks et al. 1995, as cited in Hawks 1994).

Research has demonstrated that there is a significant positive relationship between spiritual health and mental health in patients with chronic diseases, i.e. the higher and stronger the level of spiritual health, the lower the level of psychological disorders. Therefore, spiritual health can be used as a predictor of mental crisis in chronic patients (Najafi et al. 2022). Private and philanthropic foundations are currently the largest supporters of research on the intersection of spirituality and mental health. This is because federal agencies in many developed countries, including the USA and UK, have yet to issue any formal requests for funding proposals. This is even though 84% of the world's population is religiously affiliated and 68% of unaffiliated individuals believe in a higher power. Surprisingly, a phenomenon as widespread as spirituality is ignored by academic psychiatry (Rosmarin et al. 2021).

Concerning economic implications of negative health impacts, the world economy loses US $1 trillion a year due to lost productivity caused by two of the most prevalent mental illnesses, depression and anxiety. In 2010, it was predicted that the global economy lost about $2.5 trillion annually due to poor mental health and decreased productivity; by 2030, that amount is expected to climb to $6 trillion (The Lancet Global Health 2020). The fact is, before 2020, median spending on mental health services in the Americas (which includes Caribbean SIDS) was only 2% of total government health spending! The percentage of the burden related to mental disorders is six times the percentage of health dollars dedicated to mental health, with the median ratio of mental illness burden to expenditure on mental health being 6.1. The burden can range from 1.8 to 72 times the reported spending (PAHO 2018).

This chapter explores the current state of research on the connections between mental health and climate change in SIDS. Some key questions addressed refer to the types of studies that have been conducted in SIDS on the effects of climate change events on mental health and what gaps guide future research on the consequences of climate change on mental health in SIDS.

## Methodology

To determine the current state of research conducted on mental health and climate change in SIDS, a literature review was undertaken. Search terms used were mental health, climate change, and small island developing states (SIDS).

The initial online database search using Google was conducted on February 6, 2024, and lasted for six months to give ample time to find and access pertinent publications since SIDS are spread across three geographical locations. Considering the relationships between mental health and climate change in SIDS was a requirement for inclusion in the search. Also included was research which could translate from non-SIDS to SIDS in keeping with the paper's goal. For e.g., Hwong et al.'s scoping review (2022) highlights the strengths and weaknesses of existing research approaches on the effects of climate change on mental health in high-income, 'developed' countries, primarily non-SIDS. However, research gaps found there would also translate into gaps in SIDS. The key mental health outcomes evaluated were psychological distress in surveys, emergency department and hospital admissions for a psychiatric disease in administrative records, and death by suicide in public health data.

Based on the inclusion criteria, a total of 38 articles were assessed. Knowledge of the current status of research on mental health and climate change in SIDS was obtained by combining details from these articles with insights from the author's personal experience living and working in Caribbean SIDS.

## Results and Discussion

Of the 38 articles identified, one focused on current research knowledge and gaps regarding mental health and well-being under climate change in SIDS (Kelman et al. 2021).

### *Influencing Factors*

The following factors were found which are linked to a climate change event's impact on mental health: (1) the local context of culture, society, economy, and development; (2) the exposure's spatial distribution; (3) the kind of meteorological event; (4) the length and intensity of the event; and (5) the expected severity and acuity of the associated effects on physical health and community well-being, (Hwong et al. 2022, as cited in Obradovich et al. 2018). These events can cause acute stress, anxiety, depression, and post-traumatic stress disorder (PTSD), i.e. have negative impacts on mental health. These factors are further elaborated below:

#### Factor 1: The Local Context of Culture, Society, Economy, and Development

SIDS are a diverse group of countries that exhibit significant differences among themselves. They comprise four atoll nations, fourteen volcanic islands, and four continental countries. As a result, their coastal ecosystems and geomorphologies vary considerably. Their economies also differ greatly, ranging from some of the least developed countries in the world to some of the countries with the highest gross domestic product (GDP) per capita., e.g. Singapore (Vousdoukas et al. 2023). The adverse effects of climate change and disasters not only affect tangible cultural and natural properties, such as buildings, monuments, or archaeological sites, but also intangible cultural heritage, such as the loss of oral tradition and languages. This jeopardizes the investment in and development of the creative sector. In addition, the migration of young persons highlights the crucial importance of preserving culture in all its aspects and passing it down to future generations. A comprehensive vulnerability index for SIDS is currently being created by the UN due to the complex and interconnected vulnerabilities that these countries face. Coastal flooding susceptibility is highly variable among SIDS due to differences in adaptive abilities. Additionally, coral communities and their natural protection exhibit great threats from elevated temperatures and ocean acidification (Vousdoukas et al. 2023).

The various sub-regions of SIDS exhibit distinct features that have influenced their approaches to cultural policies. While not every SIDS has a cultural policy, the majority are working on creating one at the moment, either as separate laws or in conjunction with other sectoral policies (UNESCO 2024). In the Caribbean

SIDS, a large portion of the population has ancestral ties to Africa and other nations, with a shared history marked by slavery, forced migration, and the coexistence of multiple ethnic groups. The Caribbean region is traditionally divided into linguistic groups, namely English-speaking, French-speaking, Spanish-speaking, and Dutch-speaking. The SIDS of the Pacific and AIS sub-regions share a rich legacy of diverse marine ecosystems. The Pacific region, comprising over 20,000 islands separated by vast stretches of open ocean, spans around 165 million $km^2$. The Caribbean region includes over 7000 islands, islets, and cays, with a total land area of approximately 239,681 $km^2$.

**Factor 2: The Exposure's Spatial Distribution**

A considerable proportion of populations, infrastructure, and assets are situated along the coastline in areas with a high ratio of coastline to land (Thomas et al. 2020, as cited in Barnett and Waters 2016). Coastal and inland flooding, coastal erosion, extreme rainfall, terrestrial and marine heatwaves are climate change-related hazards for tropical SIDS. Anthropogenic climate change has an impact on these risks and is expected to continue to increase as global temperatures rise (Thomas et al. 2020). The encroachment of saltwater onto land is increasing day by day, which is contaminating freshwater sources and causing damage to agricultural land. This, in turn, is affecting the livelihoods of people who depend on these resources for their survival. With the rise in ocean acidity, freshwater salinification, invasive species, and evapotranspiration, the local food and water supplies are also changing, which can have severe health impacts on people living in SIDS (IPCC 2020, as cited by Kelman et al. 2021).

Although most SIDS are in the tropical cyclone (TC) belt and are frequently impacted by TCs, there are SIDS such as Bahrain, Belize, and Suriname that hardly experience TCs at all. Moreover, some SIDS are exposed to long-period swell waves, for instance, Pacific SIDS, while others mostly experience short-period wind sea waves, e.g. Caribbean SIDS. Tides are mainly microtidal in SIDS (Vousdoukas et al. 2023). Many of these islands are situated in the tropical and subtropical regions and a lot of them are considered as large Ocean States, which means that their oceanic territories are much bigger than their landmasses. Due to these vast exclusive economic zones, the marine resources of these nations are highly vulnerable to any disturbances in the ocean (Thomas et al. 2020).

SIDS are in control of approximately 30% of the world's oceans and seas, which presents significant opportunities as part of the 'blue economy'. This is vital for maintaining food security, which is inextricably linked to traditional knowledge and practices, fighting climate change, and protecting biodiversity (UNESCO 2024).

By spatially and temporally linking individuals' reports of mental health difficulties to the environmental exposures that they experienced, it was found that short-term weather, multiyear warming, and TC exposure each relate to worsened mental health outcomes (Obradovich et al. 2018).

### Factor 3: The Kind of Meteorological Event

Climate change-related events for tropical SIDS can be categorized into three types: acute, subacute, and changes lasting to the end of this century and beyond, i.e. chronic. Acute events last for days and for SIDS include hurricanes, floods, and short-duration heat waves. Subacute events last for months or years, e.g. long-duration heat waves. Chronic environmental changes last until the end of this century and beyond, such as higher temperatures, sea level rise, and a permanently altered and potentially uninhabitable physical environment (Palinkas and Wong 2020).

### Factor 4: The Length and Intensity of the Event

Effects due to climate change-related events can be classified as short, medium, or long-term. Short-term refers to < 1-year post-exposure to the event, medium as 1–5 years post-exposure, and long-term > 5 years or ongoing exposure (Hwong et al. 2022).

### Factor 5: The Expected Severity and Acuity of the Associated Consequences for Physical Health and Community Wellbeing (Dependent on 1–4)

Climate change can result in several types of events that cause different degrees of economic losses associated with property damage, loss of income and employment opportunities, and reduced economic productivity, particularly in agriculture and fisheries. It also poses threats to health and well-being, including injuries and deaths, the spread of vector-borne and respiratory illnesses, and heat-related stress. Climate change leads to population displacement, loss of attachment to the natural environment, social conflict, and inter-group violence, each of which significantly impacts mental health. The timing and circumstances surrounding climate-related events, as well as the specific associations between an event and its psychological effects, have significant implications for the development and implementation of policies and practices aimed at preventing and treating mental health problems related to climate (Palinkas and Wong 2020).

The impact of climate change on mental health can occur either directly with immediate effect, or indirectly in the short term during extreme events, or indirectly in the long term. Direct effects of climate change on mental health include stress-related and trauma-related consequences of an acute event, such as a hurricane or flood. Indirect effects refer to changes in mental health that are related to physical health factors such as increased ambient temperature, higher levels of pollen, dust, or pollution, as well as community-level effects that arise from economic damages, conflicts over scarce natural resources, displacement, or migration due to loss of habitable land. Chronic indirect mental health effects can explain eco-anxiety or climate grief, helplessness, worry, and fear of rapid climate change (Hwong et al.

2022, as cited in Albrecht et al. 2007). These events can have a significant impact on the mental health of a population and may lead to the development of psychiatric conditions like Post Traumatic Stress Disorder (PTSD), mood disorders such as depression and anxiety, increased rates of suicide and substance abuse, and more aggressive behaviour (Cianconi et al. 2020).

## *Barriers to Research Progress*

The main research gaps found on the consequences of climate change on mental health in SIDS, along with suggestions to remedy them are discussed in the following paragraphs, using the five influencing factors framework linked to a climate change event's impact on mental health in SIDS.

### Gap 1: Lack of Studies on Public Mental Health Interventions in SIDS

There are only a few studies that focus on public mental health interventions for climate change. In disaster management it is necessary to approach the study of mental health outcomes following a weather event from a population-level perspective. This includes examining factors such as how public policy responses, like funding changes for mental health services, can affect outcomes. It also involves monitoring the use of outpatient and crisis mental health services in the community after a weather event, as well as preparing and responding to the event at the community level. More healthcare providers and first responders need to receive mental health training to better manage these situations.

It is crucial to recognize the impact of environmental factors on mental health outcomes and implement community-based mental health interventions with individual-level interventions (Tselebis and Pachi 2022 as cited in Alegría et al. 2022). Based on this transformational shift, it is possible to address social and psychological needs at the community level. This approach can help reduce the number of people accessing hospital clinics, where staff may feel overwhelmed in dealing with the structural and social issues that contribute to patients' mental health needs (Tselebis and Pachi 2022).

Effective and efficient monitoring and evaluation (M&E) and more use of Trauma Signature Analysis (TSIG) are needed in SIDS to help support the role of primary health care professionals and predict needs for mental health and psychosocial support. The functions of primary mental health care are early diagnosis and intervention, evaluation, treatment planning, referral, monitoring, and prevention at the individual and community level (Tselebis and Pachi 2022).

There are three main differences between M&E for traditional management purposes and climate change adaptation planning. (1) Climate change-related time intervals and the corresponding results of adaptation are likely to be longer than those of conventional management initiatives. (2) There is uncertainty surrounding the specific nature and extent of climate change, particularly at the local level. This uncertainty can impact monitoring results and their interpretation. (3) A changing climate means traditional methods of measuring change may not work, requiring a moving baseline.

TSIG analysis is defined as an evidence-based approach that looks at the connections between the physical and psychological effects of exposing a population to a disaster, extreme event, or complicated situation and those relationships; to deliver prompt, practical advice for efficient disaster behavioral health (DBH) support that is tailored to the event's unique characteristics (Shultz et al. 2016, as cited in Shultz and Neria 2013). TSIG analysis looks at disaster survivors' exposures to empirically documented risk factors for psychological distress and common mental disorders (Shultz et al. 2016 as cited in Shultz et al. 2013). After determining the hazards, losses, and life changes, a disaster's unique 'signature' of exposure risks may predict needs for mental health and psychosocial support (MHPSS). Performing TSIG analysis in "real-time" during or after a disaster can guide on-scene MHPSS response (Shultz et al. 2016 as cited in Shultz and Neria 2013). Mental health researchers, climatologists and/or environmental scientists would need to collaborate on research that focuses on addressing this gap.

## Gap 2: Insufficient Research on Climate Change and Vulnerable and Marginalized Groups in SIDS

More research is necessary to comprehend how climate change disproportionately affects vulnerable and marginalized groups' mental health to enhance prevention, planning, response, and adaptation tools and efforts. Vulnerable populations include the homeless, the mentally ill, children, and the elderly (Coverdale et al. 2018, as cited in IPCC 2014). Conducting a thorough assessment of mental health symptoms and accurately classifying subclinical and clinical outcomes, while comparing them to appropriate populations, can significantly enhance the quality of these future mental health studies. Mental health researchers need to take the lead or collaborate on research that focuses on reducing this gap.

## Gap 3: Unknown Consequences of the Longitudinal Effects of Climate Change on Mental Health in SIDS

To determine the longitudinal effects of climate change on mental health, population-based databases and meteorological data can be used. Obradovich et al. (2018) linked mental health data with meteorological data and empirical tools from climate econometrics literature to examine the historical relationship between climatic variations and human mental health.

To determine the longitudinal effects of climate change on mental health, I would suggest utilizing the Framingham heart study (FHS) as a guide to the implementation of this research in the three SIDS regions. "Framingham is the essence of successful epidemiological research, productive of insights and applications… [and] the prototype and model of the cohort study" (Oppenheimer 2005, as cited in Susser 1987). The aim of the FHS was to investigate how coronary artery disease manifests in a normal or unselected population and identify risk factors for the disease's development by clinical and laboratory testing as well as long-term follow-up of this population (Oppenheimer 2005). It is a multigenerational cohort study that has multiple longitudinal observations. Repeated samples over several decades have allowed for studies describing temporal patterns in risk factors. It incorporated a large community-based sample and sophisticated statistical modelling and has combined integrated clinical epidemiology with translational research (Tsao and Vasan 2015).

In this context of mental health and climate change and given the many gaps in research in SIDS previously mentioned, there are many scenarios that could be investigated in a systematic way. Mental health outcomes and variables to be measured would need to be clearly defined. Mental health researchers and epidemiologists would need to collaborate on research that focuses on tackling this gap.

### Gap 4: Inadequate Research on Mental Health and Well-Being Impacts Caused by Gradual Changes Due to Climate Change in SIDS

More research needs to be conducted on mental health and well-being impacts focusing on gradual changes due to climate change, e.g. a warmer ocean and atmosphere, rising sea levels, ocean acidification, ecosystem changes, and changes in land and freshwater. There is little SIDS-related literature on self-harm, suicide, conflict, violence, and abuse which may be linked to gradual changes due to climate change. Globally, neuropsychiatric disorders are a major cause of death, and their prevalence has increased by more than 30% in recent years (Serafini et al. 2023). Due to their chronic and incapacitating nature, mental health illnesses as major depressive disorders (MDD) and suicidal behavior (SB) have emerged as diseases of critical public health concern among these conditions with high socioeconomic cost and health burden (Serafini et al. 2023). Mental health researchers need to take the lead or collaborate on research that focuses on narrowing this gap.

### Gap 5: Limited Research Conducted on the Impact of Climate Change-Related Migration on the Mental Health and Overall Well-Being of Persons Living in SIDS

Relocating or migrating can have significant impacts on mental health and well-being due to loss of place attachment, ancestral connections, and identities. This can lead to the erosion of belief systems, family ties, and local and cultural knowledge (Kelman et al. 2021 as cited in Singh et al. 2020). Places often mentioned as potential candidates for significant forced migration due to rising sea levels are the Federated States of Micronesia, Kiribati, Maldives, Marshall Islands, and Tuvalu, i.e. Pacific SIDS. In 2023 this became relevant for the Cayman Islands, i.e. a Caribbean SIDS.

The Rising Tides Departing Shores Action Plan was created to initiate this crucial dialogue on relocation with the Cayman Islands and UK governments.

Migration is a challenging experience, especially when it is not voluntary. This experience can become even more stressful due to factors such as inadequate social support, poor health systems, insufficient livelihoods, economic hardship, discrimination, and limited access to housing, education, social services, and healthcare. Possible outcomes are reduced self-esteem, difficulty adjusting to the new environment, and higher incidence of depression, phobias, and schizophrenia (Kelman et al. 2021 as cited in Henssler et al. 2020).

Once more, mental health researchers need to work together on research that focuses on narrowing this gap.

## Gap 6: Lack of Information on Physical Health and Well-Being Impacts of Climate Change in SIDS

It would be beneficial to conduct more assessments of mental health and well-being following natural disasters, to gather information on the physical health and well-being impacts of climate change. This includes identifying infectious and non-communicable diseases that are sensitive to climate change.

Assessments aimed at understanding the impact of climate change on human health can play a vital role in raising awareness and promoting acceptance of the importance of mental health and well-being. These assessments can also help in adapting to the changing weather patterns. Globally, diseases such as dengue fever, malaria, cholera, and the effects of undernutrition are some of the major indicators of the physical health impacts of climate change (Kelman et al. 2021 as cited in Watts et al. 2021).

Education initiatives, as said before, are key to overcoming this gap. Again, mental health researchers need to take the lead or collaborate on research that focuses on addressing this gap.

## Gap 7: Lack of Research on How Climate Events Affect People with Pre-existing Mental Health Conditions in SIDS

It is essential to better understand how climate events affect people with pre-existing mental health conditions or those who are predisposed to mental health conditions, the very populations seen in and around the margins of psychiatric services.

It is evident in the literature that the understanding of mental health varies between ‘developed’ and ‘developing’ countries, also, between and within SIDS. According to the WHO, mental health is defined as a condition of well-being in which a person is aware of their own potential, is able to handle everyday stress, is able to work well, and is able to give back to their community. I suggest one use a more comprehensive definition for mental health, i.e. a dynamic internal balance that allows people to apply their skills in a way that is consistent with society’s universal values. The state of internal equilibrium is influenced by various aspects of mental health, including basic cognitive and social skills, the capacity to identify, express, and regulate one's own emotions as well as empathize with others, flexibility in coping with negative life events, the ability to function in social roles, and a harmonious relationship between the body and mind (Galderisi et al. 2015).

The interpretation of 'society's universal values' and 'what leads to internal equilibrium' varies between and within SIDS. These are two variables that should be considered, among others, when doing any scientifically sound research on mental health and climate change in SIDS.

### Gap 8: Lack of Research on the Use of Effective Music and Spiritual Interventions as Therapeutical Options

Music interventions are a potential alternative for depression, instead of medications (Leubner and Hinterberger 2017). Studies to validate whether the use of music would be an effective intervention to significantly reduce the incidence of depression caused by climate change in SIDS have not been conducted and would be extremely useful. Music can be used as a tool to mitigate the current mental health crisis.

Mental health practitioners in SIDS often have a limited range of therapeutic options for patients, especially when dealing with the youth demographic. Pharmaceutical options are usually cost prohibitive and often present undesirable risks of complications due to side effects and social considerations. Providing practitioners with a clinically verified set of non-pharmaceutical options will: reduce the cost of therapy, increase the range of patient demographics that can be reached, improve options for public conversations on the topic and reduce negative stigma associated with mental illness.

In a meta-analysis of twenty-six studies, the experimental group that received a music intervention showed a significant decrease in depression levels over time compared to a control or comparison group (Leubner and Hinterberger 2017). It was observed that the reduction in mood disorders was consistent across all age groups. Music has proven to be a potential treatment option for mood disorders that has not been fully explored yet. The positive effects of music-driven interventions were noticeable after their completion. The intervention had positive effects on various aspects of the participants' lives. These included social aspects such as an increased motivation to participate in life again, as well as improvements in psychological status such as strengthened self-confidence and resilience to withstand stress. The intervention also showed promising outcomes in improving frequently occurring comorbidities such as anxiety, sleeping disorders, confidence, and self-esteem, which are often associated with depression. Research has shown that music can be an effective treatment option that significantly improves depression symptoms and quality of life, regardless of age (Leubner and Hinterberger 2017).

Previously it was mentioned that there is a significant positive correlation between spiritual health and mental health in patients with chronic illnesses, i.e. the higher and more robust the level of spiritual health, the lower the level of psychological disorders (Najafi et al. 2022). By ignoring the spiritual dimension of health, one is ignoring a crucial means of empowering individuals and populations to achieve 'healthy physical, mental, and spiritual development' (Agenda 21, Chapter 6.23, as cited in Vader 2006). Spiritual health can be used as a predictor of mental crisis in chronic patients (Najafi et al. 2022). It has been observed that persons who have a strong spiritual foundation are better able to manage their illnesses. Therefore, receiving support from spiritual or religious sources and establishing a connection with a higher power can be advantageous in improving the overall quality of life, managing mental health conditions, and reducing the severity of symptoms in patients with chronic diseases (Najafi et al. 2022). Our mental and social moods are largely influenced by our pursuit of spiritual significance, awareness of our own potential, and effort to fulfill that potential. Ultimately these elements also have an impact on our physical health. For us to live a healthy and meaningful life, we must find inner peace, actualization, and connection (Hawks et al. 1995).

**Need for Capacity-Building**

In addition to the 8 research gaps mentioned above on the consequences of climate change on mental health in SIDS, 2 key areas also need addressing. They are (1) the lack of trained mental health professionals and (2) inadequate health systems, services, training and infrastructure in SIDS, as further described below:

**The lack of trained mental health professionals in SIDS**: The MA in Clinical and Community Psychology at St George's University, Grenada is the only programme offering Community Psychology in the *English-speaking* Caribbean, and in disaster management, it is essential to have a community approach. Puerto Rico, Cuba, and the Francophone territories may well offer community psychology programmes (Dr. K. Moore, coordinator-Franciscan Institute Trauma Team, personal communication, Feb 12, 2024).

It must be emphasized that if a climate change-related event created unmet 'survival needs,' then the thought of conducting any mental health research within a SIDS would be irrational. One of many such examples was the impact of tropical storm Erika on the SIDS Dominica. August 2015, Erika within hours brought fifteen inches of rain and flash floods which claimed lives, washed away roads and bridges, and destroyed houses and entire villages. Approximately 60% of roads in the country became inaccessible. Public services were disrupted, and the youth were unable to attend school for months.

There was no water and electricity for months and teams were sent to assist persons experiencing mental trauma caused by this climate event. It should be noted that the magnitude of damage caused was by a tropical storm and not by a category five hurricane which hit Dominica 2 years later, i.e. Maria. To date, they are still recovering from both events (Dr. K. Moore, coordinator-Franciscan Institute Trauma Team, personal communication, Feb 12, 2024).

To increase the number of trained mental health professionals, residencies and fellowships must provide mentorship to trainees, allowing them to explore the relationship between climate change and psychiatric outcomes (Hwong et al. 2022).

Medical students, residents, faculty, and the wider psychiatric community should all be educated about the relevant issues (Friedrich 2017). To achieve this, medical schools could consider hiring climatologists and environmental scientists as faculty members. Alternatively, they could work with equally trained professionals or consultants who are not currently affiliated with a university or research institute, but who are experienced in conducting research and educating in this area.

- **Inadequate health systems, services, training and infrastructure in SIDS**
  There is a pressing need for the enhancement of health systems, services, training, and infrastructure. This should include improving mental health and well-being services. The approach should be to strike a balance between integrating local and indigenous perspectives and practices on health. In addition, efforts should be made to overcome deep-rooted stigmas associated with mental health and well-being difficulties. Fear and misunderstanding often lead to prejudice against individuals with mental illness and addiction, even among healthcare providers. This prejudice and discrimination create significant obstacles to diagnosis and treatment. Education initiatives, as already said, are crucial to overcome this huge barrier. As mentioned previously, before 2020, median spending on mental health services in the Americas (which includes Caribbean SIDS) was only 2% of total government health spending (PAHO 2018). This has to be increased to match the extent of the existing problem.

## *Multivariate Statistical Analysis for Addressing Existing Research Gaps*

To account for the many variables that exist between and within all SIDS that impact greatly on experiencing advancement in mental health, I would strongly recommend the use of a multivariate approach in tackling identified research gaps. Five factors linked to a climate change event's impact on mental health in SIDS were previously highlighted. Another factor that must be examined is resource capability (e.g. physical infrastructure and availability of funds to do research). Gender, age, and differences in education must also be used to tailor policies and practices to prevent and treat climate-related mental health problems as accurately as possible.

By considering multiple variables simultaneously, multivariate analysis would provide a more holistic and thorough understanding of complex datasets. The use of multivariate and/or univariate statistical analysis would depend on the objective(s) of the research study. Clearly defined objectives based on the previously identified gaps would guide the appropriate data collection and subsequent statistical data analysis techniques to be used (Fig. 21.1).

If one objective is to help support the role of primary health care professionals and predict needs for mental health and psychosocial support, then the TSIG analysis approach which was used in Haiti in 2016, can serve as a model for the Caribbean Community and Common Market (CARICOM).

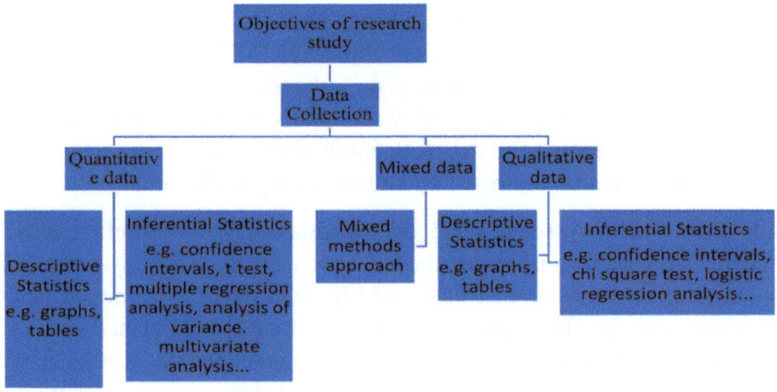


**Fig. 21.1** Examples of appropriate data collection and statistical data analysis techniques to be used

## Conclusions

Having reviewed the available literature and from firsthand knowledge gained from living and working in Caribbean SIDS, it is evident that there are significant research gaps on the connection between mental health and climate change in SIDS, and that more studies are needed to address these gaps.

For there to be noteworthy improvements in tackling mental health issues due to climate change in SIDS, there must be more action by all relevant stakeholders. This includes especially the medical and mental health professions, all SIDS governments, civil society, the private sector, international organizations and environmental scientists/climatologists/trained consultants and researchers using an evidence-based decision-making approach to address this huge problem of mental health and climate change in SIDS. Also, there must be a statistically significant improvement in the current expenditure on mental health which is woefully inadequate compared to the current mental illness burden in all SIDS.

More education initiatives and help are needed from the international community to support trained professionals living in SIDS. Studies vary greatly in samples, methodologies, measures and outcomes, leading to findings that are often inconsistent and difficult to compare, thus limiting conclusions which can be made across studies. Thus, there is a need for reliable, consistent measurements of climate events, analytical geographic units, and mental health outcomes across research projects.

Greater use of multivariate analysis is recommended which can address identified research gaps and achieve significant advancements in addressing mental health issues associated with climate change in SIDS. This will allow for major gains in mental health by accounting for the myriad factors that vary within and between each SIDS.

This paper has some limitations. More time is needed to find other studies to support the research findings from Kelman et al. (2021). Since the author is not currently working at the University of the West Indies, it was difficult at times to get needed references. A colleague at Clemson University assisted greatly with this constraint and authors of some references used in this paper.

Despite these obstacles, this paper provides comprehensive information on current research knowledge and gaps regarding mental health and climate change in SIDS and recommendations as to how to address them.

**Acknowledgments** I wish to thank Ms Petra Bridgemohan, co-author of the National Eye Study of TT (NESTT) and Dr Karen Moore, child psychologist, TT, for their insightful editing comments on this paper. Furthermore, I express my gratitude to God for leading me to write this paper and my mum who supported me while writing it.

## References

Addressing Stigma (n.d.). Available at: https://www.camh.ca/en/driving-change/addressing-stigma. Accessed 20 Feb 2024

Agenda 21 (n.d.). Available at: https://sustainabledevelopment.un.org/content/documents/Agenda21.pdf. Accessed 11 Feb 2024

Albrecht G, Sartore GM, Connor L et al (2007) Solastalgia: the distress caused by environmental change. Australas Psychiatry 15(suppl 1):S95–98

Alegría M, Zhen-Duan J, O'Malley IS, DiMarzio KA (2022) A new agenda for optimizing investments in community mental health and reducing disparities. Am J Psychiatry 179:402–416

Association of Caribbean States (2015) The Challenge of small Island developing states: Barbados, Mauritius, Samoa and beyond. Available at: https://www.eird.org/americas/news/the-challenge-of-small-island-developing-states-barbados-mauritius-samoa-and-beyond.html. Accessed 11 Feb 2024

Barnett J, Waters E (2016) Rethinking the vulnerability of small island states: climate change and development in the Pacific Islands. In: Grugel J, Hammett (eds) The palgrave handbook of international development. London: Palgrave MacMillan, pp. 731–48

Cianconi P, Betro S, Janiri L (2020) The impact of climate change on mental health: a systematic descriptive Review. Front Psych 11(74):1–15

Constitution of the world health organization. Available at: https://www.who.int/about/accountability/governance/constitution. Accessed 5 Feb 2024

Coverdale J, Balon R, Beresin E, Brenner A, Guerrero A, Louie A, Roberts L (2018) Climate change: a call to action for the psychiatric profession. Acad Psychiatry 42:317–323

COVID-19 pandemic triggers 25% increase in prevalence of anxiety and depression worldwide. Available at: https://www.who.int/news/item/02-03-2022-covid-19-pandemic-triggers-25-increase-in-prevalence-of-anxiety-and-depression-worldwide. Accessed 12 Feb 2024

Cutting edge: small Island developing states: cultural diversity as a driver of resilience and adaptation. Available at: https://www.unesco.org/en/articles/cutting-edge-small-island-developing-statescultural-diversity-driver-resilience-and-adaptation. Accessed 16 Feb 2024

Friedrich M (2017) Medical community gathers steam to tackle climate's health effects. J Am Med Assoc 317(15):1511–1513

Galderisi S, Heinz A, Kastrup M, Beezhold J, Sartorius N (2015) Toward a new definition of mental health. World Psychiatry 14(2):231–233

Hawks S (1994) Spiritual health: definition and theory. Well Persp 10:3–13

Hawks K, Hull M, Thalman R, Richins P (1995) Review of spiritual health: definition, role, and intervention strategies in health promotion. Am J Health Promot 9(5):371–378

Henssler J, Brandt L, Müller M, Liu S, Montag C, Sterzer P and Heinz A (2020) Migration and schizophrenia: meta-analysis and explanatory framework. Eur Arch Psychiatry Clin Neurosci 270:325–335
Hwong A, Wang M, Khan H, Chagwedera D, Grzenda A, Doty B, Benton T, Alpert J, Clarke D, Compton W (2022) Climate change and mental health research methods, gaps, and priorities: a scoping review. Lancet Planet Health 6(3):E281–E291
IPCC (2014) Synthesis report, summary for policymakers. Contribution of working groups I, II and III to the fifth assessment report of the intergovernmental panel on climate change. Core Working Team Pachauri RK and Meyer LA, editors. IPCC, Geneva, Switzerland
IPCC (2020) IPCC special report on climate change, desertification, land degradation, sustainable land management, food security, and greenhouse gas fluxes in terrestrial ecosystems. Geneva, Intergovernmental Panel on Climate Change
Kelman I, Ayeb-Karlsson S, Rose-Clarke K, Prost A, Ronneberg E, Wheeler N, Watts N (2021) A review of mental health and wellbeing under climate change in small island developing states (SIDS). Environ Res Lett 16:1–13
Leubner D, Hinterberger T (2017) Reviewing the Effectiveness of music interventions in treating depression. Front Psychol 8(1109):1–21
Mental health matters (2020) Lancet Glob Health 8(11):E1352
Monitoring and evaluation in climate change adaptation. Available at: https://coastadapt.com.au/how-to-pages/monitoring-and-evaluation-climate-change-adaptation. Accessed 25 Feb 2024
Najafi K, Khoshab H, Rahimi N, Jahanara A (2022) Relationship between spiritual health with stress, anxiety, and depression in patients with chronic diseases. Int J Afr Nurs Sci 17:1–5
Obradovich N, Migliorini R, Paulus M, Rahwan I (2018) Empirical evidence of mental health risks posed by climate change. Proc Natl Acad Sci USA 115(43):10953–10958
Oppenheimer G (2005) Becoming the Framingham study 1947–1950. Am J Public Health 95(4):602–610
PAHO (2018) The burden of mental disorders in the region of the Americas. Available at: https://iris.paho.org/handle/10665.2/49578. Accessed 5 Feb 2024
Palinkas L, Wong M (2020) Global climate change and mental health. Curr Opin Psychol 32:12–16
Rosmarin D, Pargament K, Koenig H (2021) Spirituality and mental health: challenges and opportunities. Lancet Psychiatry 8:92–93
Serafini G, Costanza A, Aguglia A, Amerio A, Trabucco A, Escelsior A, Sher L, Amore M (2023) The role of inflammation in the pathophysiology of depression and suicidal behavior: implications for treatment. Med Clin North Am 107:1–29
Shultz JM, Neria Y (2013) Trauma signature analysis: state of the art and evolving future directions. Disaster Health 1(1):4–8
Shultz JM, Cela T, Marcelin LH, Espinola M, Heitmann I, Sanchez C, Jean Pierre A, Foo CY, Thompson K, Klotzbach P, Espinel Z, Rechkemmer A (2016) The trauma signature of 2016 Hurricane Matthew and the psychosocial impact on Haiti. Disas Health 3(4):121–138
Singh P, Charan D, Kaur M, Railoa K and Chand R (2020) Place attachment and cultural barriers to climate change induced relocation: lessons from Vunisavisavi village, Vanua Levu, Fiji. In: Filho WL (ed) Climate change impacts and adaptation strategies for coastal communities. Cham, Springer
Small island developing states' Available at: https://sustainabledevelopment.un.org/topics/sids/list. Accessed 13 Feb 2024
The road to recovery in dominica: rebuilding and resilience after cyclone Erika. Available at: https://www.preventionweb.net/blog/road-recovery-dominica-rebuilding-and-resilience-after-cyclone-erika. Accessed 14 Feb 2024
Thomas A, Baptiste A, Martyr-Koller R, Pringle P, Rhiney K (2020) Climate change and small Island developing states. Annu Rev Environ Resour 45:1–27
Tsao C, Vasan R (2015) The Framingham heart study: past, present and future. Int J Epidemiol 44(6):1763–1766
Tselebis A, Pachi A (2022) Primary mental health care in a new era. Healthcare 10(10):1–4

Vader J (2006) Spiritual health: the next frontier. Eur J Pub Health 16(5):457

Vousdoukas M, Athanasiou P, Giardino A, Mentaschi L, Stocchino A, Kopp R, Menéndez P, Beck M, Ranasinghe R, Feyen L (2023) Small Island developing states under threat by rising seas even in a 1.5 °C warming world. Nat Sustain. 6:1552–1564

Watts N et al (2021) The 2020 report of the Lancet Countdown on health and climate change: responding to converging crises. Lancet 397:129–170

What is climate change? Available at: https://www.epa.gov/climatechange-science/frequently-asked-questions-about-climate-change#climate-change. Accessed 6 Feb 2024

# Chapter 22
# Assessing Urban Heat Island Dynamics: Vulnerability, Health Impacts, and Climate Resilience in the South Twenty-Four Parganas District, India

**Sudarshana Sinha, Ankhi Banerjee, and Archana Patnaik**

**Abstract** This study explores Urban Heat Island (UHI) dynamics by integrating Land Surface Temperature (LST) analysis with socio-economic vulnerability and public health assessments. UHI effects, exacerbated by urbanization and loss of green spaces, raise urban temperatures, impacting both environmental and human health. While prior studies focus primarily on temperature variations, this research addresses a gap by examining the interplay between UHI patterns, demographic vulnerabilities, and the health impacts of heatwaves. The research focuses on South Twenty-four Parganas, West Bengal, India, a region with diverse land use types. It seeks to identify the factors contributing to UHI, analyze how socioeconomic and infrastructural factors influence sensitivity to UHI, and assess the health impacts of UHI and heatwaves. Multiple thermal indices such as LST, Temperature Condition Index (TCI), and Relative LST were used to analyze UHI intensity, along with detailed assessments of vegetation (NDVI), waterbodies (NDWI), and built-up areas (NDBI). Primary survey data from vulnerable blue-collar workers further illuminated the socio-economic dimensions of UHI impacts. The findings demonstrate that environmental variables, in conjunction with socioeconomic differences, exacerbate vulnerabilities, particularly in densely populated places. Heat exposure significantly elevates health risks, including cardiovascular and respiratory issues. The study emphasizes the need to implement complete mitigation strategies involving urban design, public health initiatives, and community participation. This integrative approach addresses major gaps in UHI research. This study amalgamates UHI dynamics with socioeconomic and health evaluations to provide insights into UHI

S. Sinha (✉) · A. Patnaik
Department of Humanities and Social Sciences, Indian Institute of Technology Kharagpur, Kharagpur, India
e-mail: sudarshanasinha1@gmail.com

A. Banerjee
Ranbir and Chitra Gupta School of Infrastructure Design and Management, Indian Institute of Technology Kharagpur, Kharagpur, India

 
W. Leal Filho and F. Wolf (eds.), *Climate Change and Health: Perspectives from Developing Countries*, Climate Change Management,
https://doi.org/10.1007/978-3-031-93177-2_22

implications across varied landscapes, and additionally policy-driven suggestions for diminution.

**Keywords** UHI · LST · Vulnerability assessment · Heatwaves · Climate resilience

## Introduction

Land use and land cover changes (LULCC) have rapidly transformed urban landscapes, altering surface and subsurface dynamics and triggering significant alterations to the thermal environment (Irshad et al. 2024). Urban heat islands (UHIs), defined as regions of urban centres exhibiting higher temperatures than adjacent suburban and rural areas, are a notable repercussion (Zhang et al. 2021). This intensifies heat waves and disrupts heat retention within the urban locale (Jang et al. 2020). The shift in the native landscapes, marked by increased grey infrastructure, exacerbates heat anomalies (Wang et al. 2021). Concretization has additionally reduced the cooling effects of blue and green (BG) areas (Lee et al. 2021), particularly in the urban core and Central Business Districts (CBDs), triggering hotspots (Jia et al. 2024). The continuous diminution of green spaces enhances heat absorption, exacerbating the heat anomaly (Huang et al. 2020), whereas the loss of blue spaces stimulates heat retention (Dai and Liu 2022). Open spaces also play an integral part in regulating these anomalies, as they alter albedo, radiation emissivity, and absorption (Ren et al. 2022). Addressing these aspects via effective urban design is essential for mitigating the thermal impact of increased urbanization and ensuring sustainable, resilient urban ecosystems.

According to the UN (2019), the global urban population is anticipated to increase from 55.3% to 68% by 2050. In India, it increased from 27.81% in 2001 to 32.42% in 2011 (Census 2011), indicating rising urbanization. This expansion alters regional energy balances, increasing sensible heat flux while decreasing latent heat flux (Huang et al. 2022a, b). Urban warming, with a projected global temperature increase of 1.1 °C (Lin et al. 2024), poses health risks, especially in low-income countries (Ascione et al. 2024; Huang et al. 2022a, b). Rising temperatures in mid and low-latitude cities could trigger higher dermatological (Halder et al. 2024) and cardio-immune disorders (Li et al. 2024a, b). While tropical regions experience high diurnal temperature ranges, the link between heat-related mortality varies (Kacker et al. 2024). A systematic review of the literature (SLR) on urban heat's behavioural effects is limited, highlighting the need for studies with a finer spatial resolution (Jonescu et al. 2024). Most SLR emphasizes summer impacts, overlooking potential winter benefits (Macintyre et al. 2021). Understanding local urban heat effects is essential for developing resilient strategies, calling for multidisciplinary research (Chen et al. 2022).

UHI intensify temperatures, contributing to heat waves that adversely affect human health (Lasky et al. 2024). Vulnerabilities owing to thermal discomfort and heat stress are influenced by age and gender (Dutta et al. 2024; Lee et al. 2021). Heat-related conditions like exhaustion (Zhao et al. 2024), strokes (Yadav et al. 2023), cramps, and dehydration (Aboulnaga et al. 2024) disrupt urban life. Prolonged heat exposure impairs physical and cognitive abilities (Tewari et al. 2023), whereas UHI is associated with respiratory and cardiovascular disorders (Cleland et al. 2023). UHI also promotes the transmission of various vectors, further degrading urban wellness (Li et al. 2024a, b). Despite its extensive use, artificial cooling hinders adaptation to natural temperatures, resulting in detrimental health consequences (Ng et al. 2016). While SLR highlights UHI's potential winter advantages (Heaviside et al. 2017), variances in urban morphology and socioecological aspects require additional investigation (Tehrani et al. 2024; Wu et al. 2024). A comprehensive analysis utilizing niche personal data is useful for understanding individual thresholds and adaptability (Bai et al. 2024).

The UHI phenomenon gravely impacts urban residents' psychological and emotional stability, triggering mood swings and physiological distress (Dai and Liu 2022; Ma et al. 2024). UHI has been correlated to an increased risk of suicide (Aghamohammadi et al. 2022), manic episodes (Medici et al. 2016), heat-induced aggression (Stevens et al. 2024), hostility (Huang et al. 2022a, b), post-traumatic stress disorder, anxiety (Ma et al. 2024), and depression (Zhou et al. 2024). While rare, instances of schizophrenia (Lee et al. 2018) and bipolar illness (Tewari et al. 2023) have been documented.

SLR indicates a sporadic rise in suicide rates during heat stress periods, especially in early summer, with males displaying greater vulnerability due to lower tolerance levels (Abrar et al. 2022). SLR reveals a sporadic surge in suicide rates during heat waves, especially during early summer, with males being more vulnerable due to lower tolerance levels (Ming et al. 2024).

SLR identified a correlation between bipolar disorder prevalence and certain temperature circumstances, particularly when temperatures rise above 24 °C. Females tend to be more prone to bipolar-like symptoms caused by increased thermal radiation and heat fluxes (Palinkas et al. 2022). Furthermore, heat stress worsens senility (Aghamohammadi et al. 2022) and dementia in the elderly, who have difficulty adapting to changing microclimates (Linares et al. 2017; Samba et al. 2016), making them more susceptible to respiratory and dermatological issues with prolonged exposure (Stewart and Mills 2021). Urban residents with psychological disorders are also at higher risk for substance abuse and alcohol consumption as coping strategies (He et al. 2022a, b). Elevated thermal levels in tropical regions strain physiological limits and impair cognitive function (Han et al. 2023), while the impact of heat stress correlates with population density (Liu et al. 2024). Age significantly influences resilience to heat stress (Cleland et al. 2023), underscoring the need for nuanced regional analyses that consider factors such as location, morphology, and composition (Yang et al. 2021a, b). Generalizations regarding the effects of thermal stress in urban areas lack depth and require specialized analysis. UHI effects prolong nocturnal heat release, triggering insomnia, restlessness, and discomfort (Kassomenos and Begou 2022). UHI exacerbates weariness, promotes sedentary behaviour (Silva et al. 2019), interrupts commuting, increases pedestrian discomfort (Yang et al. 2020), and reduces outdoor comfort (Ren et al. 2023). It diminishes interest in recreational activities (Sanusi and Bidin 2020) and productivity (Ebi et al. 2021). Heat-related fatalities highlight UHI's detrimental impact on health and well-being, both indoors and outdoors (Heaviside et al. 2016; Taylor et al. 2018).

SLR stated that most studies focus on spatiotemporal variance in LST temperature, without integrating multi-variate ecological indicators, socioeconomic characteristics, and health-related dimensions, particularly within specific urban locales. There is a notable gap about the relationship between urban morphology, LULC, and socio-economic disparities that influence variance in LST, as well as how UHI exacerbates vulnerabilities across different populations. Furthermore, limited research explores the combined effects of UHI and heatwaves on public health and well-being, particularly in terms of long-term mitigation strategies addressing both immediate and prolonged health risks. This study addresses these gaps through an integrated methodology that encapsulates UHI patterns, evaluates regional risks, and evaluates health effects while determining feasible mitigation methods. The impact of UHI is profoundly dependent on geographical location, landscape ecology, socioeconomic framework, and daily activities. Consequently, broad observations frequently need to capture the subtleties intricate to each locale, necessitating region-specific research. Research questions entail:

1. What are the causative factors of UHI in the region, and how can LST fluctuations be effectively mapped?
2. How do socio-economic, infrastructural, and environmental variables impact influence the region's sensitivity to UHI?
3. What are the implications of UHI and heatwaves on health and well-being? What mitigation strategies are effective in minimizing these impacts?

The paper aims to:

1. Identify and analysis of regional UHI patterns and their relationship with pertinent variables.
2. Evaluate regional and demographic vulnerability, focusing on socio-economic, infrastructural, and environmental variables.
3. Evaluate the physiological, psychological, and socio-economic health effects of UHI and heatwaves.

Ultimately, this study seeks to provide a comprehensive understanding of the interplay between urbanization, UHI dynamics, and their multifaceted effects on both environmental and human health. This paper is divided into several sections for clarity. The introduction provides an overview of the topic. The following section details the materials and methods, including the analytical indices, study area, and data presented. Sections three and four present the research results and a discussion of the findings, respectively. The conclusion summarizes the outcomes and offers relevant recommendations.

## Materials and Methods

The methodological framework is as follows: Thermal anomalies and the regional UHI are then identified and analyzed using several thermal indices including Land Surface Temperature (LST) analysis, Relative LST (RLST), and the Temperature Condition Index (TCI). Regional vulnerability is assessed through landscape ecology, focusing on built-up areas, vegetation, and water bodies. Individual vulnerability is evaluated via primary surveys, offering deeper insight. Finally, a perception study investigates the perceived effects of heatwaves and UHI on health and well-being. This approach integrates both quantitative and qualitative methods for a comprehensive analysis.

### *Analysis of Thermal Conditions*

In this investigation, thermal indices such as LST, TCI, and RLST were employed to assess the thermal conditions of the study area. LST, widely recognized as a key determinant of the thermal environment, reflects the exchange of thermal energy between surface and subsurface layers (Waseem and Khayyam 2019) and has been positively correlated with anthropogenic activities, particularly the expansion of grey infrastructure. LST data were extracted from the MODIS dataset, with daytime values selected, reprojection to the appropriate coordinate system, and conversion from Kelvin to Celsius.

TCI, used to measure vegetation stress on LST, normalizes LST values by comparing them to maximum and minimum observed LSTs (Wang et al. 2019). TCI values range from 0 to 1, indicating the degree of vegetation-induced thermal stress.

$$TCI = \frac{(LST_{\mathrm{Max}} - LST)}{(LST_{\mathrm{Max}} - LST_{\mathrm{Min}})}$$

RLST illustrates the spatial distribution of thermal environments across various land-use types. The equation used for RLST (Cai et al. 2019) is:

$$RLST = \frac{LST}{LST_{\mathrm{Mean}}}$$

Areas with RLST values exceeding 2 °C were identified as high-temperature zones, or regional heat islands.

**Built-up Area Index**

Normalized Difference Built-up Index (NDBI), is an effective remote sensing algorithm to quantify urbanization. The NDBI precisely evaluates the intensity of built-up regions by evaluating reflectance in the Near-Infrared (NIR) and Shortwave Infrared

(SWIR) spectral bands. Its demonstrated capacity to detect and delineate urban areas makes it an essential tool for evaluating the pattern of urbanization.

$$NDBI = \frac{(SWIR - NIR)}{(SWIR + NIR)}$$

**Vegetation Index**

The Normalized Difference Vegetation Index (NDVI) is a vital metric to monitor vegetation health. It operates by contrasting the Near-Infrared (NIR) and visible red spectral bands, based on the premise that healthy vegetation absorbs visible light while reflecting a substantial amount of near-infrared light.

$$NDVI = \frac{(NIR - RED)}{(NIR + RED)}$$

**Waterbody Index**

The locale's waterbody status was determined using Normalized Difference Waterbody Index (NDWI) analysis, an integral remote sensing metric for delineating water bodies. NDWI leverages water's distinctive reflectance in the NIR and SWIR bands to differentiate it from other land cover types. This technique reduces atmospheric and sensor calibration issues, allowing for accurate waterbody identification.

$$NDWI = \frac{(NIR - SWIR)}{(NIR + SWIR)}$$

## *Brief Description of the Study Area*

South Twenty-four Parganas, a prominent district in West Bengal, India, is well-suited for studying the UHI phenomenon due to its diverse landscape, rapid urbanization, and significant population growth. Spanning latitudes 22°30′45″ N to 20°29′ N and longitudes 89°4′56″ E to 88°3′45″ E, the district is geographically bounded by Kolkata to the north, the Bay of Bengal to the south, and the Bidya, Matla, and Hooghly rivers to the west and east. This area offers a complex mix of urban, peri-urban, urban, rural, and mangrove forest (Sundarbans) environments, making it an ideal site for examining how varied land use impacts UHI intensity.

The district has experienced substantial population growth, increasing by 18.17% between 2001 and 2011, outpacing West Bengal's average growth rate of 13.8%. Urban expansion has been particularly notable, with a 9.85% rise in urban population and an additional 241.96 km$^2$ added to urban areas. Urbanization rates are especially high in Budge Budge-I (73.81%) and Thakurpukur Maheshtola (48.82%), highlighting concentrated urban growth, while Diamond Harbour exhibits the highest population density at 5673.86 persons/km$^2$. These dynamics underscore the rapid urbanization processes that are critical drivers of UHI in the region.

The diverse landscape and varying degrees of urbanization in South Twenty-four Parganas create significant spatial variability in thermal conditions. Areas with the highest UHI intensity (UHII) were identified and selected for focused study. These locations, characterized by intense urbanization, population density, and rapid development, provide critical insight into the localized effects of UHI and its broader implications for human health, particularly in densely populated and vulnerable urban zones (Fig. 22.1).

The data presented in this paper were sourced from primary surveys conducted in government hospitals and through satellite imagery analysis. Satellite images were used to identify areas with the highest Urban Heat Island Intensity (UHII), obtained from the USGS Explorer and MODIS datasets for 2023, with only images below 3% cloud cover included. These images underwent PAN sharpening and noise reduction before analysis.

Subsequently, primary surveys were conducted in these urban areas. A purposive stratified random sampling approach was employed to select participants, focusing on individuals in blue-collar occupations or the informal sector who typically spend 6–7 h exposed to direct sunlight. The demographic profile of respondents ranged from 20 to 50 years of age. The questionnaire was divided into sections: the first collected personal data, followed by sections assessing respondents' perceptions of

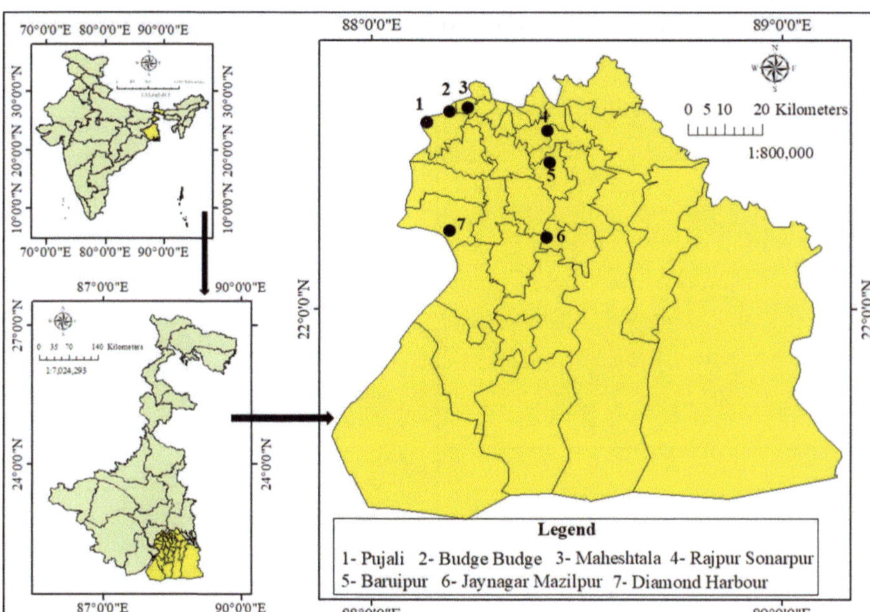


**Fig. 22.1** Location map of the study area

climate change, particularly heatwaves, and the effects on their health, well-being, work, and mental health. The final section gathered observations on climate change over the past five years and future expectations. Responses were analyzed using a five-point Likert scale, ranging from "strongly disagree" (1) to "strongly agree" (5).

# Results

## *LST and Thermal Condition Analysis*

The LST within the study area was meticulously analyzed across three transitional periods: Summer (April, May, June), Transitional Period (October, November), and Winter (December, January, February, March), (refer to Fig. 22.2). The analysis employed a fishnet technique at 10-m intervals to capture LST data throughout the study area during these months.

Following data collection, a refinement process was conducted to eliminate null values, yielding a subset of points having recorded temperatures. Among these, points with consistently high temperatures were identified, and the RLST was calculated. Particular attention was given to identifying a centroid where temperatures exceeded the surrounding areas by more than 3.47 °C.

A 9-km buffer zone was constructed around the centroid, classifying it as a regional UHI. The TCI was computed monthly, with significant fluctuations in January (0.68), February (0.72), March (0.82), April and May (0.85), June (0.71), October (0.80),

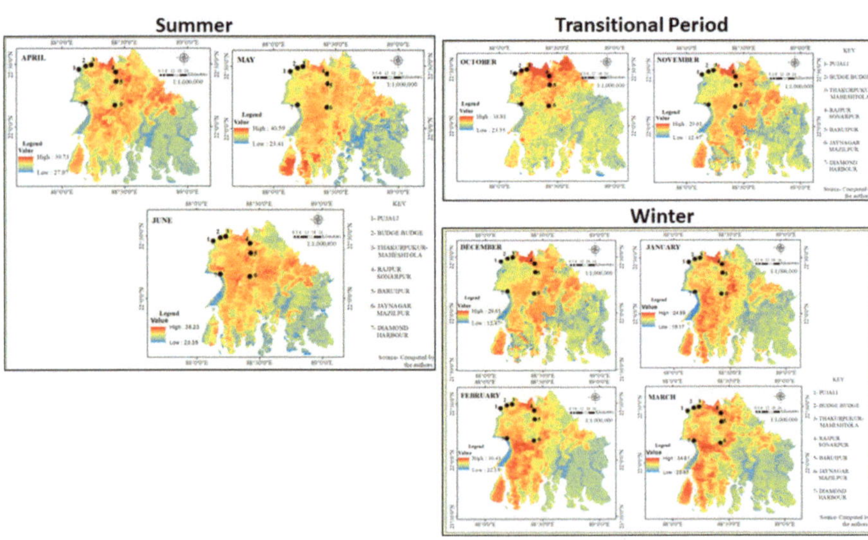


**Fig. 22.2** Seasonal variation in the LST

November (0.69), and December (0.74). This analysis underscores the region's dynamic thermal environment over time.

## *Area Vulnerability Index*

The Area Vulnerability Index is an important indicator for assessing a region's vulnerability to environmental stresses, notably UHI effects (see Fig. 22.3). A comprehensive analysis entails assessing several indices, namely the NDBI, NDVI, and NDWI, that would furnish insight into the area's infrastructure, vegetation, and waterbody distribution. The NDBI analysis depicts a considerable predominance of built-up areas closer to the regional UHI, with values ranging from 0.58 to − 0.56. This high concentration of infrastructure, driven by the region's proximity to Kolkata (a megalopolis), contributes to a higher LST, reflecting elevated thermal loads and reduced thermal balance. The NDVI analysis indicates a limited vegetation cover, with values ranging from 0.35 to − 0.10. This scarcity exacerbates the region's vulnerability to heat, as it hampers natural cooling processes, such as evapotranspiration and shading. Similarly, the NDWI analysis reveals a deficiency of waterbodies, with values between 0.37 and − 0.44, further reducing the area's cooling capacity, as waterbodies function as natural thermal regulators. These findings collectively indicate a likely upward trend in LST within the region, potentially leading to an intensification of the UHI effect in the near future. The integrated insights from NDBI, NDVI, and NDWI analyses provide a comprehensive understanding of the region's thermal dynamics and vulnerability, facilitating targeted interventions to mitigate the negative effects of increased urbanization and environmental stress.

## *Individual Vulnerability Index*

The Individual Vulnerability Index was meticulously developed using several sub-indicators, thoughtfully categorized into three overarching themes: family size, dwelling conditions, and the structural quality of residences. Analysis of the data reveals that a significant proportion of respondents reported moderate family sizes, typically comprising 5–8 members, accounting for 61% of the sample population.

In terms of dwelling arrangements, 53% of households indicated having 2–3 rooms, while 26% reported having only 1–2 rooms, and 21% reported having more than 3 rooms in their residences. Regarding the quality of housing, 61% of respondents described their residences as being in moderate condition, 11% indicated their homes were of good quality, and 28% reported living in substandard conditions.

Further analysis highlights that asbestos and concrete were the predominant materials used in the construction of respondents' homes. Additionally, a notable factor contributing to vulnerability was the limited ability to invest in advanced cooling

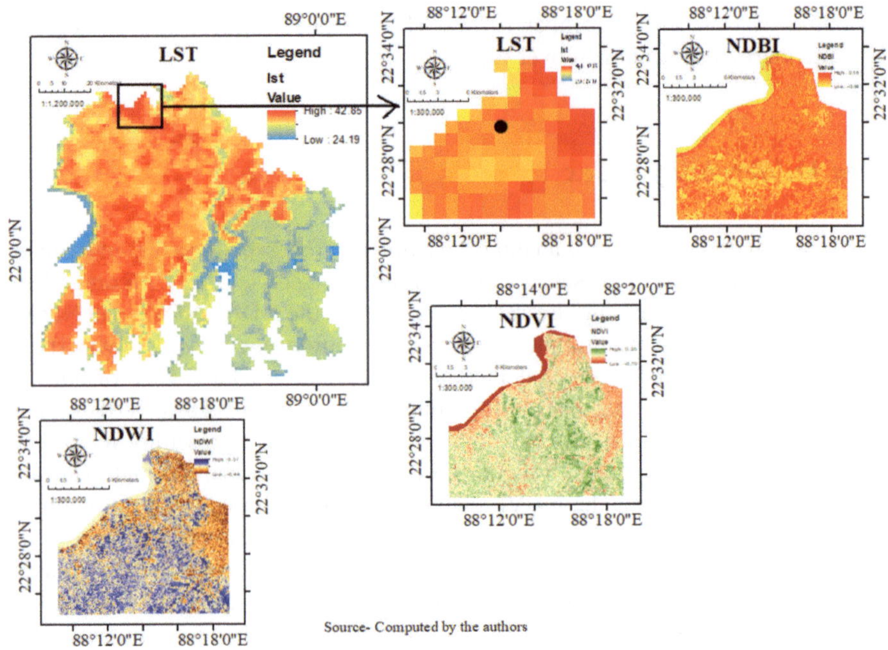


**Fig. 22.3** LST, NDWI, NDBI, NDVI analysis

systems. Due to financial constraints, most respondents relied on ceiling or table fans, as the cost of air conditioning was prohibitive.

This comprehensive analysis highlights various factors contributing to individual vulnerability, including family dynamics, housing conditions, and economic limitations. These insights are critical for informing targeted interventions and policy measures aimed at reducing vulnerabilities and improving overall well-being within the community.

## *Impact on Health and Wellbeing*

When respondents were asked about the intensity of physical discomforts arising from climate change, the majority cited various dermatological issues, with 61% reporting an increase in skin rashes and allergies, and 43% experiencing heat exhaustion and cramps. Additionally, 39% reported respiratory and cardiopulmonary disorders, while 50% identified cardiovascular ailments. A significant number of respondents experienced heightened breathlessness (61%), excessive sweating (59%), heat-related fever (57%), and diarrhoea (37%) due to heatwaves. In more severe cases, respondents reported incidents of heatstroke in their communities, necessitating urgent medical attention at government hospitals.

Respondents were also asked to assess their discomfort levels regarding mental health, lifestyle, and well-being using a five-point scale (see Table 22.1). A considerable portion of the population reported increased fatigue due to heatwaves, with notable occurrences of stress and panic attacks, often leading to insomnia and further deteriorating their physical and mental health. Cognitive impairment and behavioral disturbances, such as increased aggression and conflicts over trivial matters, were also reported. Furthermore, many respondents expressed that heatwaves had adversely affected their livelihoods, particularly those in the informal sector, where reduced clientele during specific hours led to decreased daily earnings. Respondents also noted a decline in work productivity due to fatigue and thermal discomfort, as well as a reduction in fitness levels, opting for sedentary activities over active pursuits during heatwaves.

The research further analyzed hospitalization data from outpatient departments of hospitals near the UHI, as depicted in the accompanying figure. A noticeable surge in hospitalizations occurred during May and June (see Fig. 22.4). A significant increase in ophthalmic conditions, induced by heatwaves, was particularly evident during these months, with many patients seeking medical attention for heat-related ocular issues. Hospitalizations due to cardiovascular conditions and asthma also saw a significant rise, aligning with established epidemiological trends linking elevated temperatures to increased vulnerability to cardiovascular complications and respiratory conditions such as asthma. This highlights the need for targeted interventions to mitigate the health risks posed by extreme heat.

Additionally, the data revealed a persistent rise in hospitalizations due to hypertension and stroke, particularly during the summer months, coinciding with recurring heatwaves. These findings underscore the sustained impact of extreme heat events on cardiovascular health, with increased incidences of hypertension and stroke-related hospitalizations.

The analysis reveals a concerning trend of increased hospitalizations during the summer months, primarily due to heat-related health issues. The surge in cases

**Table 22.1** Intensity of discomfort due to heatwaves as perceived by the respondents

| Discomforts | (In %) | | | | |
|---|---|---|---|---|---|
| | 1 | 2 | 3 | 4 | 5 |
| Increase in Fatigue | 4 | 21 | 29 | 43 | 17 |
| Increase in stress and panic attacks | 7 | 17 | 23 | 42 | 11 |
| Insomnia | 4 | 11 | 31 | 45 | 9 |
| Increased levels of aggression | 9 | 17 | 26 | 31 | 17 |
| Reduction in work productivity | 3 | 5 | 14 | 54 | 24 |
| Has affected the livelihood | 11 | 21 | 29 | 28 | 11 |
| Alteration in fitness and lifestyle | 5 | 12 | 23 | 33 | 27 |

*Source* Data obtained from the primary survey and computed by the authors

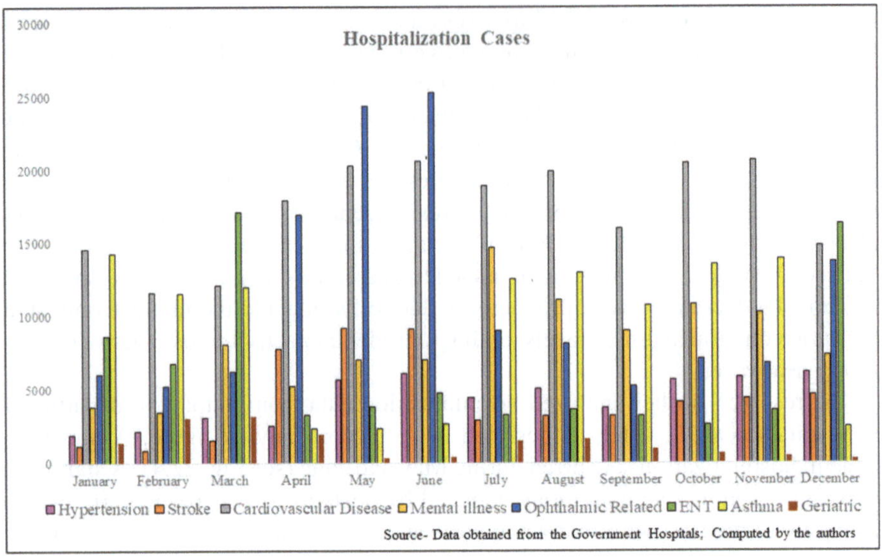


**Fig. 22.4** Month-wise hospitalization cases

related to ophthalmic conditions, cardiovascular complications, asthma exacerbations, hypertension, and stroke underscores the urgent need for proactive measures. Addressing these challenges requires a multifaceted approach, including public health interventions, urban planning strategies, and community outreach initiatives to build resilience and protect vulnerable populations from the harmful effects of extreme heat events.

## Discussion

The analysis of the study area reveals critical links between environmental factors, vulnerability indices, and health outcomes, emphasizing the profound impact of extreme heat events. The Area Vulnerability Index and LST data underscore the region's susceptibility to the UHI effect, exacerbated by dense built-up areas, reduced vegetation, and limited waterbodies. These environmental stressors, combined with socioeconomic disparities, intensify vulnerabilities in certain demographic groups.

The Individual Vulnerability Index highlights the role of family size, housing conditions, and economic constraints in shaping susceptibility to heat-related health issues, such as respiratory, cardiovascular, and dermatological conditions. The rise in hospitalizations during summer months, particularly for conditions like asthma, hypertension, and stroke, underscores the urgent need for targeted public health interventions.

Addressing these challenges requires integrating scientific evidence into urban planning, public health, and community engagement strategies. A holistic approach promoting resilience, equitable access to resources, and adaptive measures can mitigate the health impacts of environmental stressors and foster healthier, more sustainable urban environments.

## Conclusion

UHIs present significant risks to environmental sustainability and public health, particularly in densely populated urban areas. Addressing these challenges requires a comprehensive approach combining urban planning, public health interventions, social support, and policy development.

- Urban planners should prioritize the creation of green spaces, such as parks and green roofs, to enhance natural cooling. Sustainable construction practices, like cool roofs and reflective pavements, can reduce heat absorption. Water management strategies, including rainwater harvesting and permeable pavements, also help regulate urban temperatures.
- Public healthcare professionals must develop heatwave preparedness plans, early warning systems, and community outreach programs. Health education initiatives should promote awareness of heat-related risks, while healthcare professionals must be trained to manage heat emergencies. Investments in housing improvements, particularly in low-income areas, are essential for enhancing thermal comfort.

Policymakers should incorporate climate resilience into urban planning, integrating green infrastructure and heat mitigation measures. A coordinated, interdisciplinary approach is vital to mitigate UHI effects, promote sustainability, and protect public health, fostering resilient and healthier urban communities.

## Novelty of the Research, Limitations and Future Scope

This research presents a novel approach to understanding UHI dynamics by integrating LST analysis with socio-economic vulnerability and public health assessments. Unlike traditional studies that focus primarily on temperature variations, this study explores the complex relationship between UHI patterns, demographic vulnerabilities, and their multidimensional health impacts, both physical and psychological. The use of multiple thermal indices LST, TCI, and RLST combined with detailed analyses of vegetation, waterbodies, and built-up areas, offers a refined and comprehensive assessment of UHI intensity.

Focusing on South Twenty-four Parganas, a region with diverse land uses, this study uniquely examines UHI effects across urban, peri-urban, and rural landscapes.

Incorporating primary survey data from vulnerable populations, it addresses critical gaps in understanding how socio-economic factors exacerbate UHI impacts. This multidisciplinary approach, alongside policy-driven recommendations, offers a substantial contribution to existing UHI research.

However, the study's temporal scope is restricted to a single year (2023), as is its initial concentration on blue-collar workers, which might exclude other vulnerable populations such as children, the elderly, and those with pre-existing health disorders. Additional research is needed to assess the long-term UHI trends, interannual variability, and mental health consequences of extended heat exposure.

Future research should also incorporate multi-year data sets for better comprehension of UHI patterns and their relationship to climate change and urban transition. For further comprehension of UHI-related health hazards, vulnerable groups, such as the elderly and those with pre-existing health disorders, should also be conducted.

## References

Aboulnaga M, Trombadore A, Mostafa M, Abouaiana A (2024) Understanding urban heat Island effect: causes, impacts, factors, and strategies for better livability and climate change mitigation and adaptation. Livable cities: urban heat Islands mitigation for climate change adaptation through urban greening. Springer, Cham, pp 283–366

Abrar R, Sarkar SK, Nishtha KT, Talukdar S, Rahman A, Islam AR, Mosavi A (2022) Assessing the spatial mapping of heat vulnerability under urban heat Island (UHI) effect in the Dhaka Metropolitan area. Sustainability 14(9):4945

Aghamohammadi N, Fong CS, Farid NDN, Ramakreshnan L, Mohammadi PA (2022) Heat and mental health in cities. In: Urban overheating: heat mitigation and the impact on health. Springer, Singapore, pp 81–107

Ascione F, Bianco N, Manniti G, Mastellone M, Tariello F, Vanoli GP (2024) The impact of heat mitigation on low-income population. In: Mitigation and adaptation of urban overheating. Elsevier, Amsterdam, pp 123–153

Bai X, Yu Z, Wang B, Zhang Y, Zhou S, Sha X, Li S, Yao X, Geng X (2024) Quantifying threshold and scale response of urban air and surface temperature to surrounding landscapes under extreme heat. Build Environ 247:111029

Cai Y, Chen Y, Tong C (2019) Spatiotemporal evolution of urban green space and its impact on the urban thermal environment based on remote sensing data: a case study of Fuzhou City, China. Urban for Urban Green 41:333–343

Census (2011) Census tables. Office of the register general & Census Commissioner, India and Ministry of Home Affairs, Government of India

Chen Y, Yang J, Yang R, Xiao X, Xia JC (2022) Contribution of urban functional zones to the spatial distribution of urban thermal environment. Build Environ 216:109000

Cleland SE, Steinhardt W, Neas LM, West JJ, Rappold AG (2023) Urban heat island impacts on heat-related cardiovascular morbidity: a time series analysis of older adults in US metropolitan areas. Environ Int 178:108005

Dai Y, Liu T (2022) Spatiotemporal mechanism of urban heat Island effects on human health—evidence from Tianjin City Of China. Front Ecol Evol 10:1010400

Dutta K, Basu D, Agrawal S (2024) Ambient temperature modelling from surface characteristics and associating urban morphology with thermal discomfort. Singap J Trop Geograp 1–10

Ebi KL, Capon A, Berry P, Broderick C, de Dear R, Havenith G, Honda Y, Kovats RS, Ma W, Malik A, Morris NB, Nybo L, Seneviratne SI, Vanos J, Jay O (2021) Hot weather and heat extremes: health risks. Lancet 398(10301):698–708

Halder B, Chatterjee P, Rana B, Bandyopadhyay J, Pande CB, Ahmed KO, Elkhrachy I, Radwan N (2024) Delineating the climate change impacts on urban environment along with heat stress in the Indian tropical city. Phys Chem Earth Parts a/b/c 136:103745

Han D, Zhang T, Qin Y, Tan Y, Liu J (2023) A comparative review on the mitigation strategies of urban heat island (UHI): a pathway for sustainable urban development. Climate Dev 15(5):379–403

He BJ, Wang J, Zhu J, Qi J (2022a) Beating the urban heat: situation, background, impacts and the way forward in China. Renew Sustain Energy Rev 161:112350

He BJ, Zhao D, Dong X, Xiong K, Feng C, Qi Q, Darko A, Sharifi A, Pathak M (2022b) Perception, physiological and psychological impacts, adaptive awareness and knowledge, and climate justice under urban heat: a study in extremely hot-humid Chongqing, China. Sustain Cities Soc 79:103685

Heaviside C, Vardoulakis S, Cai X-M (2016) Attribution of mortality to the urban heat island during heatwaves in the West Midlands, UK. Environ Health 15(27):1–12

Heaviside C, Macintyre H, Vardoulakis S (2017) The urban heat Island: implications for health in a changing environment. Curr Environ Health Rep 1–11

Huang H, Li Y, Zhao Y, Zhai W (2022a) Analysis of the impact of urban summer high temperatures and outdoor activity duration on residents' emotional health: taking hostility as an example. Front Public Health 10:955077

Huang HC, Zhou XH, Qu YC, Zhang H (2022b) Spatial effects of climate warming on health risks due to urban heat exposure: a case study Of China. DYNA 97:162–168

Huang HC, Deng X, Yang HL, Zhou XH, Jia Q (2020) Spatiotemporal mechanism underlying the effect of urban heat Island On cardiovascular diseases. Iran J Public Health 49:1455–1466

Irshad Z, Hassan M, Akbar S, Farooq M, Chishtie FA (2024) Spatiotemporal changes in LULC and associated impact on urban Heat Islands over Pakistan using geospatial techniques. Urban Clim 57:102112

Jang J, Lee W, Choi M, Kang C, Kim H (2020) Roles of urban heat anomaly and land-use/land-cover on the heat-related mortality in the national capital region of South Korea: a multi-districts time-series study. Environ Int 145:106127

Jia S, Wang Y, Liang TC, Weng Q, Yoo C, Chen W, Ding X (2024) Multiscale estimation of the cooling effect of urban greenspace in subtropical and tropical cities. Urban For Urban Green 128390

Jonescu EE, Ramanayaka CE, Olatunji OA, Uylaki TJ (2024) Understanding the impact of urban heat islands on crime: insights from temperature, population density, and green canopy cover. Crime Sci 13(1):15

Kacker K, Srivastava P, Mukherjee M (2024) Heat stress risk at an intra-urban level: a case study of Delhi, India. Build Environ 264:111897

Kassomenos P, Begou P (2022) The impact of urban overheating on heat-related morbidity. In: Urban overheating: heat mitigation and the impact on health, pp 39–80

Lasky E, Costello S, Ndovu A, Aguilera R, Weiser SD, Benmarhnia T (2024) The health benefits of reducing micro-heat islands: a 22-year analysis of the impact of urban temperature reduction on heat-related illnesses in California's major cities. Sci Total Environ 949:175284

Lee YY, Md Din MF, Iwao K, Lee YH, Anting N (2021) Impact of thermal behaviour of different environmental conditions on ambient environment and thermal discomfort In Malaysia. Indoor Built Environ 30:520–534

Lee S, Lee H, Myung W, Kim EJ, Kim H (2018) Mental disease-related emergency admissions attributable to hot temperatures. Sci Total Environ 616:688–694

Li F, Yigitcanlar T, Li W, Nepal M, Nguyen K, Dur F (2024a) Understanding urban heat vulnerability: scientometric analysis of five decades of research. Urban Climate 56:102035

Li X, Tang J, He BJ (2024b) Highly heterogeneous heat-related health impacts in china and implications for sustainable urban heat governance. Adv Sustain Syst 8(5):2300538

Lin Z, Xu H, Hu X, Liu Z, Yao X, Zhu Z (2024) Characterizing the seasonal relationships between urban heat island and surface energy balance fluxes considering the impact of three-dimensional urban morphology. Build Environ 265:112017

Linares C, Culqui D, Carmona R, Ortiz C, Díaz J (2017) Short-term association between environmental factors and hospital admissions due to dementia in Madrid. Environ Res 152:214–220

Liu D, Zhou R, Ma Q, He T, Fang X, Xiao L, Hu Y, Li J, Shao L, Gao J (2024) Spatio-temporal patterns and population exposure risks of urban heat island in megacity Shanghai, China. Sustain Cities Soc 108:105500

Ma X, Lin J, Liu A, Nian X, Hong B (2024) Psychophysiological responses of the elderly to outdoor thermal stress: a focused analysis. Urban Clim 57:102094

Macintyre HL, Heaviside M, Cai X, Phalkey R (2021) The winter urban heat Island: impacts on cold-related mortality in a highly urbanized European region for present and future climate. Environ Int 154:106530

Medici CR, Vestergaard CH, Hadzi-Pavlovic D, MunkJorgensen P, Parker G (2016) Seasonal variations in hospital admissions for Mania: examining for associations with weather variables over time. J Affect Disord 205:81–86

Ming Y, Liu Y, Liu X, Tian Z (2024) Demographic disparity in diurnal surface urban Heat Island exposure across local climate zones: a case study of Chongqing, China. Sci Total Environ 923:171203

Ng CFS, Boeckmann M, Ueda K, Zeeb H, Nitta H, Watanabe C, Honda Y (2016) Heat-related mortality: effect modification and adaptation in Japan from 1972 to 2010. Glob Environ Chang 39:234–243

Palinkas LA, Hurlburt MS, Fernandez C, De Leon J, Yu K, Salinas E, Garcia E, Johnston J, Rahman MM, Silva SJ, McConnell RS (2022) Vulnerable, resilient, or both? A qualitative study of adaptation resources and behaviors to heat waves and health outcomes of low-income residents of urban heat Islands. Int J Environ Res Public Health 19(17):11090

Ren J, Shi K, Li Z, Kong X, Zhou H (2023) A review on the impacts of urban heat islands on outdoor thermal comfort. Buildings 13(6):1368

Ren J, Yang J, Zhang Y, Xiao X, Xia JC, Li X et al (2022) Exploring thermal comfort of urban buildings based on local climate zones. J Clean Prod 340:130744

Samba H, Guerchet M, Ndamba-Bandzouzi B, Mbelesso P, Lacroix P, Dartigues JF, Preux PM (2016) Dementia-associated mortality and its predictors among older adults in Sub-Saharan Africa: results from A 2-year follow-up in Congo (the EPIDEMCA-FU study). Age Ageing 45(5):681–687

Sanusi R, Bidin S (2020) Re-naturing cities: impact of microclimate, human thermal comfort and recreational participation. In: Climate change, hazards and adaptation options: handling the impacts of a changing climate, pp 545–562

Silva RPM, Barros CLM, Mendes TT, Garcia ES, Valenti VE, de Abreu LC, Garner DM, Espindola FS, Penha-Silva N (2019) The influence of a hot environment on physiological stress responses in exercise until exhaustion. PLoS ONE 14:e0209510

Stevens HR, Graham PL, Beggs PJ, Ossola A (2024) Associations between violent crime inside and outside, air temperature, urban heat island magnitude and urban green space. Int J Biometeorol 68(4):661–673

Stewart ID, Mills G (2021) The Urban Heat Island. Elsevier, Amsterdam

Taylor J, Wilkinson P, Picetti R, Symonds P, Heaviside C, Macintyre HL, Davies M, Mavrogianni A, Hutchinson E (2018) Comparison of built environment adaptations to heat exposure and mortality during hot weather, West Midlands Region, UK. Environ Int 111:287–294

Tehrani AA, Veisi O, Delavar Y, Bahrami S, Sobhaninia S, Mehan A (2024) Predicting urban heat island in European cities: a comparative study of GRU, DNN, and ANN models using urban morphological variables. Urban Clim 56:102061

Tewari K, Tewari M, Niyogi D (2023) Need for considering urban climate change factors on stroke, neurodegenerative diseases, and mood disorders studies. Comput Urban Sci 3(1):4
UN (2019) World population prospects 2019: highlights, vol 11. United Nations Department for Economic and Social Affairs, New York, NY, p 125
Wang Y, Yi G, Zhou X, Zhang T, Bie X, Li J, Ji B (2021) Spatial distribution and influencing factors on urban land surface temperature of twelve megacities in China from 2000 to 2017. Ecol Ind 125:107533
Wang K, Li T, Wei J (2019) Exploring drought conditions in the three river headwaters region from 2002 to 2011 using multiple drought indices. Water 11(2):190
Waseem S, Khayyam U (2019) Loss of vegetative cover and increased land surface temperature: a case study of Islamabad, Pakistan. J Clean Prod 234:972–983
Wu T, Wang Z, Xu Q (2024) Spatio-temporal heterogeneity of the urban heat effect and its socio-ecological drivers in Yangzhou City, China. Land 13(9):1470
Yadav N, Rajendra K, Awasthi A, Singh C, Bhushan B (2023) Systematic exploration of heat wave impact on mortality and urban heat island: a review from 2000 to 2022. Urban Clim 51:101622
Yang J, Shi B, Xia G, Xue Q, Cao SJ (2020) Impacts of urban form on thermal environment near the surface region at pedestrian height: a case study based on high-density built-up areas of nanjing city In China. Sustainability 12:1737
Yang J, Yang Y, Sun D, Jin C, Xiao X (2021a) Influence of urban morphological characteristics on thermal environment. Sustain Cities Soc 72:103045
Yang Z, Chen Y, Guo G, Zheng Z, Wu Z (2021b) Characteristics of land surface temperature clusters: case study of the central urban area of Guangzhou. Sustain Cities Soc 73:103140
Zhao H, Fang Y, Xu X (2024) Quantifying morphology evolutions of urban heat islands and assessing their heat exposure in a metropolis. Sustain Cities Soc 102:105244
Zhang Y, Balzter H, Li Y (2021) Influence of impervious surface area and fractional vegetation cover on seasonal urban surface heating/cooling rates. Remote Sens 13:1263
Zhou W, Wang Q, Li R, Zhang Z, Kadier A, Wang W, Zhou F, Ling L (2024) Heatwave exposure in relation to decreased sleep duration in older adults. Environ Int 183:108348

MIX
Papier aus verantwortungsvollen Quellen
Paper from responsible sources
FSC® C105338

If you have any concerns about our products,
you can contact us on
**ProductSafety@springernature.com**

In case Publisher is established outside the EU,
the EU authorized representative is:
**Springer Nature Customer Service Center GmbH**
**Europaplatz 3, 69115 Heidelberg, Germany**

Printed by Libri Plureos GmbH
in Hamburg, Germany